Illustrated Special Relativity
Through Its Paradoxes:
A Fusion of Linear Algebra, Graphics, and Reality:

Text by John de Pillis and José Wudka
Illustrations and animations by John de Pillis

Standard Edition

2

Contents

This book is dedicated to my wife,
Susan de Pillis, whose support over the years
allows me to truly understand why authors write dedications.

I. A First Pass

Preface *presents introductory remarks.*

Chapter 1 *offers an intuitive overview of some of the paradoxes in special relativity. Detailed graphical analysis follows in Chapters (9)-(14).*

Chapter 2 *concerns moving clocks (that always run slow) and moving objects (that always shrink in the direction of motion).*

Chapter 3 *introduces inertial frames as one-dimensional spaces (X−axes), each with a clock at every point. Parallel rulers or trains on parallel tracks traveling at constant relative speeds are mnemonic models for frames.*

Chapter 4 *develops the canonical mathematical model (a two-dimensional spacetime graph) for tracking and recording the one-dimensional movements of parallel inertial frames.*

Preface

(0.1) Exposition and Paradoxes

The Nature of Exposition

(0.1.1) **There are two flavors of scientific exposition** — the qualitative and the quantitative. We provide both. The qualitative narrative gives us a story, a description of <u>what</u> happens. The quantitative exposition also tells us what happens, but adds the <u>how</u> and <u>why</u> of these happenings.

If we only want to know *what* happens, then we are content with descriptions and assertions. We need no explanation. We are happy to learn that rods in motion shrink in the direction of motion, or that moving clocks always run slower than identical stationary clocks. But we need not know why.

But if we also want to know *why* things happen, then we must use observed measurements along with mathematical structures to represent physical objects and phenomena.[1]

(0.1.2) **Why must we use mathematics in our quantitative analysis?** This is a question that confounded Albert Einstein, who wrote, ([13], pg. 28)

> *"How can it be that mathematics, being after all a product of human thought independent of experience, is so admirably adapted to the objects of reality?"*

Einstein was not alone in wondering why mathematics proved necessary to explain Reality. Physicist Eugene Wigner [33] and mathematician R. W. Hamming [19], were also puzzled over the Unreasonable Effectiveness of mathematics.

(0.1.3) To see how mathematics, a chess game of the mind, proves its "un-

---

[1]It may be only apocryphal, but it has been said that when Stephen Hawking was writing his book, *A Brief History of Time* [20], he was advised by his editor that each formula he used would cut the sales by half.

reasonable" effectiveness, note how often in this book we will take the following INTELLECTUAL ROUND TRIP:

(i)   Consider a natural phenomenon in the "real" world,

(ii) Represent that phenomenon as a mathematical model,

(iii) Run a mathematical analysis on that model, and, finally,

(iv) Return the results of the mathematical analysis back to the original "real" world.

Without this round trip from Reality to mathematics and back to Reality, we would not know, for example, that

(0.1.4)   • Moving clocks must run slow (*Figure* (1.10.2), *pg.* 26) and moving rods must shrink in the direction of motion (*Figure* (1.10.4), *pg.* 27).

• The front end of a moving train is always younger than its back end (*Appendix* (C), *pg.* 359).

• The assumption of perfectly rigid bodies implies that time can be reversed in the sense that a future event would affect the present ((14.2.4), *pg.* 177).

NOTE: A stark example of this intellectual round trip between Reality and mathematics is the set of Maxwell's equations (*Chapters* (17)-(22)).

. . . .

The successes of (0.1.4), drawn from graphs and formulas, show how mathematics can model Reality and take us to truths about the physical world that lie well beyond the apprehension of our five humble input ports, or senses. But beware!

1.*sight*
2.*smell*
3.*hearing*
4.*taste*
5.*touch*
6.$\begin{bmatrix} in \\ progress \end{bmatrix}$

(0.1.5)   **The Occasional Illusion:** Although we will be applauding the success of mathematical modeling throughout this book, we must also be mindful that we often grasp Reality with a slippery hand. For example, in (23.6), we show how Descartes applied a false hypothesis to obtain a true result — Snell's Law. Fermat, who disputed Descartes' premise, applied an alternate (true) hypothesis to arrive at the same result ((23.6.2), *pg.* 323).

As indicated above, this book offers both qualitative and quantitative expositions since the reader may benefit from one style or the other — or both.

The Nature of Paradox

(0.1.6) Paradoxes are phenomena or arguments that challenge our intuition to a such a degree that we are tempted to declare the statements to be outright contradictions to what we "certainly" know is true.

Part of the fun is to acknowledge these first impressions of disbelief, and then to resolve them through the force of logic and mathematics, and thereby clarify the contours of Reality. These are characteristics of non-intuitive assertions or paradoxes which we now define:

(0.1.7)

> **DEFINITION:** *A* PARADOX *is a statement or proposition that is* <u>*seemingly*</u> *contradictory or absurd, but, in reality, expresses a truth without contradiction.*

(0.1.8) How are paradoxes explained? As events will reveal, paradoxes often arise from the ambiguity of spoken language and reluctance to abandon "obvious" intuitive truths.

For example, our intuition that tells us that the earth is flat, that the sun moves around the earth, and that atoms, if they exist at all, must be relatively solid — they can not consist mostly of empty space.

More generally, we might say that paradoxes arise when we fail to apply CRITICAL THINKING[2] which includes the *questioning and evaluating of basic assumptions.*

---

[2]*Critical thinking* is not the same thing as criticizing. The declaration, "I don't like those shoes you are wearing!" is not necessarily a product of critical thinking which must be based on *clear identification of assumptions* and their logical consequences.

(0.2) Organization of this Book

- **Preface:** *This chapter*

SR[3] **QUALITATIVE MATERIAL** (*what* the paradoxes are)

- **Chapter (1)** *Introduction to the Paradoxes:* This is an introductory overview of six paradoxes in special relativity without full explanations or proof.

- **Chapter (2)** *Clocks and Rods in Motion:* It is stated, without extensive proof, that moving clocks always run slow and moving objects always shrink in the direction of motion.

SR QUANTITATIVE MATERIAL (questions of *why* and *how.*)

- **Chapter (3)** *The Algebra of Frames:* The inertial frame, $\mathcal{F}$, the basic setting for special relativity, is a choice of an origin, a coordinate system and clocks at every point in space, which together allow a unique specification of the location and time of each event. Without loss of generality, we can restrict the three space coordinates to one space coordinate x (1.3.1). Examples of inertial frames experiencing no forces with a clock at each point would be a railroad car or a ruler moving at constant speed. Distance x and time t at that point x defines an ordered pair, $[x, t]$, the set of which is a *two-dimensional vector space* (*Section* (3.2)). Finally, we present the six rules for frames (*Section* (3.4)).

- **Chapter (4)** *The Graphing of Frames:* Since an inertial frame is mathematically modeled as a two-dimensional vector space (*Ch.* (3)), each frame can also be represented as a two-dimensional *spacetime graph*. Each ordered pair $[x, t]$ can then be pictured as a point on the spacetime graph that has coordinate axes X for distance and T for time.
  The tools of linear algebra now come into play: If we have two frames, $\mathcal{F}^A$ and $\mathcal{F}^B$, say, moving parallel to each other with a constant relative speed (*imagine two railroad cars on parallel tracks*), then to peek into one frame from another along a virtual *line-of-sight*, is to establish a relationship or pairing between the observer point $[x, t]_A$ in frame $\mathcal{F}^A$ and the observed point $[x', t']_B$ in frame $\mathcal{F}^B$. The set of all pairings, $[x, t]_A \rightarrow [x, t]_B$, defines a

---

[3]**SR** is an abbreviation for *Special Relativity.*

mathematical *function* that also turns out to be, in mathematical terminology, a *linear transformation* ((A.4.1b), *pg.* 343).

- **Chapter (5) *Galilean Transformations:*** The standard tools of linear algebra allow us to create a unique 2×2 matrix that totally reflects the properties of the corresponding linear transformation $[x, t]_A \to [x, t]_B$ described in Chapter (4). What does this matrix look like? Since Galileo assumed that there was only one universal clock by which all observers in all frames can simultaneously synchronize their watches (*an intuitive idea behind railroad timetables, and one that we see in spy movies all the time*), it follows, in a non-obvious way, that the speed of light must be infinite. Assuming infinite speed of light, we arrive at a unique 2×2 matrix called the Galilean transformation matrix that allows us to view between the pair of inertial frames along line-of-sight corridors ((5.4.1a), *pg.* 90).

- **Chapter (6) *Constant c in Spacetime:*** By 1905, c, the speed of light in a vacuum, was determined not to be infinite, but to have a value of 186,000 miles per second. Einstein's fundamental assumption is that this single value of c must be the <u>same</u> for all observers regardless of the speed of either the light-source or the observer (*No matter how fast you chase after a photon of light, its speed is still the constant c*). With this additional assumption, spacetime graphs are called *Minkowski diagrams*. With the aid of Minkowski diagrams, we learn that two events occurring simultaneously in one frame can never occur simultaneously in another frame that is moving relative to the first frame. That the constant speed of light destroys simultaneity (*Figure* (6.2.3)) is anything but clear from just the reality of our five senses.

- **Chapter (7) *Lorentz Transformations:*** Since the speed of light c is *not* infinite and is measured to be constant by *all* observers regardless of motion, we are able to use the geometry of the Minkowski diagram to derive the linear transformation $[x, t]_A \to [x, t]_B$ that pairs an observer at $[x, t]_A$ in $\mathcal{F}^A$ viewing an observer at $[x', t']_B$ in $\mathcal{F}^B$ through a virtual line-of-sight corridor that runs between the two (parallel) frames. This pairing, the *Lorentz linear transformation*, has a corresponding 2×2 matrix representation called the *Lorentz matrix* ((7.2.2), *pg.* 109). With this non-Galilean matrix in hand, we find that speeds no longer add in the "usual" way that accords with our experience. We show this to be true analytically through a matrix analysis (*Section* (7.4)) and then geometrically using areas (*Section* (7.5))

- **Chapter (8)** *The Hyperbola of Time-Stamped Origins:* Consider infinitely many parallel frames, each moving with a different relative speed. For fixed time t_0 in each frame, plot the respective origins in the Minkowski *X-T* diagram. The resulting curve is a well-defined hyperbola. Different fixed times t_0 produce different hyperbolas. (*This hyperbola reduces to a horizontal line in Gaussian/Newtonian physics.*) Each moving frame $\mathcal{F}$ gives rise to a tangent line (*of the t_0 hyperbola*) which then corresponds to all the t_0 simultaneous events in frame $\mathcal{F}$.

- **Chapter (9)** *The Accommodating Universe:* The theory of special relativity holds that particles and signals can never travel faster than the speed of light. However, conditions are described under which an intergalactic traveler (skateboarder) apparently does travel faster than light. The analysis reveals how this "contradiction" is a consequence of the (mis)use of language.

- **Chapter (10)** *The Length-Time Comparison Paradoxes:* Let two inertial frames be in parallel constant motion relative to each other. Then observers in each frame see the clocks in the other (*moving*) frame run slower than the clocks of their own (*stationary*) frame. Similarly, identical rods in the other (*moving*) frame are shorter than the rods in their own (*stationary*) frame. This chapter answers the questions: How can clocks in each frame simultaneously run slower than clocks in the other frame? How can identical rods in each frame simultaneously be shorter than rods in the other frame?

- **Chapter (11)** *The Twin Paradox:* On the day of their birth, one twin sets off in space in a very fast rocket while the other twin stays on Earth. After 25 Earth years, the traveling twin's space ship turns around and takes another 25 years to return to Earth. The paradox is that the Earth-bound twin has aged 50 years while the space ship twin, upon return, has aged only 30 years. On one hand, all moving clocks run slow, so naturally, the moving space ship clock slows down and reads only 30 years instead of 50 years on its return to Earth. On the other hand, it is the Earth clock that is moving as observed from the space ship. If so, then why doesn't the (*moving*) Earth clock run slower than the (*stationary*) space ship clock?

- **Chapter (12)** *The Train-Tunnel Paradox:* Viewed from the tunnel, a moving train inside the tunnel avoids the shock of simultaneous lightning bolts that strike the tunnel entrance and the exit. Viewed from the train, the (*moving*) tunnel becomes shorter than the train. Do the (*simultaneous?*) lightning bolts now strike the "too-long" train?

- **Chapter (13)** *The Pea-shooter Paradox:*   Our intuition tells us that if a train is going at $u = 50$ mph and a ball on the train is thrown forward at $v = 20$ mph, then an observer standing outside the train will see the ball traveling at $u+v = 20+50 = 70$ mph. Not so. The actual (Lorentzian, or relativistic) sum of speeds, is always $u \oplus v = (u+v)/(1+uv/c^2)$ which, for small speeds, is imperceptibly less than the familiar combined speed $u + v$. There are two ways to derive $u \oplus v$: analytically through matrix calculation (7.4) and graphically through addition of areas (7.5).

- **Chapter (14)** *The Bug-Rivet Paradox* is one of the least explored paradoxes of special relativity. The assumption of absolute rigidity — where objects can never be stretched or compressed — implies that *the future affects the present!* Using a principle demonstrated by the Slinky toy (14.3), special relativity tells us that a speeding rivet (which, from a stationary frame, is seen to be shortened), reverts to its at-rest length after its speed is reduced to zero. In our example, (14.1), a certain collision is designed to bring the rivet speed to zero. However, perfect rigidity implies that the rivet will "sense" the impending collision and will start to reach its at-rest length even *before* the collision occurs.

ENERGY and MASS

- **Chapter (15)** *E=mc²* This celebrated equation is almost an afterthought once the connection is made between the pea-shooter paradox, Chapter 13, and conservation of momentum.

THE MATHEMATICS of WAVES and LIGHT

- **Chapter (16)** *The Nature of Waves:*   We develop the mathematical model of propagated waves since light exhibits the properties of a propagated wave under certain circumstances.

MAXWELL'S EQUATIONS (*a unification of theories of electricity, magnetism, and optics*)

- **Chapter (17)** *Maxwell's Mathematical Toolkit:*   At the very outset of his seminal paper in special relativity [12], Einstein acknowledges the work of Clerk Maxwell and his unifying work in

which he developed just four equations ((20.7.3), *pg.* 280) that
unified the theories of electricity, optics, and magnetism. In this
chapter, we see how observable (one-dimensional) lengths and (two-
dimensional) areas are interpreted mathematically as three-dimensional
vectors, or arrows (See *Vectors Model Reality* (17.5), *pg.* 226).

- **Chapter (18)** *Electric and Magnetic Fields:* This is an overview of electricity and magnetism basics including Coulomb's Law
  (18.2.4b)

- **Chapter (19)** *Electricity and Magnetism. Gauss' Laws:* After developing the notion of flux, we state Gauss' Law for both
  electricity (19.3.1) and magnetism (19.4.1).

- **Chapter (20)** *Towards Maxwell's Equations:* After discussion
  of the Biot-Savart Law (20.1), the Ampère Maxwell Law (20.3),
  along with the laws of Faraday (20.5) and Lentz (20.6), we finally
  present the elegant four equations that Maxwell fashioned (20.7)
  to embrace all the laws previously stated.

- **Chapter (21)** *Electromagnetism. A Qualitative View:* We
  offer a general description of the geometry of electromagnetic wave
  propagation.

- **Chapter (22)** *Electromagnetism. A Quantitative View:* An
  analytic explanation of topics in the previous chapter are presented.
  We see how Maxwell calculated c, the speed of light. (*Actually, he
  calculated the speed of electromagnetic waves, not knowing at the
  time that light was one example of electromagnetism*). He did this
  by merely computing $c = \sqrt{\varepsilon_0 \mu_0}$, the square root of the product
  of ε_0, the permittivity (18.2.5a) from Coulomb's law, and μ_0, the
  permeability (20.2.2) from the Biot-Savart Law. This is an amazing
  mathematical prediction.

FINAL THOUGHTS

- **Chapter (23)** *Epilogue. Final Thoughts:* We find an overview
  of Einstein's Annus Mirabilis of 1905 (23.2), a comparison of the
  Galileo/Newton relativities with that of Einstein/Lorentz (23.3), a
  note on the reliability of Conventional Wisdom, and a description of
  a feud between Descartes and Fermat, each of whom derived Snell's
  Law (23.6). The difference is that Descartes falsely assumed that
  light would gain speed as it entered water, while Fermat correctly
  asserted that light would slow down. Yet they both arrived at the
  same conclusion.

APPENDIX

- **Appendix (A)** *Linear Algebra Overview.* A brief exposition of vector spaces, functions, and linear transformations along with their matrix representations, eigenvalues and eigenvectors.

- **Appendix (B)** *Hyperbolic Functions.* Graphical properties of the hyperbola $f(x) = 1/x^2$ become the mathematical foundation for Section (7.5), pg. 112, *Addition of Speeds via Areas* in the special case that the speed of light $c = 1$. Properties of hyperbolas and their areas are central to developing a graphical resolution of the pea-shooter paradox.

- **Appendix (C)** *Deconstructing a Moving train.* A railroad car at constant speed is a physical object. When modeled mathematically, it becomes an inertial frame (*a ruler with a clock at each point* ((1.5.1), *pg.* 16). What is not obvious in the "real" physical reality, is that observers on a platform will see that the front of the train is younger than the rear of the rain. The Minkowski diagram ((C.2.1), *pg.* 360) justifies this conclusion.

(0.2.1) **Acknowledgments.** Lectures from this book were presented in the winter of 2005 at the University of Padua, Italy, with the generous support of Professoressa *Maria Morandi Cecchi*, Dipartimento Matematica Pura ed Applicata. The authors benefited from conversations with physicist *Sid Fiarman* and with Professors *Jack Denny* of Arizona State University, *Al Kelley* of the University of California, Santa Cruz, Professors *Fred Metcalf, Larry Harper, David Ellerman,* and *John Baez* of the University of California, Riverside, Professor *John Mallinckrodt*, California State Polytechnic University, who commented on the paradoxes, *Dr. Joan Kleypas* of the National Center for Atmospheric Research in Boulder, Colorado, and *Richard Reeves,*. Special thanks are owed to *Jerry Alexanderson* and particularly to *Don Albers* of the Mathematical Association of America who was consistently supportive of this book with its color illustrations, and the animations.

(1) Introduction to the Paradoxes

(1.1) Aristotle vs. Galileo

(1.1.1) ARISTOTLE (384-322 B.C.) believed that a stone fell toward the earth because the stone and the earth were both in the "Earth" category among the four basic elements, Earth, Air, Fire, and Water. According to this reasoning, smoke, which consists of air and fire, wants to be closer to the sky (air) and further from an unlike element (earth). Hence, earthly objects move naturally toward the earth while "airy" objects move naturally upward. The assumed fifth element, the heavenly substance he called the QUINTESSENCE along with an assumed PRIME MOVER accounted for the "perfect" circular and uniform motion of the heavens. Finally, Aristotle asserted that motion of any object required a continuing force on that object as long as motion persisted.

As if to simplify the discussion, ST. THOMAS AQUINAS (1225-1274) later declared that something is in motion when it is both *what it is already* and *something else that it is not yet*. (*Apparently, new explanations do not always lead to clarity.*)

In a complete departure from the convoluted explanations of Aristotle and Aquinas, GALILEO GALILEI (1564-1642) claimed that a force was *not* necessary to maintain constant, straight-line motion. This property is in evidence even today where it can be observed that an object in space will drift along forever without external forces.

Galileo's thesis was that a force was necessary only to produce a *change* in speed, that is, an acceleration or deceleration. For example, it is the force of friction that brings a rolling ball to a stop on a flat surface.

Aristotle and Galileo disagreed on another point. Where Aristotle said a heavy body would fall faster than a lighter one (*Doesn't a rock fall faster than a feather?*), Galileo claimed that two bodies, when dropped from the same height, always fall at the same rate, when acted upon by the force of gravity alone. (*The effect of friction due to air resistance must be negligible, which is not the case with a falling feather.)*

(1.1.2) FIGURE: *Galileo's Thought Experiment for Falling Bodies*

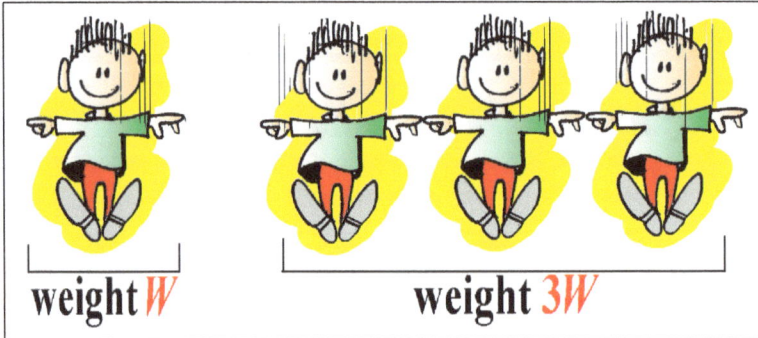

A *body of weight* **W** *(left panel) falls at the same rate as the heavier body of weight* **3 W** *(right panel) since it can be considered as three independent bodies, each of weight* **W**.

If Aristotle were correct in claiming that heavier bodies fall faster than lighter ones then once the three individuals of Figure (1.1.2) touch fingers, forming a collective weight $3W$, they would start to fall faster. And once their fingers separated, the speed of each individual, being only W units, would slow down.

(1.2) Frames of Reference

A Frame of Reference is a way of specifying the position of each event with respect to a given point, and the time at which the event occurred.

For example, while directing a tourist from a train station to her hotel room, we need to tell her how many meters east she should walk, then how many north, and how many stories up the building she should go; we can also tell her shell be there at 4 PM. So we imagine three lines, one east-west, one north-south and one up-down, and a clock at very point; That is, from our location (the origin), we can specify any point and any time — that is, any event — by providing 4 numbers.

In sum, the chosen set of lines and clocks represent a reference frame. Using this frame we can tell the where and when of any event. The corresponding 4 numbers are the coordinates of that event with respect to that reference frame

Here is a more formal definition:

(1.2.1)

> **DEFINITION:** *A* FRAME OF REFERENCE, *$\mathcal{F}$, is a choice of an origin, a coordinate system and clocks at every point in space, which together allow a unique specification of the location and time of each event.*

(1.3) Straight-Line Trajectories in 3-Space

(1.3.1) Representing 3D Motion as 1D Motion: If the motion of a particle is only along a straight line in a space with X, Y, and Z-axes, then we can choose our X-axis, which is also a straight line, to exactly *coincide* with that straight line path in 3-space. As illustrated by Figure (1.3.2), the particle traveling along this X-axis has spacetime coordinates of the form $[x, 0, 0, t]$ — the Y and Z coordinates are fixed at 0 while the x and t values vary with time. Hence, the coordinates $[x, t]$ contain the same information about the particle's motion as does $[x, 0, 0, t]$.

(1.3.2) FIGURE: *1D Representation of 3D Straight-Line Trajectory.*

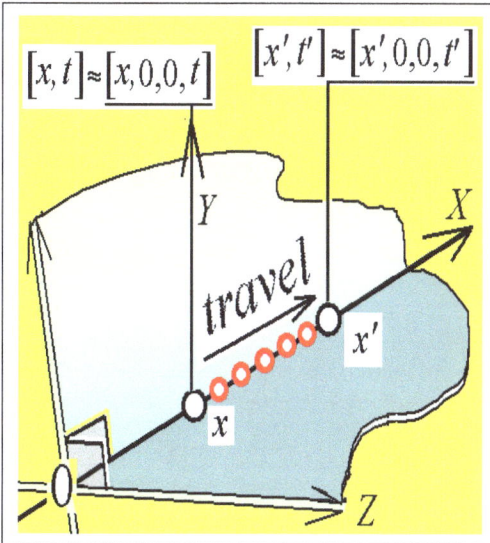

Without loss of generality, we choose the X-Y-Z axes so that the X-axis falls on the straight-line trajectory of an object in 3-space. Two positions (x coordinates) are shown of an object traveling on the X-axis at times t_0 and t_1. The two positions are economically represented, without loss of information, by the ordered pairs, $[x, t]$ and $[x', t']$ — no y or z values are needed.

(1.3.3) FIGURE: *When Pictures Fail*

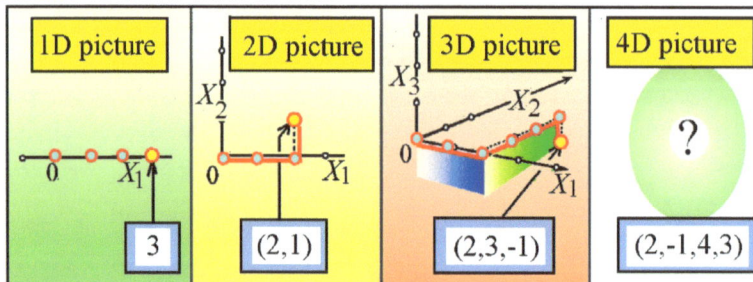

1D picture	2D picture	3D picture	4D picture
3	(2,1)	(2,3,-1)	(2,-1,4,3)

Although time is involved, Figure (1.3.2) does not show a separate time axis, T which would necessarily add an extra dimension to the 3-dimensional graph. Such a complete spacetime diagram, or frame of reference $\mathcal{F}$ (1.2.1) with coordinates $[x, y, z, t]$, would require four space dimensions to plot which, in our 3-dimensional universe, we cannot do.

But if the trail of a uniformly moving particle in 3-space can be represented as having only one space dimension (and one time dimension), i.e., we need only two fixed coordinates, as in (1.3.2), then we can represent all the information of this object's motion in a 2-dimensional spacetime graph.

(1.4) Galilean Relativity

(1.4.1) It was Galileo who first formulated a principle of relativity where speeds are measured relative to a frame of reference (1.2.1). This principle can be observed, for example, when a passenger on an airplane (*frame of reference*) traveling at 600 mph relative to the ground (*another frame of reference*) throws a ball vertically in the air. The ball travels on a vertical up-and-down path *relative to the passenger's frame of reference*. The

ball does not, as intuition and the illustration might suggest, snap toward the back of the airplane at 600 mph.

Galileo considered two ships (frames of reference (1.2.1)), each of which has an experimenter below deck in a cabin without windows. Each ship is moving with constant speed, which could be zero, relative to land. Galileo then asserted that there is no experiment that the experimenters can devise that would reveal whether or not their respective ships are in motion. Moreover, all their experiments within the ship (e.g., balls rolling down tilted planes, collisions of billiard balls) would produce identical results.

(1.4.2) Using the model of two ships, each of which is a separate frame of reference, Galileo stated his relativity principle as follows:

> Shut yourself up with some friend in the main cabin below decks on some large ship, and have with you some flies, butterflies, and some other small flying animals...hang up a bottle that empties drop by drop into a wide vessel beneath it...have the ship proceed with any speed you like, so long as the motion is uniform and not fluctuating this way and that...The droplets will fall...into the vessel beneath without dropping toward the stern, although, while the drops are in the air the ship runs many spans...the butterflies and the flies will continue their flights indifferently toward every side, nor will it ever happen that they are concentrated toward the stern, as if tired out from keeping up with the course of the ship.

The point of this quote is that passengers on the ship can not determine whether the ship is moving (with constant speed) relative to the shore unless they view the shore from their cabin. Within the cabin frame of reference, the ship speed is zero, while in a frame anchored to the shore, the ship may have a non-zero speed. Speed or velocity is therefore relative to the frame from which it is measured.

This quote can be found in a 1953 translation by Stillman Drake [9] and in the more recent 2004 book by Roger Penrose [27], *The Road to Reality*.

(1.5) Special Relativity: A First Pass

Briefly put, SPECIAL RELATIVITY is the study of physical frames of reference (1.2.1) — coordinate systems with clocks at each point — with the additional condition that the frames move at constant speed relative to each other. Inertial frames experience no forces or acceleration and objects or observers within each frame are fixed at their respective positions. The PRINCIPLE OF RELATIVITY asserts that all laws of physics, including those of electromagnetism and quantum mechanics, are equally valid in each inertial frame. In sum,

(1.5.1)

> DEFINITION: *An* INERTIAL FRAME OF REFERENCE *is a frame of reference* [(1.2.1), pg.13] *which is non-accelerating.*

In an *inertial* frame, whose objects experience no forces, a freely-moving ball on a horizontal friction-free table would remain stationary.

The theory of special relativity is based on the following assumptions:

(1.5.2) TWO BASIC PREMISES (POSTULATES)_____

> **(1.5.2a)**   The laws of physics for mechanics (for objects), electromagnetism, and quantum theory, are the same and equally valid for all inertial frames of reference (points of view) (1.2.1).
>
> **(1.5.2b)**   The speed of light in a vacuum, denoted c, is the same for all observers, regardless of the speed of its source, or the speed of the observers.

(1.5.3) **Assumptions, postulates, axioms, or premises,** like those of (1.5.2), are, by definition, fundamental properties, observations, or "starting points" that are given without proof. Scientists and mathematicians hold to the importance of clearly identifying these basic and unproven assumptions. This is so that we can better understand their consequences — theorems and properties that necessarily follow. Chapter **??** on the scientific method and chapter **??** on mathematical logic expand on this principle.

. . . .

(1.5.4) **Assumption (1.5.2b) is, admittedly, anti-intuitive.** Although you can run after a *material* object, like a train or a ball in flight, to make its speed appear slower (relative to you), you can not do this with *electromagnetic* light "particles" or photons.

If you release photons from a light source, then you will always measure their speed to be **186,282 miles per second**, or **2.99792458 × 10^8 meters per second.**[1] regardless of your motion or the motion of the light source.

(1.5.5) **F**IGURE: *Chasing a Photon from a Light Beam*

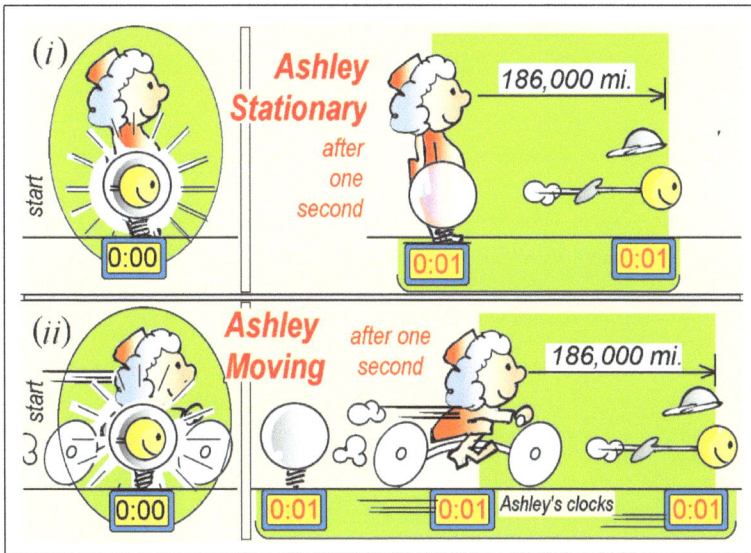

At time ⬚ 0:00 *seconds, Ashley is synchronized with the emitted photon at the light source. One second later, Ashley's clock reads* 0:01 *. Ashley will <u>always</u> find herself separated from the emitted photon by a distance of*

---

[1]We approximate these values with $c \approx 186,000$ miles/hr $\approx 3 \times 10^8$ meters/sec.

186,000 *miles, whether she is*
    (*i*)   *stationary relative to the light source or*
    (*ii*) *moving relative to the light source.*

....

(1.5.6) **The Importance of Maxwell's Work.**   In listing the two postulates of special relativity (1.5.2), one might think they are independent of each other. In fact, if the laws of electromagnetism, i.e.,
Maxwell's equations, are valid in all inertial frames, as is implied by
(1.5.2a), then we must accept the consequences of these equations,
one of which is (1.5.2b), namely, c, the speed of all electromagnetic
waves in a vacuum (*and the speed of light in particular*) is the same
in all frames from which it is measured. (*See Section* (22.4), *pg.* 303,
to see how Maxwell calculated c *using just two laboratory measurements.*)

(1.6) A Symmetry Principle

If frame $\mathcal{F}^A$ measures the relative speed of frame $\mathcal{F}^B$ to be rightward,
at, say, 30 miles per hour, then is it possible for frame $\mathcal{F}^B$ to measure
the relative speed of frame $\mathcal{F}^A$ to be leftward at 20 miles per hour?

Such a lack of symmetry is not possible if all inertial frames are equal
in the sense that no frame is any more special or unique than any
other frame. Thus, if $\mathcal{F}^A$ sees frame $\mathcal{F}^B$ moving rightward at 30
miles per hour, then frame $\mathcal{F}^B$ must see frame $\mathcal{F}^B$ moving leftward
at 30 miles per hour, or at -30 miles per hour.

We state this symmetry principle more formally:.

(1.6.1)

THE SYMMETRY PRINCIPLE FOR RELATIVE SPEEDS. *Inertial frames
$\mathcal{F}^A$ and $\mathcal{F}^B$ are moving parallel to each other with constant
speed. If observers in frame of reference $\mathcal{F}^A$ measure the speed
of frame $\mathcal{F}^B$ to be v in the positive direction, then conversely,
observers in $\mathcal{F}^B$ will measure the speed of frame $\mathcal{F}^A$ to be $-v$,
i.e., observers in $\mathcal{F}^B$ measure the speed of $\mathcal{F}^A$ to have the same
magnitude but the negative (opposite) direction.*

(1.7) Lorentzian Relativity

(1.7.1) NOTE: *At this point, our exposition is purely descriptive — we do not explain <u>how</u> these results arise. Proof of why moving clocks run slow, and why moving rods shrink, will come later in Theorem (2.3.3).*

(1.7.2) Lorentzian Relativity _____

The following results, (1.7.2a) and (1.7.2b), are inevitable consequences of the two assumptions noted in (1.5.2), along with the symmetry principle for speeds (1.6.1). Suppose a clock and an object are fixed in frames that are observed to be moving relative to an observer in another frame. Then

(1.7.2a) The moving clock is always *observed* to run slower than stationary clocks. (*This effect is too small to be observed until the clock reaches speeds close to the speed of light.*) Within its own frame, a stationary observer does not see an identical clock running any slower.

(1.7.2b) The length of the moving object physically contracts in the direction of its motion as measured by the stationary observer. Within its own frame, a stationary observer does not measure an identical rod to be any shorter.

A further consequence of these two results concerns the unique rôle of the speed of light in a vacuum, namely,

(1.7.2c) The speed of light in a vacuum is an upper limit for the speed of all matter and all signals in the Universe — no material object or wave can achieve a speed greater than the speed of light in a vacuum.

(1.8) The Ubiquitous Shrinkage Constant

(1.8.1) In (1.7.2a) and (1.7.2b), the clock and the object are fixed in frames that are observed to move at a specific speed v. The observed "shrinkage" in length and time is quantified by the speed-dependent constant, σ_v which we define next in (1.8.2).

(1.8.2)

> **DEFINITION:** *The speed-dependent* **Lorentz contraction factor** *is defined to be*
>
> $$\sigma_v \overset{def}{=} \sqrt{1 - (v/c)^2} < 1 \quad \text{whenever speed } v < c.$$

(This value of σ_v will be justified in (7.1.3b), pg. 106.)

Specifically, suppose clocks are SYNCHRONIZED so that the clocks at the origin $x = 0$ in each frame read time $t = 0$ (*A more formal definition will be found at (3.4.5a), pg. 62*). Then the moving clock will always tick at a rate $\sigma_v \leq 1$ times the rate of the stationary clock. Similarly, if d_0 is the observed rod length in a stationary frame, then observers in the stationary frame observe the length of the moving rod to be $\sigma_v d_A = d_0$.

(1.8.3) || **NOTE:** ***One proof of the fact that moving rods shrink, and moving clocks slow down*** is found in Theorem (2.3.3), pg. 49. This theorem uses properties of Figure (2.3.2) which are based only on the Pythagorean Theorem.

The properties of length contraction and time dilation are summarized in the following table.

(1.8.4) **TABLE:** ***At-Rest and Speed-v Lengths and Times***

	Observed at rest	Observed at speed v
Clock time	(t_0) *hrs.*	$(\sigma_v \cdot t_0)$ *hrs.*
Rod length	(d) *ft.*	$(\sigma_v \cdot d)$ *ft.*
	$\sigma_v = \sqrt{1 - (v/c)^2}$	

|| **NOTE:** *Objects do not shrink when measured in their own frames and clocks do not slow down when time is measured in their own frames.*

(1.8.5) **The online animation**[2] of Lorentz frames shows that in relativistic physics, a ruler contracts to 80% of its at-rest length when it travels at relative speed $0.6\,c$ (*Note that $\sigma_v = 0.8$ when $v = 0.6c$ in* (1.8.2)). The online animation of Galilean frames shows that in classical Galilean physics, a ruler traveling at any speed enjoys no length contraction.

The following examples are products of the contraction formula (1.8.2).

(1.8.6) Example: *Any Degree of Contraction is Achievable.* From (1.8.2), we see that the number $\sigma_v = \sqrt{1 - (v/c)^2}$ can be as close to zero as we like as long as speed, v, is close enough to c, the speed of light in a vacuum. (*More formally, we write "$v \to c$ implies $\sigma_v \to 0$."*) From Table (1.8.4), any rod of length d can have an arbitrarily small observed length, $(\sigma_v \cdot d)$, if its speed, v, is close enough to c.

As a special case of (1.8.6), we have

(1.8.7) Example: *The Ever-Shrinking Tunnel.* A tunnel or corridor has an at-rest length of 100 miles. Therefore, from (1.7.2b) or (1.8.6), a traveler moving through the tunnel who sees the tunnel flying by at speed v can make that tunnel shrink in the direction of travel to any small non-zero length, namely, $(\sigma_v \times 100)$ miles. From the traveler's point of view, the moving tunnel's length will be multiplied (shrunk) by the Lorentz contraction factor, $0 < \sigma_v = \sqrt{1 - (v/c)^2} \leq 1$ which can be made as small as we like if v is taken close enough to c.

However, from the point of view of an outsider who is *not* moving relative to the tunnel, the tunnel's length remains at the fixed length of 100 miles.

When we compare similar observations and measurements of an observer *outside* a tunnel with those of a moving skateboarder *inside* the tunnel (*see Figure* (1.9.4c)), we generate the Accommodating Universe Paradox in Section (1.9).

---

[2]http://www.special-relativity-illustrated.com

(1.9) Paradox: The Accommodating Universe

(1.9.1) **Inertial frame** $\mathcal{F}^E$ — contains the "dumbell" configuration of
(*i*)   the Earth at the point $x = 0$,
(*ii*) Star XYZ at point $x = d_E$ and
(*iii*) a tunnel or corridor with at-rest length d_E in frame $\mathcal{F}^E$ that
connects Earth and Star XYZ.

Inertial frame $\mathcal{F}^S$ — whose X-axis is parallel to that of $\mathcal{F}^E$ and
contains a skateboard and a skateboarder fixed at point $x = 0$, the
origin of $\mathcal{F}^S$. The origin of frame $\mathcal{F}^S$ is fixed to the skateboard
and travels through the tunnel from Earth to Star XYZ at speed v
relative to frame $\mathcal{F}^E$.

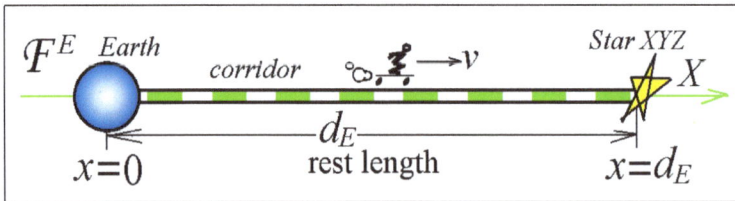

From (1.6.1), the skateboarder in frame $\mathcal{F}^S$, will see the corridor
traveling leftward at speed $-v$. From properties of the ubiquitous
shrinking factor (1.8.4), the skateboarder will measure the corridor to
have a length of only $\sigma_v d_E$. This means ***as speed v approaches the
speed of light, the length of the corridor becomes arbitrarily
small from the skateboarder's point of view.***[3]

What does this mean?

(1.9.2) **The Universe** is   <u>ACCOMMODATING</u>   in the sense that the faster
the skateboarder races toward the star, the shorter will be the length
$\sigma_v d_E$ of the corridor from the skateboarder's point of view in $\mathcal{F}^S$.

---

[3]The same argument implies that any two points in the entire universe become
arbitrarily close to each other from a traveler's point of view if the speed is close
enough to c.

view of stationary wall from Earth frame $\mathcal{F}^E$

view of moving wall from skateboard frame $\mathcal{F}^S$

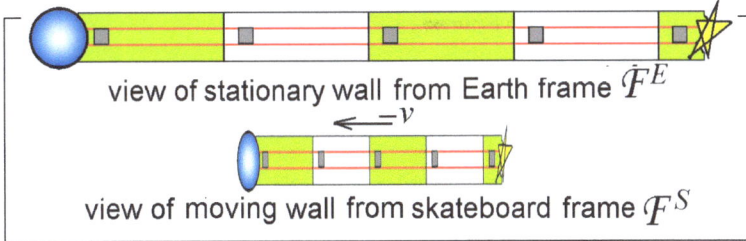

Hence, the skateboarder, traveling at speed v along a shortened corridor will reach Star XYZ in an arbitrarily short time.

To set the stage, we will need one more definition.

(1.9.3)

> **DEFINITION:** *A* LIGHT YEAR *is defined as the distance in a vacuum that light travels after a one-year journey. Similarly, a* LIGHT SECOND, LIGHT MINUTE, *etc. is the distance traveled by light in a second, or a minute, etc.*

Statement of the Paradox

Suppose the length of the corridor between Earth and Star XYZ is $d_E = 150$ light years (1.9.3) as seen from Earth. This means a photon requires 150 years to travel from Earth to Star XYZ. Yet, as mentioned in (1.9.2) above, for sufficiently high speeds, the skateboarder can travel through the corridor from Earth to Star XYZ in less than 150 years.

Question: How can the skateboarder travel faster than light?

(1.9.4) Analysis: In both the Earth's frame of reference $\mathcal{F}^E$ and the skateboarder's frame $\mathcal{F}^S$ (1.9.1), the time of travel of any object traveling at speed w over distance $dist$ is

(1.9.4a) $$time = \frac{dist}{w}.$$

Using (1.9.4a) in frame $\mathcal{F}^E$,

(1.9.4b)

• <u>Skateboarder time in $\mathcal{F}^E$</u> required to cover distance d_E at speed v is $\boxed{d_E/v}$ seconds.

• <u>Photon time in $\mathcal{F}^E$</u> required to cover distance d_E at speed c is $\boxed{d_E/c}$ seconds.

(1.9.4c)

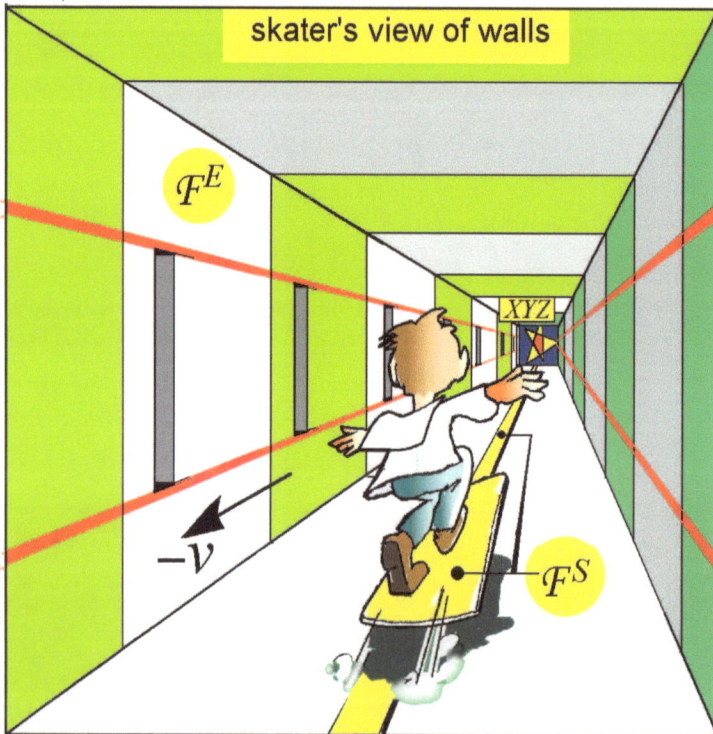

skater's view of walls

• <u>Skateboarder time in $\mathcal{F}^S$</u> required to cover distance $\sigma_v d_E$ at speed v is $\boxed{\sigma_v d_E/v}$ seconds.

• <u>Photon time in $\mathcal{F}^S$</u> required to cover distance $\sigma_v d_E$ at speed c is $\boxed{\sigma_v d_E/c}$ seconds.

(1.9.5) Skateboarder Travels Faster Than Light: *A Specious Argument.* Consider the ratio of the times required, for the skateboarder and for a photon, to cover the distance between Earth and Star XYZ.

(1.9.5a)
from (1.9.4c),   skateboarder time $= \sigma_v d_E/v$
from (1.9.4b),   photon time $= \quad d_E/c$

which implies $\dfrac{skateboarder\ time}{photon\ time} = \dfrac{\sigma_v d_E/v}{d_E/c} = \dfrac{\sigma_v}{(v/c)}$.

To fix ideas, set the skateboarder's time equal to 1% of the photon's time. Then calculation of the ratio $\frac{\sigma_v}{(v/c)}$ of (1.9.5a) yields the result

(1.9.5b) $\dfrac{skateboard\ time}{photon\ time} = \dfrac{\sigma_v}{(v/c)} = 0.01\ implies\ (v/c) \approx 0.99995.$

(1.9.6) **Something is Horribly Wrong.** We have just shown in (1.9.5b) that if the skateboarder travel time is 99% less than the photon travel time (*for the same Earth* $\rightarrow$ *Star XYZ journey*), then the skateboarder speed v is also *less than* the photon speed c. How can both statements be true at the same time?

> **Why Argument (1.9.5) is Specious:**
> *An exercise in critical thinking.*

(1.9.7) **CRITICAL THINKING:** *An example.* The error of (1.9.6) lies not in the calculations, but in the interpretation of their meaning. Recall that we used the following assumed "truths."

    (1.9.7a) Moving lengths observed to have speed v, shrink by a factor of σ_v (*from (1.7.2b)*).

    (1.9.7b) Neither matter nor information (electromagnetic waves) can move faster than the speed of light (*from (1.7.2c)*).

    (1.9.7c) Clocks in Earth frame $\mathcal{F}^E$ and the skateboarder's frame $\mathcal{F}^S$ run at the same rate. (*In (1.9.5a), we chose the clock of $\mathcal{F}^S$ to measure the skateboarder's journey time and the clock of $\mathcal{F}^E$ to measure the photon's journey time.*)

In the list of (1.9.7), we started by accepting that (1.9.7a) and (1.9.7b) are true — moving lengths do shrink and nothing travels faster than light. Given these assumptions, we then concluded that (1.9.7c) must be false — when we compare travel times of two objects (*skateboarder and photon*) we can not use clocks from two different frames. So, if magnitudes of lengths depend on the reference frame, so will the length of time intervals.

Chapter (9) offers a full analysis of (1.9.4b), (1.9.4c), and (1.9.5b).

(1.10) Paradox: Time and Distance Asymmetry

Moving Clocks Run Slow

The paradox of Time and Distance Asymmetry comes in two parts:

(1.10.1) **I. The Clocks on a Train.** In both panels of Figure (1.10.2), the car (frame $\mathcal{F}^B$) is traveling to the right at speed $0.66c$, or 0.66 times the speed of light, as seen from the platform (frame $\mathcal{F}^A$). Each panel shows two identical clocks, one in the frame of the moving train and the other in the frame of the stationary platform.

(1.10.2) **F**IGURE: *Motion Makes Clocks Run Slow*

- LEFT PANEL OF (1.10.2): *A snapshot at time $t = 0$ shows the clock of the moving car and a stationary platform clock simultaneously synchronized to read* $\boxed{0:00}$ *milliseconds.*

- RIGHT PANEL OF (1.10.2): *The same clock of the car in the moving frame, $\mathcal{F}^B$ reads only* $\boxed{0:30}$ *milliseconds while, at the same moment,* *another stationary platform clock in frame $\mathcal{F}^A$ registers* $\boxed{0:40}$ *milliseconds.*

The Ubiquitous Contraction Factor: From Definition (1.8.2), when $v = 0.66c$, the Lorentz contraction factor is $\sigma_v = 0.75$ (within two-place decimal accuracy). Thus, the moving car records $\boxed{0:30}$ = $\sigma_v \cdot \boxed{0:40} = 0.75 \cdot \boxed{0:40}$ at the moment that time $\boxed{0:40}$ is read by the clock on the stationary platform.

> **Moving Rulers Shrink in the Direction of Motion**

(1.10.3) A Quantitative View of (1.7.2b) *A Ruler on a Moving car.* In both panels of Figure (1.10.4), the car travels rightward at $0.66c$, or 0.66 times the speed of light. Each panel shows two identical rulers, one on the moving train and the other on the stationary platform.

(1.10.4) FIGURE: *Motion Shrinks Distance*

- LEFT PANEL OF (1.10.4): *The car, at rest, is 4 units wide as measured by identical rulers, one on the car and the other on the fixed platform.*

- RIGHT PANEL OF (1.10.4): *The car is now in motion with a speed of $0.66c$ as it passes a second point on the platform. This snapshot shows the platform ruler measures the moving car width as only 3 units.*

. . . .

The Ubiquitous Contraction Factor: From (1.8.2), the Lorentz contraction factor $\sigma_v = 0.75$ when $v = 0.66\,c$. Hence, the ruler on the moving car contracts in the direction of motion from 4.0 units, as measured in its stationary state, to $3.0 = \sigma_v \cdot 4.0 = 0.75 \cdot 4.0$ units as measured while the car is moving at constant speed $0.66\,c$.

Statement of the Paradox

It can be argued that the assertion that motion alone will cause clocks to slow (*Figure* (1.10.2)) and rulers to shrink (*Figure* (1.10.4)) is a paradox in itself.

But we wish to go further: We agree that all rulers and clocks are identical when observed at rest. Table (1.8.4) tells us that, in all cases, at observed speed v moving clocks run slow by a factor of σ_v and moving objects shrink (in the direction of motion) by the same factor σ_v. The (apparent) contradiction of asymmetry is this:

(1.10.4a)   *From the railway platform* point of view (*Figure* (1.10.2)), the moving clock on the railway car runs slower than the stationary platform clock by a factor of σ_v. Also, (*Figure* (1.10.4)), the moving ruler (rod) attached to the railway car is shorter than the identical ruler on the platform by a factor of σ_v.

(1.10.4b)   *From the railway car* point of view, the moving clock is the one on the platform. Hence, it is now the platform clock that runs slower than the railway car clock by a factor of σ_v. Similarly, the moving ruler is now the one on the platform. Hence, it is now the platform ruler that is shorter than the railway car ruler by the same factor σ_v.

. . . .

(1.10.5) **How can (1.10.4a) and (1.10.4b) both be true?** That is, how can each observer — one in the railway car and one on the platform — see each other's clock run slower than his own? And how can each observer see the other's (identical) ruler as smaller than his own?

Comments on How the Paradox is Resolved

Spoken language is the culprit with respect to this paradox. We use the verb "see" as in (1.10.5), when *an* observer in one frame measures the time in another frame — we imagine only one person doing the seeing.

However, in Figures (2.4.2) and (2.4.3), pg. 51, it will be shown that to measure lengths of moving rods and rates of moving clocks, a team of *two* observers in the stationary frame is always required. With this configuration of observers, the paradox of mutual shrinkage and mutual slowing of moving clocks is resolved.

(1.11) Paradox: The Traveling Twin

Statement of the Twin Paradox

(1.11.1) **A Qualitative View.** A pair of twins carry identical clocks. One twin travels into space at a speed close to that of light. After the return to Earth, the traveling twin is biologically many years younger than the Earthbound twin.

animated
online

Since observers in each frame see the moving clocks of the other frame running slow (*see* (1.7.2a) *and Figure* (1.10.2)), the paradox, or apparent contradiction, is this:

- If each twin sees the moving clock of the other twin running more slowly than his/her own clock, how can one of the twins, the traveling one, end up being younger than the Earthbound twin?

In this case, <u>travel into the future</u> — by the traveling twin — is possible.

(1.11.2) **A Quantitative View.** The following four panels, (1.11.2a)-(1.11.2d), illustrate the twin paradox from the Earthbound twin's point of view when the speed of the rocket ship is $0.8c$ (80% the speed of light). After 50 years, as measured by the earthbound clock, the traveling twin returns, having aged only 30 years biologically.

(1.11.2a) FIGURE: *The Take-Off:*

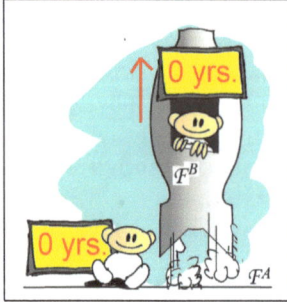

At the moment of takeoff, two identical clocks, the Earth clock in frame $\mathcal{F}^A$ and the spaceship clock in frame $\mathcal{F}^B$ are synchronized to read $t = 0$ years. The time unit of each clock is in "years."

(1.11.2b) FIGURE: *In Flight*

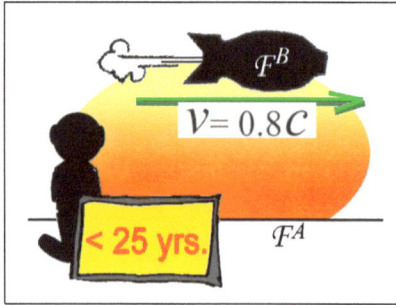

For the next twenty five years, as timed by the Earth clock in $\mathcal{F}^A$, the spaceship in frame $\mathcal{F}^B$ races away from Earth, traveling at a constant speed of $0.8\,c$ (*8/10 the speed of light*).

(1.11.2c) FIGURE: *Transfer to the Returning Spaceship*

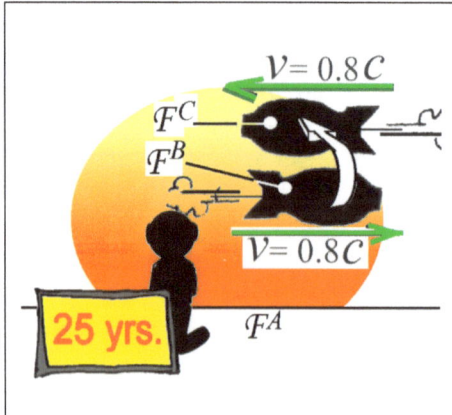

At the moment the Earth clock reads 25.0 years, the traveling twin instantly switches from frame $\mathcal{F}^B$, the outbound ship, to frame $\mathcal{F}^C$, an incoming spaceship going in the opposite direction. The incoming ship is traveling at a constant speed of $0.8\,c$ (*8/10 the speed of light*).

(1.11.2d) **F**IGURE: *Return to Earth: The Reunion*

According to the Earth clock in $\mathcal{F}^A$, the return journey of frame $\mathcal{F}^C$ containing the traveling twin, takes another 25 years. Therefore, the full round trip has taken 50 years. On the other hand, the spaceship clock, which is identical to the Earth clock, will have ticked off only 30 years!

Informal Resolution of the Paradox

(1.11.3) *The Earthbound Twin Point of View.* The Earthbound twin in frame $\mathcal{F}^A$ always sees the clock of frame $\mathcal{F}^B$ (the spaceship) running slow since it is moving relative to the Earth frame $\mathcal{F}^A$ with speed $0.8\,c$. Specifically, the Lorentz contraction factor, $\sigma_v = 0.6$ (1.8.2) guarantees that the 50 years on the stationary Earth clock correspond to a total of $\sigma_v \times 50 = 0.6 \times 50 = 30$ years on the moving spaceship clock — 15 years are registered on the spaceship clock in each direction. There is no paradox here.

(1.11.4) *The Spaceship Twin Point of View.* The spaceship twin in $\mathcal{F}^B$ also sees the clock on Earth in $\mathcal{F}^A$ running slow since it is moving at relative speed $0.8\,c$. The same Lorentz factor, $\sigma_v = 0.6$, guarantees that the 30 years of the spaceship clock (*see* (1.11.3)), correspond to $0.6 \times 30 = 18$ years on the (relatively moving) Earth clock — 9 years are registered on the slower Earth clock in each direction. (*See Table* (1.11.5).)

Now here is the surprising part. After the "instantaneous" change of frames (*from outgoing $\mathcal{F}^B$ to incoming $\mathcal{F}^C$*), the traveling twin in frame of reference, $\mathcal{F}^C$, now sees the Earth clock reading 41 years — a full 32 years later than the Earth clock reading of 9 years he had just seen from frame $\mathcal{F}^B$. (*See Table* (1.11.6).)

This "surprise" is due to the fact that during the moment of spaceship transfer, the traveling twin is no longer in an *inertial* frame since he suffers an extreme acceleration (change of speed) which affects the way clocks tick during that moment. See (11.5), pg. 147 for details.

Here is a tabulated summary of answers (1.11.3) and (1.11.4) of the twin paradox which is illustrated in Panels (1.11.2a)-(1.11.2d):

(1.11.5) TABLE: *Two Phases of the Spaceship Clock Journey as Seen from the Stationary Earth Clock:* $\sigma_v = 0.6$

	Leg	View from Stationary Earth
(1)	Outbound	*The stationary Earth clock registers 25 years. The moving spaceship clock runs slow and registers only $0.6 \times 25 = 15$ years.*
(2)	Homebound	*The stationary Earth clock registers 25 more years (for a total of $25 + 25 = 50$ years). The moving spaceship clock runs slow and registers $0.6 \times 25 = 15$ more years (for a total of $15 + 15 = 30$ years).*

(1.11.6) TABLE: *Three Phases of the Moving Earth Clock as Seen from the Stationary Spaceship:* σ_v =0.6

	Leg	View from Stationary Spaceship
(1)	Outbound	*The stationary spaceship clock registers 15 years, as was shown in (1.11.5). The moving Earth clock runs slow and registers only $0.6 \times 15 = 9$ years.*
(2)	Instantaneous turn-around	*After passing from outgoing frame $\mathcal{F}^B$ (where the Earth clock reads 9 years) to the inbound frame $\mathcal{F}^C$ (where the Earth clock is seen to read 41 years), the Earth clock is seen by the traveling twin, now in $\mathcal{F}^C$, to have gained 32 years.*
(3)	Homebound	*The stationary spaceship clock registers 15 more years (for a total of $15 + 15 = 30$ years). The moving Earth clock runs slow and registers $0.6 \times 15 = 9$ more years (for a total of $9 + 32 + 9 = 50$ years).*

But *why* does the spaceship twin see a time-jump on the moving Earthbound clock as described in (1.11.6)(2)? Once spacetime diagrams, known as Minkowski diagrams (6.1.1), pg. 97, are understood, the facts of this phenomenon are rendered accessible through Figure (11.3.1), pg. 143.

(1.12) Paradox: The Train in the Tunnel

Statement of the Paradox

(1.12.1) **Train-in-Tunnel Paradox:** *A Qualitative View.* When at rest, a train and a tunnel have the same length, d. This is illustrated in Figure (1.12.1a).

The following panels illustrate a pair of lightning strikes at each end of the tunnel and describe two *different* perceptions — one from Bernie who sits on the tunnel in frame, $\mathcal{F}^A$, and the other from the train's point of view in frame $\mathcal{F}^B$.

(1.12.1a) Figure: *Tunnel and Train (length d) at Rest*

(1.12.1b) Figure: *Lightning Avoids the Short Train:*

From Bernie's point of view in frame $\mathcal{F}^A$, the tunnel is stationary while the train in frame $\mathcal{F}^B$ is in motion. According to (1.10.4), pg. 27, the train shrinks in the direction of motion. (*Note how the wheels have become elliptical.*) Hence, the two lightning bolts that are simultaneous in Bernie's frame, $\mathcal{F}^A$, never hit the shorter train.

(1.12.1c) FIGURE: *Does Lightning Hit the Too-Long Train?*

Relative to the stationary frame $\mathcal{F}^B$ (the train), the frame, $\mathcal{F}^A$ (the tunnel), is observed to be moving. The moving tunnel $\mathcal{F}^A$ is seen from the train's frame $\mathcal{F}^B$ to shrink in the direction of motion (1.7.2b). If the (shrunken) tunnel is always shorter than the train, then, shouldn't the train be electrocuted by the "simultaneous" lightning strikes at the ends of the tunnel?

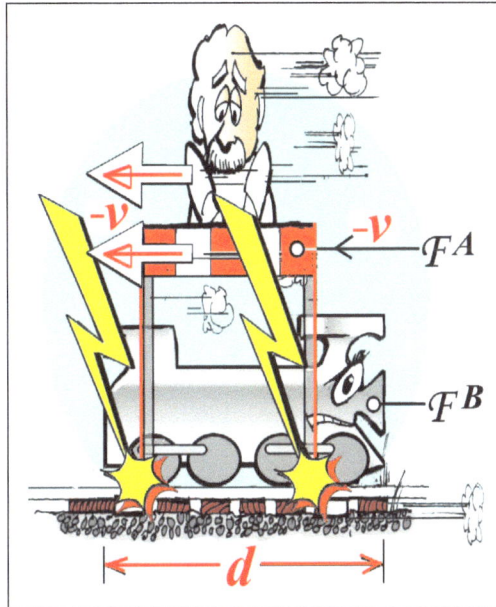

(1.12.2) **Summary:** From the viewpoint of the tunnel's frame, $\mathcal{F}^A$, the shortened train in frame $\mathcal{F}^B$ avoids the lightning (*see Figure* (1.12.1b)). Yet, in a change of viewpoint from which the train is stationary, we see the moving, shortened tunnel where "simultaneous" lightning strikes hit the too-long train (*see Figure* (1.12.1c)).

Question: How does the train escape electrocution in one frame but not in another?

Informal Resolution of the Paradox

The apparent inconsistency between the figures is explained by the fact that the double lightning strike in Figure (1.12.1b) is *simultaneous* whereas, in Figure (1.12.1c), the lightning strikes are not simultaneous. (*In* (A.1.1), *we first noted this loss of simultaneity.*) In Figure (1.12.1c), the right bolt of lightning strikes *ahead* of the "too-long train" and, a little later — only after the tail of the train has entered the tunnel — does the left bolt of lightning decide to strike. In both frames of reference, the train avoids electrocution.

How do we know this to be true? These sequences are read from the spacetime Minkowski diagrams in Section (C.2).

(1.13) Paradox: The Pea-Shooter

(1.13.1) **Relation to $E{=}mc^2$.** The relativistic addition of speeds, illustrated in this example, lies at the heart of the celebrated mass-energy equation, $E = mc^2$ (*A full exposition appears in Chapter* (15).)

A Brief Preview: In our everyday experience, the speed of an object — like a bullet — is equal to the speed v of the gun plus the speed w of the bullet relative to the gun. The total speed is the ordinary sum $v + w$ which is very accurate *provided* the speeds are much smaller than c.

However, to calculate accurately with very high speeds v and w, we need a more general relativistic "sum," $v \oplus w$, which will be defined in (1.13.3). Hence, if mass m has speed v with consequent momentum $[m]\,v$, then an addition of speed w produces a new momentum $[m](v \oplus w)$ instead of the classical momentum $[m](v + w)$. This implies that mass m must increase with increasing speed (*Section* (15.2)). Finally, the increase in mass is found to be due to (equivalent to) the energy required to produce the increase in speed (*Section* (15.3)).

Statement of the Paradox

(1.13.2) **The Pea-Shooter Paradox:** *A Quantitative View*

Bernie and Ashley are stationary relative to each other — they are both in the same frame of reference $\mathcal{F}^A$. From their common point of view (frame), the pea, in moving frame $\mathcal{F}^C$, is seen to have a leftward speed of $v_0 = 0.8c$ or 0.8 times the speed of light.

(1.13.2a) FIGURE: *A Shared Frame of Reference*

(1.13.2b) FIGURE: *From Bernie's Frame of Reference*

Bernie, in frame $\mathcal{F}^A$, sees Ashley's wagon in frame $\mathcal{F}^B$ speeding leftward at $0.9c$. Ashley sees the pea in frame $\mathcal{F}^C$ moving with speed $0.8c$, yet Bernie sees the same pea (propelled by the wagon) moving leftward with speed $0.988c$, a speed that is *less than* $1.7c = 0.8c$ *(pea speed seen by Ashley)* $+ 0.9c$ *(wagon speed seen by Bernie)*.

Informal Resolution of the Paradox

Consistent with our personal experience, speeds are additive. That is, when a train passes in front of us at 30 mph while a passenger walks toward the front of the train at 4 mph, then we expect to see the passenger moving at

(1.13.2c)   30 mph + 4 mph = 34 mph

relative to us.

To discuss the "true addition of speeds," we use $\oplus$, a different plus sign, to write the assertion that

(1.13.2d) $0.8c$ (mph) $\oplus$ $0.9c$ (mph) = $0.988c$ (mph)

which is *not* the value $0.8c + 0.9c = 1.7c$ we would expect by applying the usual arithmetic sum used in (1.13.2c).

(1.13.3)

> DEFINITION: **The true addition of speeds, $v \oplus w$.** *Given c is the speed of light in a vacuum. Then for speeds v and w, the sum, $v \oplus w$, is called the* LORENTZ SUM *of speeds v and w and is defined by*
>
> (1.13.3a) $v \oplus w \overset{def}{=} \dfrac{(v+w)}{1+(vw/c^2)}.$

The defining equation (1.13.3a) shows that for low speeds, such as $v = 30$ mph and $w = 4$ mph, the quantity (1.13.2c) is a close approximation to the true sum, $v \oplus w$ of (1.13.3). In fact, whenever we have speeds v and w whose product $v \cdot w$ is tiny relative to c^2, we have from (1.13.3)

(1.13.4) $v \oplus w = \dfrac{(v+w)}{1+\underbrace{(vw/c^2)}_{negligible}} \approx v + w.$

Justification for this unlikely "sum" of speeds (1.13.3a) will be developed in (7.3.1a).

(1.13.5) ‖ NOTE: *Letting $c = 1$. It is often more convenient to take $c = 1$ from the outset so that instead of dealing with speeds of the form v/c and w/c, we can work with "normalized" speeds, $-1 < v, w < 1$.*

(1.13.5a)   EXAMPLE: **One way to force c to equal one** is to introduce a new measure of length that equals the distance light travels in one second. Since $c = 2.997925 \times 10^8$ meters/second (or $186,282.4$ miles/hour), we can define

(1.13.5b)   *One* KABOOGLE $\overset{def}{=} 2.997925 \times 10^8$ meters.

Then $c = 1$ kaboogle/second.

(1.14) Paradox: The Bug and Rivet

(1.14.1) **The Bug-Rivet Paradox.** *A Qualitative View:* Two vertical parallel metal plates in inertial frame $\mathcal{F}^A$ have lefthand surfaces that are separated by a distance d. The plates have aligned holes large enough to allow a rivet shank to pass through, but not large enough for the rivet head.

(1.14.2) FIGURE: **Setup for the Bug-Rivet Paradox:**

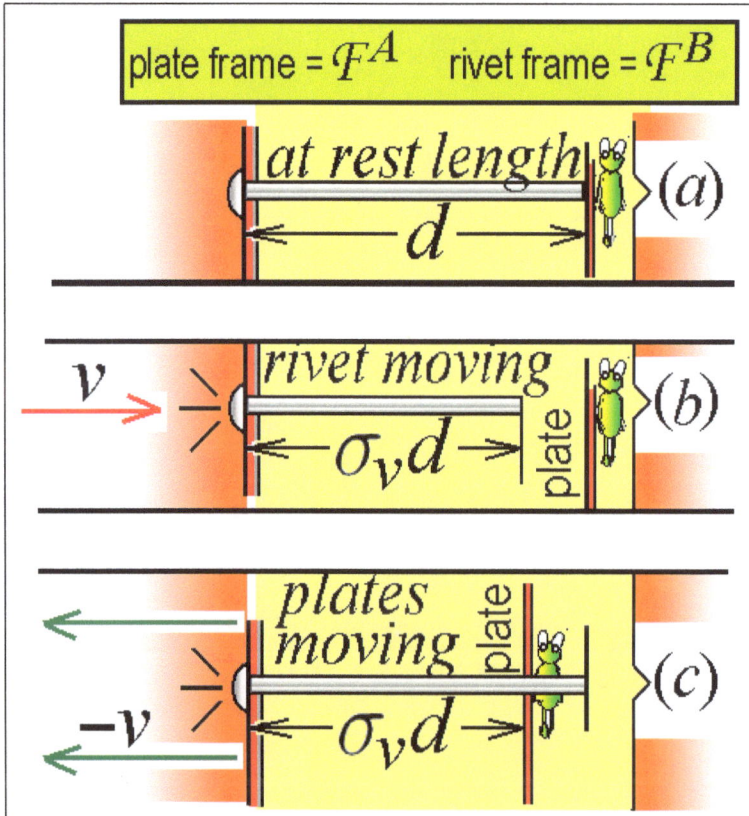

- **Fig. (1.14.2), Band (a):** The at-rest length of the rivet shank d equals the at-rest distance between the left surfaces of the parallel plates. We ignore all momentum issues and assume the mass of the rivet is miniscule compared to that of the plates.

- **Fig. (1.14.2), Band (*b*):** From the (stationary) plates' point of view in $\mathcal{F}^A$, the rivet's frame, frame $\mathcal{F}^B$, moves at constant speed v. The length of its moving shank, as measured from frame $\mathcal{F}^A$, is therefore less than d. Once the rivet head collides with the surface of the left metal plate, the rivet head stops abruptly but, as seen from frame $\mathcal{F}^A$, the shank length eventually increases to d (*see the X-T spacetime diagram in Figure* (1.14.2)).

- **Fig. (1.14.2), Band (*c*):** From the (stationary) rivet's point of view in $\mathcal{F}^B$, i.e., we are riding on the rivet, the plates in frame $\mathcal{F}^A$ are in leftward motion, traveling at constant speed $-v$. The distance between the two moving plates, as measured from the stationary rivet frame, $\mathcal{F}^B$, is therefore less than d. Once the (moving) lefthand plate collides with the (stationary) rivet head, the left plate stops.

> **The Paradox Question:** *Is the bug squashed when the rivet head is stopped by the left surface of the lefthand plate?*

If configuration (b) of Figure (1.14.2) prevails, then the answer is no — the rivet length is less than d at the moment of collision and the rivet point never reaches the bug.

If configuration (c) of Figure (1.14.2) prevails, then the answer is yes — the distance between the two plates is less than d, the length of the rivet shank. Hence, the rivet point will reach the bug that is hiding behind the righthand plate. Subsequently, the distance between the two plates increases to its at-rest value.

(1.14.3) **The resolution of the paradox** lies in the fact that the length of the rivet must expand <u>beyond</u> its at-rest length before it finally returns to its at-rest length. (*Proof and justification will be found in Chapter* (14) *which is illustrated by a property of the Slinky® toy.*) The situation is similar to a stretched spring relaxing and returning to its unstretched length.

Otherwise, if the rivet is assumed to be absolutely rigid, then the following pair of <u>false</u> statements must be valid:

(1.14.4)

Absolute rigidity implies:

• *The moment the head stops (as it collides with the plate), the tip also stops — simultaneously. This is equivalent to saying the information, "I've hit the plate" is transmitted to the end of the rivet at a speed greater than c. See (1.7.2c).*

• *The <u>future</u> affects the <u>present</u> in the sense that in one frame, the rivet tip "knows" before the event that the rivet head will collide with the plate. See (14.4.4)*

(1.15) Exercises

(1.15.1)
Exercise
Using the fact that c, the speed of light, is $186,000$ miles per second, show that the time parameters of Figure (1.10.2) require a platform length of 744 miles.

(1.15.2)
Exercise
The Skateboarder's Necessary Speed, *v*. At what speed, v, must the skateboarder of (1.9.4c) travel to reach StarXYZ in (a) one hour? (b) one day? (c) one year? (d) $1\frac{1}{2}$ years?

(1.15.3)
Exercise
Slow Speeds 1. Compute the value of $(vw)/c^2$ for the speeds $v = 30$ miles per hour and $w = 40$ miles per hour. (See Appendix **??** for the conversion of "miles per hour" to "miles per second.")

(1.15.4)
Exercise
Slow Speeds 2. With the values of v and w of Exercise (1.15.3), how close is the "intuitive" speed $v + w$ to the true speed of $v \oplus w$ given by (1.8.2)?

(2) Clocks and Rods in Motion

(2.1) The Perfect Clock

We start with an informal definition of a clock:

(2.1.1)

> DEFINITION: *A CLOCK is a device that ticks, pings, or presents some discernable signal or readout at regular intervals.*

(2.1.1a) **Starting the Clock.** The clock is started when an initial photon passes the clock and triggers the clock's first tick or ping as required in Definition (2.1.1). Although $\boxed{0.00}$ is a natural starting time, the initial starting time can be set arbitrarily.

(2.1.1b) **Setting the Tick Rate.** A second passing photon defines the length of the time interval that will be repeated for all subsequent ticks or pings — the clock "remembers" the time interval between the first and second photons and it waits for this interval to pass before it issues each subsequent tick.

(2.1.2) **Defining the One-Second Ticking Rate.** Figure (2.1.3) shows a clock with properties of (2.1.1a) and (2.1.1b) whose ticking rate is being set to one-second intervals.

(2.1.3) FIGURE: *Setting the One-Second Tick Rate of a Clock.*

animated online

- LEFT PANEL OF (2.1.3): *The photon passes over the clock at point* 0 *and simultaneously sets off the clock's first tick. The initial time may be set at* $\boxed{0.00}$ *seconds* (2.1.1a).

- MIDDLE PANEL OF (2.1.3): *At the 93,000 mile mark, the photon strikes the mirror and starts its return journey to the clock at the origin, 0.*

- THIRD PANEL OF (2.1.3): *On completion of the 186,000 mile round trip, the photon passes the clock, setting the second tick which the clock records as* $\boxed{1.00}$*, or 1-second. From this point on, the clock ticks at exactly the rate of these first two ticks. The time interval between ticks is permanently set (2.1.1b) and is defined (as follows) to be* one second.

(2.1.4)

> **DEFINITION:** ONE SECOND *is that time interval required for a photon of light to travel 186,000 miles in a vacuum. Any clock that remains stationary, and is constructed according to Figure (2.1.3), is said to read the* PROPER TIME *of its own frame of reference.*

(2.1.5) **How time can be identified with distance.** More accurately, the time of a *photon traveling in a vacuum* can be correlated to a precise *distance* once we rephrase Definition (2.1.4) as follows:

> *Whenever light (in a vacuum) travels a distance of 186,000 miles, then exactly one second of time will have passed.*

(2.1.6) **In our relativistic setting,** where c is constant for all observers, we constructed a clock (2.1.2), (2.1.3) whereby *time* is a secondary quantity derived from two primary quantities, *distance* and c, the *speed of light in a vacuum.* For example, in (2.1.5), *time* is defined as a certain *distance* covered by a *speeding* photon. By contrast, in the Newtonian, nonrelativistic setting, *speed = distance/time* is a secondary quantity derived from the two primary quantities, *distance* and *time.*

(2.1.7) **Why is this clock of (2.1.2), (2.1.3) "perfect?"** Assumption (1.5.2b) tells us that the speed of light does not change, even if its source and observer are moving relative to each other. Moreover, since all frames of reference share the same yardstick to mark distances along the direction of the photon's motion, the same 1-second interval of proper (2.1.4) time will be recorded within all inertial frames of reference.

As we shall see in Section (2.2) if we follow the construction described in Figure (2.1.3), then we can produce as many identical clocks as we like that will simultaneously read the same proper time within a single frame, $\mathcal{F}$.

(2.2) Synchronizing Clocks within a Single Frame

Informally, we may define two clocks in a single frame $\mathcal{F}$ (1.2.1) as being *synchronized* if they show the same time simultaneously.

Moreover,

(2.2.1)

> DEFINITION: *Any two clocks that are constructed so as to tick at exactly the same rate when observed at rest are called* IDENTICAL *clocks.*

(2.2.2) What We May Not Do: *Invalid Synchronizing Methods.*

- We cannot shout, mail, or wire the readings of one clock to another remote clock since any such message takes time to deliver — any time lag renders the reported time as stale and invalid.

- Also, it does no good to remove the distance problem by bringing all the clocks to one place in order to synchronize them because eventually, we have to move them back to their original positions. As we have noted (1.7.2a), motion itself will slow the clock down and so, will distort its readings.

(2.2.3) What We May Do: *A Valid Synchronizing Method.* We know from (2.1.5) and Exercise (2.4.1) that if a photon travels a total distance of d miles, then $t = d/186,000$ seconds will have passed.

Each clock is constructed so that it starts to tick (at one-second intervals, say) as soon as it is hit by a photon. Before a clock starts ticking, we set an initial time according to its distance from a common photon emitter at the *Origin Clock* positioned at $x = 0$ with a setting of $t = 0$. Once a single burst of simultaneously emitted photons is sent into space, the Origin Clock starts ticking and the other non-Origin or clocks, Cohort Clocks, will start ticking at the instant of collision with one of these photons.

. . . .

To fix ideas, Figure (2.2.4) shows how to synchronize three cohort clocks at respective positions $x = -186,000\,miles$, $x = 186,000\,miles$, and $x = 279,000\,miles$. The general definition will be given in (2.2.5).

(2.2.4)   FIGURE: *Synchronizing Three Cohort Clocks*
              with the Origin Clock

animated
online

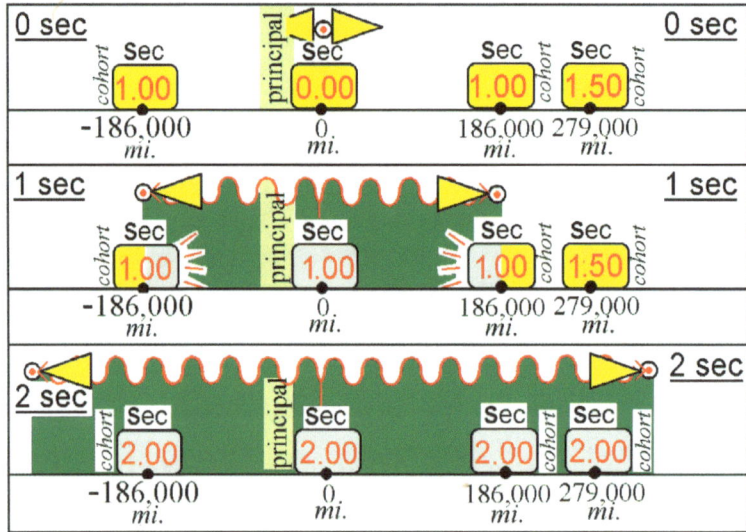

0 sec   *The Origin Clock will eject photons in the plus and minus X directions at exactly* 0.00 *seconds. As per (2.1.1a), the three cohort clocks will start to tick only when their start mechanisms collide with a photon.*

        A photon from the Origin Clock at $x = 0$ will take exactly one second to reach each of the two cohort clocks that are positioned at $x = -186,000\,miles$ and $x = 186,000\,miles$. A photon will require $1\frac{1}{2}$ seconds to reach the cohort clock situated at $x = 279,000\,miles$. As allowed by (2.1.1b), before they start ticking we may preset the cohort clocks with initial readings 1.00 *seconds and* 1.50 *seconds, respectively.*

1 sec   *It is one second later according to the Origin Clock. At this instant,*

photons collide with the start mechanism of the two cohort clocks where the initial readings had been set to $\boxed{1.00}$ *sec. The cohort clock at position* $x = 279,000\,mi$ *will have to wait for its collision photon 1/2 second from now, that is, when all (three) running clocks will be reading* $\boxed{1.50}$ *seconds.*

2 sec *All three cohort clocks have been started by their photon triggers. The three clocks are now synchronized with the Origin Clock and with each other.*

We generalize the algorithm of Figure (2.2.4) which is valid for three cohort clocks, and define the synchronization of N stationary clocks within a frame of reference.

(2.2.5)

> **DEFINITION:** *Given a functioning Origin Clock at* $x = 0$ *and* N *cohort clocks, at points* $x = d_1$, $x = d_2$, ..., $x = d_k$, ..., $x = d_N$. *Then if clock* k *is initially set at time* d_k/c *(2.1.1a) and is started by a photon emitted from the Origin Clock in accordance with (2.1.1b), then the* N *cohort clocks and the Origin Clock are said to be mutually* SYNCHRONIZED *within their shared frame of reference.*

(2.2.6) **The age of points.** Since every point in a frame has its own clock, and all the clocks are synchronized to simultaneously read the same time, it will be useful to think of a point in a frame as having the **same age** as every other point in that frame. In terms of Definition (2.1.4), the *proper time* at any point is equivalent to the *age* of that point.

(2.3) Moving Clocks Run Slow, Moving Rods Shrink

(2.3.1) **NOTE:** In the following thought experiment, Ashley's Clock in $\mathcal{F}^A$ and Bernie's clock in $\mathcal{F}^B$ are **synchronized** in the sense that their clocks simultaneously read $t = 0$ at the precise moment the photon starts its journey from the light source on the floor to the ceiling. This ensures that the respective clock readings, t_A and t_B, coincide with the <u>changes</u> in time, Δt_A and Δt_B, since the release of the photon.

. . . .

(2.3.2) Figure: ***The <u>Vertical</u> Photon Path in $\mathcal{F}^B$ is a <u>Diagonal</u> Path in $\mathcal{F}^A$***

Ashley's frame $\mathcal{F}^A$ contains a horizontal bar with the marks ① and ②. Bernie's frame $\mathcal{F}^B$ contains a light bulb that is affixed to the floor of a platform that can move left or right. D is the distance from the floor of $\mathcal{F}^B$ to the horizontal bar of $\mathcal{F}^A$.

- Left panel of (2.3.2): *In Bernie's frame $\mathcal{F}^B$, the light bulb is vertically beneath the moving mark ① at time $t = 0$, the precise moment it releases a photon that travels vertically toward the leftward-moving ceiling. At time t_B in $\mathcal{F}^B$, after traveling D units vertically, the photon hits the leftward-moving ceiling at the mark ②. Since distance = speed × time,*

(2.3.2a) $\boxed{D = c\,t_B = \textit{vertical distance from floor to ceiling}}$

   where c is the speed of light and t_B is the time (as measured in $\mathcal{F}^B$) required for the photon to travel D units vertically from floor to ceiling.

- Right panel of (2.3.2): *Ashley's frame $\mathcal{F}^A$ is fixed and Bernie's rolling platform (including the light bulb) has rightward speed v. Since distance = speed × time,*

(2.3.2b) $\boxed{d_A = v \cdot t_A = \textit{ the at-rest distance in } \mathcal{F}^A \textit{ between } ① \textit{ and } ②}$

   where t_A is the time in $\mathcal{F}^A$ for the light bulb to travel rightward at speed v from a point vertically below ① to a point vertically below ②.

The path of the photon from floor to ceiling (that was a vertical line in $\mathcal{F}^B$) is seen by Ashley to be a diagonal line in $\mathcal{F}^A$. Since distance = speed × time,

(2.3.2c) $\boxed{c\,t_A = diagonal\ distance}$

where c is the speed of the photon and t_A is the $\mathcal{F}^A$ time required for the photon to travel the diagonal path from floor to ceiling.

Based on the right triangle lengths of Figure (**??**), we show why moving clocks run slow and why moving rods shrink.

(2.3.3)

> **T**HEOREM: **Moving Clocks Run Slow, Moving Rods Shrink.** *Inertial frame $\mathcal{F}^B$ has speed v relative to frame $\mathcal{F}^A$. Then clocks in the moving frame $\mathcal{F}^B$ run at a rate that is $\sigma_v \leq 1$ (1.8.2) times the rate of clocks in $\mathcal{F}^A$. That is, if clocks of the two frames are synchronized (3.4.5a) at their origins (at time $t_A = t_B = 0$), then when the moving clock time t_B runs slower than all stationary clocks in $\mathcal{F}^A$. Specifically,*
>
> **(2.3.3a)** $t_B = \sigma_v\,t_A$
>
> *where the shrinking constant $\sigma_v = \sqrt{1 - v/c^2}$ is given by (1.8.2).*
> *If a rod in the moving frame $\mathcal{F}^B$ has at-rest length d_A as measured in the direction of motion, then it has length*
>
> **(2.3.3b)** $d_B = \sigma_v\,d_A$
>
> *as measured from the stationary frame $\mathcal{F}^A$.*

PROOF: *Moving Clocks Slow Down.* Combining paths of the right triangle, we see that the three lengths of Figure (2.3.2) listed in (2.3.2a), (2.3.2b), and (2.3.2c), correspond to sides of a right triangle. The Pythagorean Theorem implies

(2.3.3c) $(c\,t_A)^2 = (D)^2 + (v\,t_A)^2$

$$= (c\,t_B)^2 + (v\,t_A)^2 \qquad\qquad from\ (2.3.2a).$$

Solving for t_B, we obtain

(2.3.3d) $\boxed{t_B = \sqrt{1-(v/c)^2}\,t_A = \sigma_v\,t_A}$

which proves (2.3.3a).

Moving Rods Shrink: We refer to Figure (2.3.2) and change our viewpoint so that frame $\mathcal{F}^B$ is the stationary frame and frame $\mathcal{F}^A$ is moving leftward relative to $\mathcal{F}^B$. In this case, points ① and ② are moving leftward past the (fixed) light bulb in $\mathcal{F}^B$.

At time $t = 0$ in $\mathcal{F}^B$, point ① is directly overhead the stationary light bulb. At time t_B in $\mathcal{F}^B$, the leftward-traveling point ② passes vertically over the light bulb. Since distance = speed × time,

(2.3.3e) $\begin{aligned} d_B &= v \cdot t_B & d_B \text{ and } t_B \text{ are measured in frame } \mathcal{F}^B \\ &= v \cdot \sigma_v t_A & \text{from (2.3.3d)} \\ &= \sigma_v \cdot d_A & \text{from (2.3.2b)} \end{aligned}$

which proves (2.3.3b). The proof is done. ∎

(2.4) Exercises

(2.4.1) Exercise Show that any time interval t' can be defined in the manner of (2.1.2) and Figure (2.1.3). Apply your construction to the definition of $t' = 1/4$ second, and $t' = 1/1,000$ second.

Hint: *Calibrate the clock by setting the reflecting mirror d' miles from the point 0 instead of the $93,000$ miles that was used in Figure (2.1.3). The distance required for the round trip of the photon is now $2d'$ miles, so the time for the round trip, the time between the very first pair of clicks, will be $t' = 2d'/c$ which, depending on the value of d', can be any time interval desired. The units of speed c and distance d' determine the units of time t', i.e., seconds, minutes, hours, etc. (See Appendix ??.)*

(2.4.2)
Exercise

Speed of a Moving Point, and the Rate of a Moving Clock:
The Two-Observer Method

Step 1: Frame $\mathcal{F}^B$ is moving rightward (*in the positive direction*) with speed v relative to the platform frame $\mathcal{F}^A$. A single mark, **x** and a clock are placed at the right hand end of the train. At time t'_0 in $\mathcal{F}^A$, the x mark on the train in $\mathcal{F}^B$ passes in front of Ashley at point x_0 on the platform (*left panel*).

Step 2: At a later time, $t_1 > t_0$ in $\mathcal{F}^A$, the mark x passes a cohort observer standing at point x_1 on the platform in $\mathcal{F}^A$ (*right panel*).

Questions: How can we use x and t coordinates in the two panels of (2.4.2) to measure

(a) **the speed of the moving point** x in $\mathcal{F}^B$ as measured from the platform frame $\mathcal{F}^A$? (*Ans.* $v = \Delta x/\Delta t = (x_1 - x_0)/(t_1 - t_0)$)

(b) **the ratio $\Delta t_B/\Delta t_A$ of the time changes** Δt_B and Δt_A of the respective moving frames $\mathcal{F}^B$ and $\mathcal{F}^A$? (*Ans.* $\Delta t_B/\Delta t_A = (t'_1 - t'_0)/(t_1 - t_0)$).

(2.4.3)   **Measuring Moving Lengths:** *The Two-Observer Method* In
Exercise   the following exercise, we place two marks on a train or rod — an **x**
             on the front endpoint and an **o** on the back endpoint.

Ashley, and infinitely many cohort observers are standing with synchronized clocks at each point x on the platform. At time t_0, there
are only two unique observers, Ashley and a cohort (*shown in the diagram*), each of whom simultaneously sees an endpoint of the train, **o**
and **x**, at their respective platform positions, x_0 and x_1.

Question: **(c)** How can we use the x and t coordinates of (2.4.3)
         to calculate the length of the moving train in $\mathcal{F}^B$ as measured in
         stationary frame $\mathcal{F}^A$. (*Ans.* Moving train length $= (x_1 - x_0)$.)

· ·

(2.4.4)
<u>Exercise</u>

Graphing the Measurements for Finding Speed, Clock Rates, and Moving Lengths: *The Two-Observer Method*

Complete the figure to the right by locating the quantities x_0, x_1, t_0, t_1, t_0', t_1' that appear in the figures and the Question sections of (2.4.2) (a), (b), and (2.4.3) (c).

Each single point of the grey area of the figure is represented by a unique coordinate pair $[x, t]$ in $\mathcal{F}^A$. The train travels rightward with speed v.

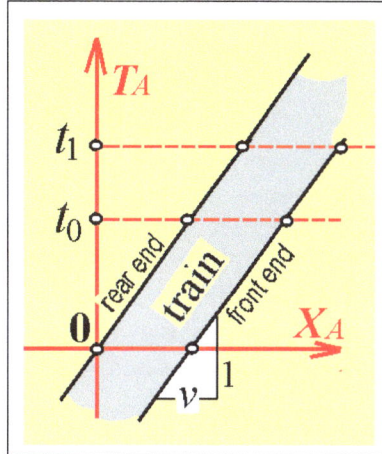

(3) The Algebra of Frames

(3.1) Inertial Frames of Reference

We shall see how a physical idea — *the inertial frame of reference* — is modeled as a mathematical concept — a set of *ordered* $[x, t]$ *pairs* or *vectors*. The following is a repeat of Definitions (1.2.1), pg. 13 and (1.5.1), pg. 16.

> **DEFINITION: (1.2.1), (1.5.1)** *A* FRAME OF REFERENCE, $\mathcal{F}$, *is a space or environment in which each point x can be located by a coordinate system and the time t_x at point x is determined by a clock at that point. An* INERTIAL FRAME OF REFERENCE *is a frame of reference which is non-accelerating, that is, free bodies move at constant speed in a straight line or are at rest.*

For example, in three-space each point is described by its space co-ordinates $[x, y, z]$ and its time coordinate t. The quartet $[x, y, z, t]$ therefore describes the state of each point in the frame of reference $\mathcal{F}$.

(3.1.1) ‖ **NOTE:** Following the simplification of Section (1.3), we choose our X-axis so as to replace the three space coordinates with just the x coordinate.

(3.1.2)   **FIGURE:** *The Ruler-Clock Structure of a Frame*

animated online

A point x on the ruler at time t on the clock is represented by the ordered pair $[x, t]$. In this figure, clocks are shown to be synchronized so at all points, all clocks simultaneously read the same time, namely, 3.41. In this case, all frame coordinates take the form $[x, 3.41]$.

(3.2) Vector Space Structure of Frames

(3.2.1) As seen in Figure (3.1.2), each point x on the ruler of the frame of reference corresponds to an ordered pair $[x, t]$ where x is the distance mark for x on the ruler and t is the time at the point x.

Do these ordered pairs have the mathematical structure of a vector space as required by Definition (A.2.1)? The following proposition says yes.

(3.2.2)

> **PROPOSITION:** *The set of all ordered pairs $\{[x, t]\}$ in frame of reference $\mathcal{F}$ forms a vector space as required by (A.2.1). Specifically, for all $[x_1, t_1], [x_2, t_2], [x, t]$ in $\mathcal{F}$*
>
> **(3.2.2a)** $\quad [x_1, t_1] + [x_2, t_2] = [x_1 + x_2, t_1 + t_2]$.
>
> **(3.2.2b)** $\quad$ *For all scalars α, $\alpha[x, t] = [\alpha x, \alpha t]$.*

PROOF: (PLAUSIBILITY ARGUMENT.)

Justifying (3.2.2a): Suppose a particle or observer from an arbitrary point on the ruler moves

(3.2.2c) $\quad$ rightward $\quad$ 8 feet, in $\quad$ 4 seconds, *then moves*
$\qquad\qquad$ leftward $\quad$ 6 feet, in $\quad$ 3 seconds.

In natural language: The particle covered a net total of $(8 - 6)$ ft. $=$ 2 ft. to the right in $(4 + 3)$ seconds $= 7$ seconds. In other words, a net total of 2 ft. were covered in a total of 7 seconds. To pass from natural to mathematical language, rewrite (3.2.2c) as

(3.2.2d)

$$
\begin{array}{r}
\text{ft. \ sec.} \\
[+8, +4] \\
+ \quad [-6, +3] \\
\hline
= \quad [+2, +7]
\end{array}
$$

where $+$ in (3.2.2d) corresponds to the phrase *"then moves"* in (3.2.2c).

The generalization of (3.2.2d) is that for all distances x_1, x_2, and all times, t_1, t_2,

(3.2.2e)　　$[x_1, t_1] + [x_2, t_2] \overset{def}{=} [x_1 + x_2, t_1 + t_2]$

which justifies (3.2.2a) since this agrees with the definition of vector addition (A.2.4b).

Justifying Definition (3.2.2b): From (3.2.2e), we have

(3.2.2f)　　$\underbrace{[x, t] + [x, t]}_{2[x,t]} \;=\; [2x, 2t],$

which gives us $a[x, t] = [ax, at]$ when $a = 2$. Using continuity, (3.2.2f) can be extended to (A.2.4c), scalar multiplication for *all* numbers $a \neq 2$. (*See Exercise* (3.5.1).)

This ends the proof. ■

(3.3)　Several Parallel Moving Frames

(3.3.1)　**The Train-as-Frame Metaphor:** We expand on the analogy drawn in (1.2.1), pg. 13 which presents a train as a model of an inertial frame. We now consider the case of several inertial frames on parallel tracks, and the way communication proceeds between pairs of frames. In fact, an observer in one train (*inertial frame* $\mathcal{F}^A$) can communicate with another frame on a parallel track only via a line-of-sight corridor that is perpendicular to both trains. That is, a passenger in one train (*frame*) can only look perpendicularly out his/her own window into the window of a parallel train (*frame*).

　　　　···

(3.3.1a) *A Railroad Model for Three Frames*

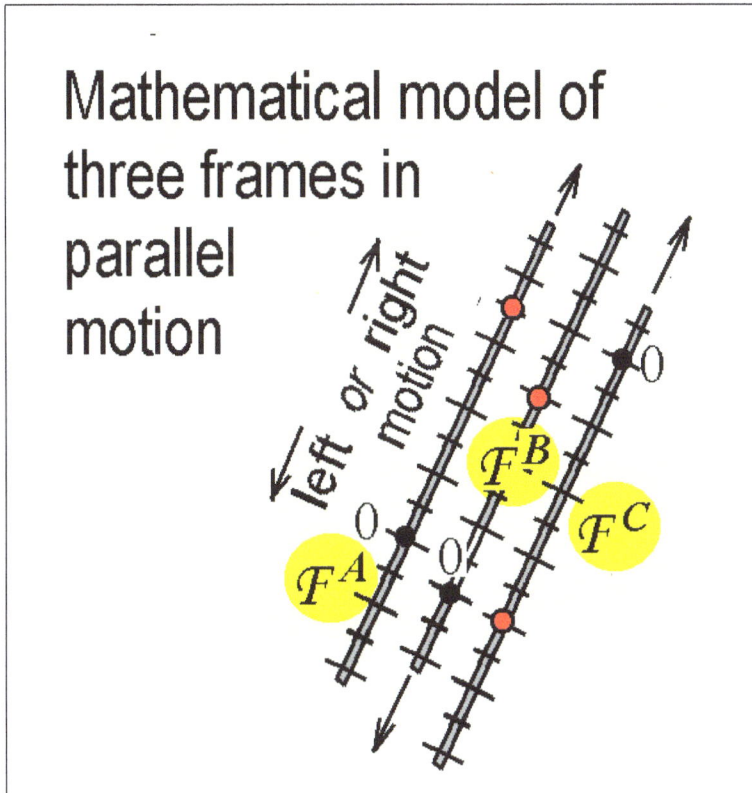

Inertial frames, $\mathcal{F}^A$, $\mathcal{F}^B$, and $\mathcal{F}^C$ are moving in left-right motion relative to each other on parallel paths. One can think of three infinite-length trains (frames) traveling on separate parallel tracks. Each moving train can be considered to be its own X-axis and the seated passengers inside the train are stationary observers, each with his/her own clock and position x.

58 *The Algebra of Frames*

(3.3.2) **The distance between the parallel tracks** is assumed to be negligible so that there is no time delay for information traveling between the moving frames (trains).

(3.4) Six Rules for Frames

There are six rules, (3.4.1)-(3.4.6), that describe the structure of frames and how communication proceeds between them.

(3.4.1) **All Frames are Physically Identical When at Rest.** All parallel frames have rulers that are identical when measured at rest and clocks that are identical in that they run at the same rate when measured at rest. As shown in Figure (3.1.2), each single frame of reference consists of a ruler (X-*axis*) with identical synchronized clocks placed at each point, x.

(3.4.2) **Like Being on Flypaper, Every Observer Remains Fixed Within His/Her Own Proper Frame (*or train or X-axis*).** Once an observer decides to move along an X-axis, like a passenger in a railroad car who decides to stroll down the center aisle, that observer is instantly assigned his/her *own* parallel X-axis, or HOME FRAME of reference, $\mathcal{F}$, relative to which he/she is, once again, stationary.

The newly-occupied <u>stationary</u> or <u>rest</u> frame is not necessarily unique. This is seen in Figure (3.3.1a) where each of the moving red dots acquires its own frame, $\mathcal{F}^A$, $\mathcal{F}^B$, or $\mathcal{F}^C$, inside of which the red dot is fixed. The entire frame (*with its fixed red dot*) is seen by an outsider to be moving. That is, observers (*red dots*) within their own proper "railroad cars" are always stationary.[1]

---

[1] We will restrict our attention to frames that have *constant* *speed* relative to each other.

(3.4.3) **Origin Observers and Clocks.**

> **DEFINITION:** *For each frame of reference, $\mathcal{F}^A$, $\mathcal{F}^B$, $\mathcal{F}^C$, ..., the one stationary observer fixed at the origin, $x = 0$, is called the* <u>ORIGIN OBSERVER</u> *of that frame. The clock at $x = 0$ is the* <u>ORIGIN CLOCK</u>. *All other observers, fixed at some $x \neq 0$, are called* <u>COHORTS</u> *of the origin observer in the frame.*

(3.4.4) **Line-of-Sight Functions Between Frames.**

In (3.3.1a), pg. 57, we represented frames as a set of trains or rulers moving on parallel tracks. There is a clock at each point of the train or ruler.

In frame (*ruler*) $\mathcal{F}^A$, Ashley is at position $x = 2$ and time $t = 3.41$, or at $[\,2,\,3.41\,]_A$ and in the parallel frame (ruler) $\mathcal{F}^B$, Bernie is at position $x = 0$ (*the origin*) and time $t = 3.00$, or at $[\,0,\,3.00\,]_B$.

As per (3.3.1), at the $\mathcal{F}^A$ end of the line-of-sight corridor (*the vertical red line in the diagram*), Ashley sees Bernie's coordinates $[\,0,\,3.00\,]_B$ while Bernie, at the opposite $\mathcal{F}^B$ end of the line-of-sight corridor, sees Ashley's coordinates $[\,2,\,3.41\,]_A$. The ***linking*** of their coordinates is mathematically expressed by a function

$$G : \mathcal{F}^A \to \mathcal{F}^B, \ (A.3.2),$$

where

$$G([\,2,\,3.41\,]_A) = [\,0,\,3.00\,]_B$$

or, in terms of the inverse function $G^{-1} : \mathcal{F}^B \to \mathcal{F}^A$,

$$G^{-1}([\,0,\,3.00\,]_B) = [\,2,\,3.41\,]_A.$$

The physical line-of-sight pairing between all coordinates of $\mathcal{F}^A$ and $\mathcal{F}^B$ can be expressed as input-output pairs of the function G as follows:

(3.4.4a)

> DEFINITION: *Frames (rulers with clocks) $\mathcal{F}^A$ and $\mathcal{F}^B$ are travel-ing parallel to each other. If coordinates $[x, t]_A$ of frame $\mathcal{F}^A$ and $[x', t']_B$ of frame $\mathcal{F}^B$ lie at opposite ends of an imaginary line that is perpendicular to both frames (rulers), then this per-pendicular line is called a* LINE-OF-SIGHT CORRIDOR *between the frames. The linked coordinates at the corridor end-points define the* LINE-OF-SIGHT FUNCTION
>
> $$G : [x, t]_A \to [x', t']_B \quad or \quad G([x, t]_A) = [x', t']_B$$
>
> *(See Definition (A.3.2) and Figure (3.4.4).)*

In sum, Figure (3.4.4) shows Ashley in frame $\mathcal{F}^A$ and Bernie in (parallel) frame $\mathcal{F}^B$ viewing each other's coordinates through functions $G : \mathcal{F}^A \to \mathcal{F}^B$ and $G^{-1} : \mathcal{F}^B \to \mathcal{F}^A$.
....

NOTE: **The Structure of G.** In Chapters (5) and (7), we shall see that line-of-sight functions (3.4.4a) will necessarily be *linear (repre-sentable as 2×2 matrices)* as defined by Definition (A.4.1). Func-tional correspondence, (3.4.4) must be constrained by physical re-quirements such as *synchronizeation* between pairs of frames, which is to say, that at time $t = 0$ in each frame, origin observers (at $x = 0$) are at opposite ends of a perpendicular line-of-sight corridor. Syn-chronization is formally defined in (3.4.5) following.

(3.4.5) All Pairs of Frames are Synchronized at $[0, 0]$.

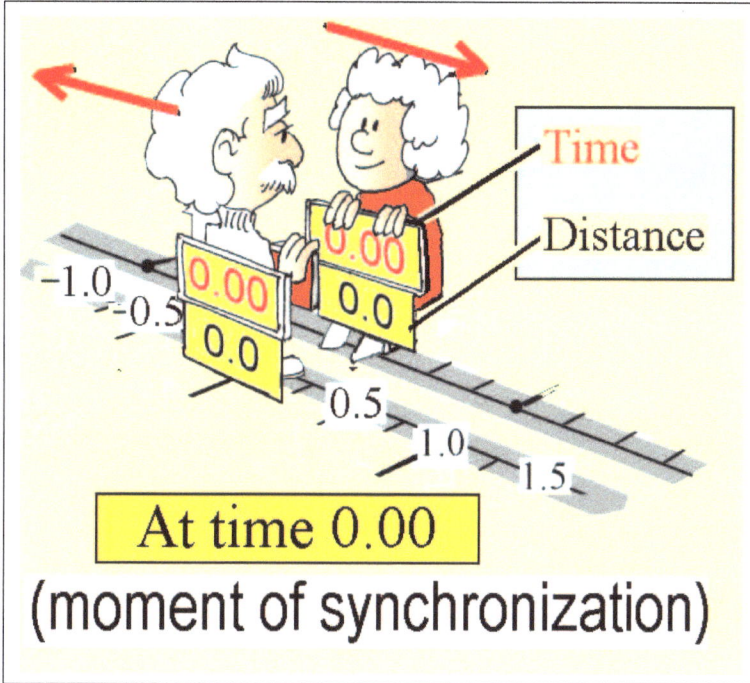

Synchronization between a pair of frames, $\mathcal{F}^A$ and $\mathcal{F}^B$, means that at $x = 0$ and time $t = 0$ in frame $\mathcal{F}^A$, the origin observer (3.4.3) looks through the line-of sight corridor at $[0, 0]_A$ in $\mathcal{F}^A$ and sees the coordinates $[0, 0]_B$ of the origin observer in $\mathcal{F}^B$ and vice versa.

We formally define synchronization in terms of the line-of-sight function $G: \mathcal{F}^A \to \mathcal{F}^B$ (3.4.4a):

(3.4.5a)

> **DEFINITION:** *Origin observers (in respective frames* $\mathcal{F}^A$ *and* $\mathcal{F}^B$*) are mutually* SYNCHRONIZED AT $[0,0]$ *if the line-of-sight function,* $G{:}\mathcal{F}^A \to \overline{\mathcal{F}^B}$*, has the property that*
>
> $$G(\vec{0}_A) \;=\; \vec{0}_B,$$
>
> *that is,*
>
> $$G([0,0]_A) \;=\; [0,0]_B,$$
>
> *where* $\vec{0}_A = [0,0]_A$ *and* $\vec{0}_B = [0,0]_B$ *are the respective origins* $x = 0$ *at time* $t = 0$ *of frames of reference,* $\mathcal{F}^A$ *and* $\mathcal{F}^B$*.*

(3.4.5b) **QUESTION:** *Are there preferred points or observers?*
Although we claim there is no preferred frame, there always is a
special, or preferred point within each frame — its origin $x = 0$.
In fact, as noted in (3.4.5a), all (*parallel*) frames must have their
respective origins $x = 0$ aligned along a perpendicular line-of-sight
corridor at local time $t = 0$. In this sense, the origin and the origin
observers (3.4.3) within each frame are special or "preferred."

(3.4.6) Symmetry Between Frames.

The following notation G_v for the line-of-sight function enhances that
of (3.4.4a) where v is the relative speed between frames.

(3.4.6a) **NOTATION:** *Suppose frame* $\mathcal{F}^B$ *has speed* v *as seen from
frame* $\mathcal{F}^A$*, and* $[x,t]_A$ *in frame* $\mathcal{F}^A$ *and* $[x',t']_B$ *in frame* $\mathcal{F}^B$ *are
physically at opposite ends of a line-of-sight corridor. We then write*

(3.4.6b) $G_v : [x,t]_A \to [x',t']_B,$

where symbol G_v *is an enhanced form of the symbol* G *(3.4.4a).*

Postulate (3.4.6c) following uses the terminology of (3.4.6b) to recast
the symmetry postulate (1.6.1) in terms of line-of-sight functions as
illustrated in Figures (3.4.4) and (3.4.5).(*A postulate* or *axiom* *is an
assertion that is accepted without proof* (1.5.3)).

(**3.4.6c**) Postulate: (*Symmetry Principle and Lines of Sight.*) *When frame $\mathcal{F}^B$ is moving with constant speed, v, relative to frame $\mathcal{F}^A$, then frame $\mathcal{F}^A$ will be observed to have speed $-v$ relative to frame $\mathcal{F}^B$. Symbolically,*

(**3.4.6d**) $G_v : \mathcal{F}^A \to \mathcal{F}^B$ *if and only if* $G_{-v} : \mathcal{F}^B \to \mathcal{F}^A.$

(**3.4.6e**)

> Proposition: **Analytic Statement of Symmetry Principle.** *The line-of-sight functions G_v and G_{-v} are inverses of each other, which is to say*
>
> (**3.4.6f**) $G_v^{-1} = G_{-v}.$

Proof: In terms of individual coordinates, we rewrite (3.4.6d) as

$[x,t]_A \xrightarrow{G_v} [x',t']_B$ *and* $[x',t']_B \xrightarrow{G_{-v}} [x,t]_A$

if and only if $[x,t]_A \xrightarrow{G_v} [x',t']_B \xrightarrow{G_{-v}} [x,t]_A$ *abbreviation of previous line*

if and only if $[x,t]_A \xrightarrow{G_{-v} \circ G_v} [x,t]_A$ *Definition (A.3.6) of composition, $\circ$*

if and only if $G_{-v} \circ G_v = I_2.$ *Definition (A.3.4)*
   Factors G_{-v} and G_v are inverses of each other,

which is to say

(**3.4.6g**) $G_{-v} = \mathrm{Inverse}(G_v)\ \ = G_v^{-1}$ *and*

(**3.4.6h**) $G_v = \mathrm{Inverse}(G_{-v}) = (G_{-v})^{-1}.$

It is Equation (3.4.6g) above that directly gives us (3.4.6f). (*Equation (3.4.6h) indirectly yields (3.4.6g). See Exercise (3.5.4).*) This proves the proposition. ∎

(3.5) Exercises

(3.5.1)   **Justifying Scalar-Vector Multiplication.** Using a continuity ar-
Exercise   gument, here is how vector addition alone (A.2.4b), pg. 336 implies
the scalar multiplication property (A.2.4c).

(3.5.1a)   Show that for all integers, n, $n[x,t] = [nx, nt]$.

(3.5.1b)   Show that for all integers n, $(1/n)[x,t] = [(1/n)x, (1/n)t]$.

(3.5.1c)   Show that for all rational numbers (fractions), $q = m/n$,
$q[x,t] = [qx, qt]$.

(3.5.1d)   Assuming the scalar product function

$$(*) \ \Phi(q) = [qx, qt]$$

is continuous in q, show that for all real numbers a we have

$$(**) \ a[x,t] = [ax, at].$$

(*Continuity means* $\lim_i \Phi = \Phi(\lim)$, *or, when* $\lim_i q_i = a$,
$\lim_i \Phi(q_i) = \Phi(\lim_i q_i) = \Phi(a)$.)

Hint: *Let* $\lim_i q_i = a$ *where* $\{q_i\}$ *is a sequence of* <u>*rational*</u> *numbers.*
Then from the continuity of Φ,

$$(***) \lim_i \Phi(q_i) = \Phi(\lim_i q_i) = \Phi(a).$$

From (*), *the rightmost term of* (***) *is* $\Phi(a) = [ax, at]$ *which
also equals its leftmost term, namely,*

$$\lim_i \Phi(q_i) = \lim_i [q_i x, q_i t] \qquad\qquad\qquad from \ (*)$$

$$= \lim_i q_i [x, t] = a[x, t] \qquad\qquad\qquad from \ (3.5.1c).$$

(3.5.2)
Exercise

The Frame for Slowing Clocks. Show how the example of Figure (1.10.2) can be expressed in terms of a spacetime diagram. That is, in each of the two panels, describe the placement of identical rulers in frame $\mathcal{F}^{car}$ of the car and in frame $\mathcal{F}^{platform}$ of the platform.

(3.5.3)
Exercise

The Frame for Shrinking Lengths. Show how the example of Figure (1.10.4) can be expressed in terms of a spacetime diagram. That is, in each of the two panels, describe the placement of identical rulers in frame $\mathcal{F}^{car}$ of the car and in frame $\mathcal{F}^{platform}$ of the platform.

(3.5.4)
Exercise

Inverse of G_v. Show that (3.4.6h) implies (3.4.6g).

Hint: *Take the multiplicative inverse of both sides of (3.4.6h) and use the fact that for any invertible function f we have $(f^{-1})^{-1} = f$.*

1

(4) The Graphing of Frames

(4.1) The Filmstrip Model of Spacetime

Our objective is to provide an intuitive way of graphing particles in motion.

In 3-dimensional space, we may represent a particle by the quartet $[x, y, z, t]$. When the particle moves, it will change its location with time, that is, the space coordinates become functions of the time coordinate t, that is, $x = x(t)$, $y = y(t)$, and $z = z(t)$. The set of these coordinates gives us a picture of the *motion* of that point as well as its position.

We shall concentrate on motion in 1-dimensional space only so that points are represented by coordinates of form $[x, t]$. (*We have seen in Section* (1.3) *that this "special case" is, in fact, sufficiently general for our purposes.*)

We already have a familiar model for representing objects that move in only one dimension — it is the filmstrip, which records snapshots of an object's position at regular time intervals.

We now expand upon this idea.

_____

(4.1.1) FIGURE: *Bernie's Motion from Ashley's Point of View:*

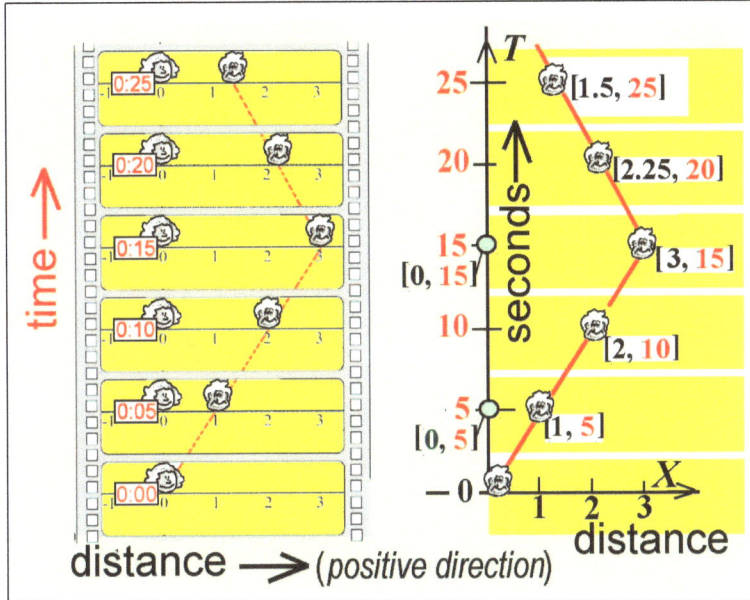

F*ive-second intervals of time correspond to each vertically advancing film-strip frame in the left half of the figure. As we follow successive filmstrip frames in the upward direction, we view the left-to-right (positive) changes in Bernie's position along the X-axis.*

(4.1.1a) The Setup for the Filmstrip of Figure (4.1.1). As indicated in Figure (4.1.1) the positive direction of distance is to the right and the negative direction is to the left — regardless of the direction in which Bernie and Ashley are facing.

Although the film strip is two-dimensional, Bernie's motion is in one-dimensional (*horizontal*) in space only. The second (*vertical*) axis tells us the time that Bernie finds himself at any position on the horizontal axis.

(4.1.1b) Extracting the *X-T* Graph (*Mathematics Models Reality*): In Figure (4.1.1), lift Bernie's path, like a decal, from the filmstrip (left hand panel) to create the two-dimensional graph (right hand panel). In this natural way, Bernie's 1-dimensional, left-to-right or right-to-left position x, say, corresponds to the horizontal X-axis reading and the time t corresponds to the vertical T-axis reading.

The right hand side of Figure (4.1.1) motivates the following definition:

(4.1.2)

> **DEFINITION:** *Given an object in a frame of reference (1.2.1) that can move in a single direction. An object's position at point x and time t is represented by ordered pairs of the form*
>
> **(4.1.2a)** $[x, t]$ *or* $\begin{bmatrix} x \\ t \end{bmatrix}$
>
> *in a graph with distance axis X and time axis T. This graph is called the* SPACETIME GRAPH (*or* SPACETIME DIAGRAM) *where each ordered pair (4.1.2a) is called an* EVENT *in spacetime.*
>
> *The path of all coordinates of a moving object over time — the set all its events — is called the* WORLDLINE *of that object in spacetime.*

For an object to move in three directions, ordered pairs (4.1.2a) are replaced with $[x_1, x_2, x_3, t]$.

(4.1.3) ‖ NOTE: **The space and time axes** are not necessarily perpendicular to each other.

(4.1.4) **Question:** Why do we use the term "event" to describe the point $[x, t]$ in a spacetime graph?

Answer: An object in spacetime, which is generally in motion, will have coordinates of the form $[x, t]$ which are generally changing in time. (*The coordinates of an object will change since t is always changing, even if the object is fixed at some point x.*) Hence, when we refer to a particular coordinate $[x, t]$ in spacetime, we are describing not only the location of an object, but the time it held that position. Specifying where and when something is corresponds to the intuitive definition of an *event*.

(4.2) Constant Velocities in Spacetime

Some of the most interesting effects in special relativity occur when we consider objects moving with constant velocity relative to one another (1.5) which we now formally define.

(4.2.1)

> **DEFINITION:** CONSTANT VELOCITY *is that motion along a straight line in space, where equal distances are covered in equal times. Therefore, any motion that veers off a straight line in space, or that changes speed along a straight line, has* NON-CONSTANT VELOCITY.

(4.2.2) ‖ **NOTE:** *The notation $\vec{v}$ that uses the overhead arrow, indicates <u>speed</u> v and <u>direction</u>. When we are only interested in the speed, a number for which no directions are involved, then we use the unornamented, symbol, v.*

The following theorem equates constant speed (a physical quantity) with worldlines that are straight (a geometric quantity).

(4.2.3) **THEOREM:** (***Constant Speed $\Leftrightarrow$ Straight Line Trajectory***) *A particle moving on the X-axis has constant speed if and only if its worldline (trajectory) in the spacetime graph (4.1.2) is a straight line.*

PROOF: (*informal*) Figure (4.2.4) shows that even though motion of a particle proceeds along a straight 1-dimensional track — the X-axis — the 2-dimensional graph of this motion in spacetime can be either curved or straight. The straight path corresponds to constant speed. Here are details.

(4.2.4) **FIGURE:** *Straight-Line Motion Along the X-Axis*

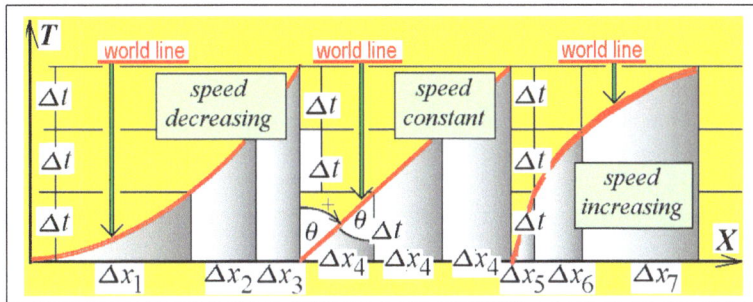

The Δt intervals on the vertical T-axis of Figure (4.2.4) are all equal. Since average speed over the Δx_ith interval is $v_i = \Delta x_i / \Delta t$, we have

- LEFT HAND WORLDLINE IN (4.2.4): *Since $\Delta x_1 > \Delta x_2 > \Delta x_3$,*

$$\frac{\Delta x_1}{\Delta t} > \frac{\Delta x_2}{\Delta t} > \frac{\Delta x_3}{\Delta t}, \qquad \textit{which means } v_1 > v_2 > v_3, \quad \textit{or v \underline{decreases}} \\ \textit{as time increases.}$$

- MIDDLE WORLDLINE IN (4.2.4): *Since all distances equal Δx_4, the average speed over each of the three distance intervals is*

$$\frac{\Delta x_4}{\Delta t} = v_4 \qquad \textit{which means speed v is \underline{constant} for all time t.}$$

- RIGHT HAND WORLDLINE IN (4.2.4): *Since $\Delta x_5 < \Delta x_6 < \Delta x_7$,*

$$\frac{\Delta x_5}{\Delta t} < \frac{\Delta x_6}{\Delta t} < \frac{\Delta x_7}{\Delta t}, \qquad \textit{which means } v_5 < v_6 < v_7, \quad \textit{or v \underline{increases}} \\ \textit{as time increases.}$$

This ends the proof. ∎

(4.2.5) **A curved trajectory** might be observed when the moving frame is not inertial (1.5.1). If the spacetime graph is not straight, the velocity is not constant which implies the object is accelerating — a force is acting on it. In the following figure, (4.2.6), the force in question is that of *gravity*. Hence, the trajectory of the falling ball is seen to have two shapes, depending on the frame of the observer.

(4.2.6) FIGURE: *One Trajectory, Two Perceptions.*

Bernie's point of view   Ashley's point of view

Bernie, from his frame, $\mathcal{F}^B$, sees the accelerating ball following a straight, vertical path. But Ashley, from her frame, $\mathcal{F}^A$, sees the ball moving to her right which causes her to observe the same ball on a downward parabolic path.

(4.3) Worldlines are Parallel to the Home Frame Time Axis

(**4.3.1**) From (3.4.2), if each observer is at rest at some fixed point x_0 in frame $\mathcal{F}^A$ say, then the worldline of that observer in the X_A-T_A spacetime graph is a straight line that is parallel to the time axis T_A. This follows since stationary observers have coordinates $[x_0, t]$ where distance x_0 is constant and time t varies.

As a special case, if the (fixed) observer is the Origin Observer (3.4.3) at $x = 0$, then his/her world line coincides with the T_A-axis itself.

Suppose an observer fixed in a home frame $\mathcal{F}^A$ sees an observer fixed in a moving frame $\mathcal{F}^B$ that has constant speed relative to $\mathcal{F}^A$. Then the spacetime graph showing both frames will show the two time axes, T_A and T_B, at some angle relative to each other. (*Due to synchronization* (3.4.5) T_A *and* T_B *intersect at common time* $t = 0$.) Also, from Figure (4.2.4), we know that in $\mathcal{F}^A$ the B-observer will trace a <u>straight</u> line at some angle; if the B-observer is at the origin in $\mathcal{F}^A$, then this straight line corresponds to the T_B-axis as seen in $\mathcal{F}^A$.

Hence, if a spacetime graph shows the relative motion of n inertial frames of reference, then there are at most n slopes that can be assigned among all the worldlines that are drawn in the graph.

. . . .

The following spacetime graph illustrates the only two slopes that are possible for all worldlines of moving objects that belong to either one of two synchronized inertial frames of reference, $\mathcal{F}^A$, or $\mathcal{F}^B$.

(4.3.2) FIGURE: *Two Frames Imply Only Two Slopes for All World-lines*

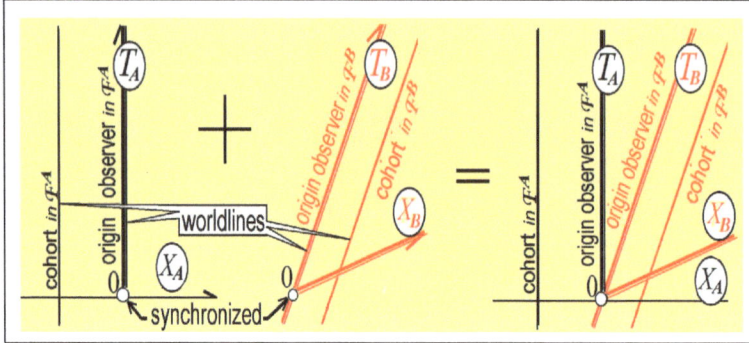

$\mathbf{T}$*hese graphs depict the worldlines of objects or observers in two synchronized inertial frames of reference, $\mathcal{F}_A$ and $\mathcal{F}_B$, that are in constant motion relative to each other. In each separate spacetime diagram, the origin observer and all her (fixed) cohorts have worldlines that are parallel straight lines — all worldlines in each frame have the same slope. The worldline of the origin observer at $x = 0$ for any t, coincides with her own time axes, T_A or T_B. When the spacetime diagrams of the frames overlap (the rightmost graph of Fig. (4.3.2)), the origins must overlap due to synchronization (3.4.5a). In sum, in the overlapping spacetime diagrams, there can only be two slopes for the worldlines of all the observers in frames $\mathcal{F}^A$ and $\mathcal{F}^B$. (Do not be distressed to see the X-axes, X_A and X_B, are not both at the same angle. As we shall see in (6.2.3), pg. 99, these axes represent respective lines of simultaneity — at time $t = 0$ — and will necessarily be drawn at different angles.)*

(4.4) Simultaneous and Static Events

There is a straightforward, graphical way to characterize *simultaneous events* and *static particles* in spacetime diagrams. First, we present a formal definition of these terms:

(4.4.1)

> **DEFINITION:** (*See Figure* **(4.4.4)**.)
> SIMULTANEOUS EVENTS in a specific frame $\mathcal{F}$ are represented as a pair of events, $[x, t_0]$ and $[x', t_0]$, in a spacetime graph with the same time coordinate t_0. A LINE OF SIMULTANEOUS EVENTS is the set of all events $[x, t_0]$ that take place at time t_0.
>
> **(4.4.1a)** **Geometrically,** *we say that a line that is parallel to the distance axis, X, is a line of simultaneous events.*
>
> A STATIONARY, *or* STATIC SET OF EVENTS *at point* x_0 *is the set of points,* $[x_0, t]$, *with a fixed space coordinate,* x_0, *and a range of the time coordinate,* t *— the observer is fixed and time varies.*
>
> **(4.4.1b)** **Geometrically**, *a stationary or static event is represented as a worldline that is parallel to the time axis, T.*

(4.4.2) ‖ **NOTE:** We have already seen in (4.1.3) that the X and T axes need not be mutually perpendicular. This non-perpendicularity is illustrated in Figure (4.4.4).

(4.4.3) ‖ **NOTE:** One can interpret simultaneous events in frame $\mathcal{F}$ that occur at time t_0 — points of the form $[x, t_0]$ — in the spacetime graph as a subset of the X-axis of $\mathcal{F}$ at time t_0 (*the X-axis at "age"* t_0 *in* $\mathcal{F}$).

‖ *Hence, lines of simultaneity at time* t_0, *are always* **parallel** *to the "usual"* X *axis at time* 0.

In sum, A stationary event has a worldline (trajectory) parallel to the T-axis. The worldlines of leftward (or rightward) traveling objects are characterized by their positive (or negative) angles with respect to the T-axis. Simultaneous events lie on a line that is parallel to the X-axis.

(4.4.4) FIGURE: *Worldlines of Simultaneous,*
Stationary, and Other Events .

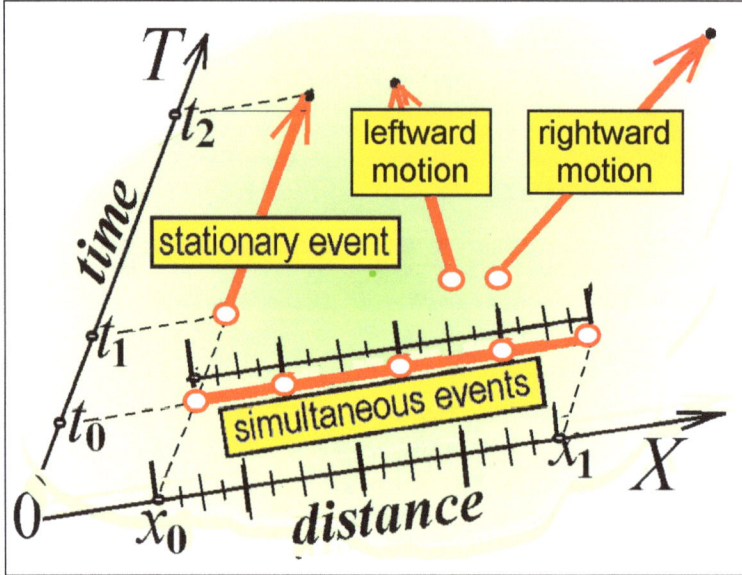

(4.4.5) EXAMPLE: Worldlines of two particles are illustrated in the following figure.

(4.4.5a) FIGURE: *Worldlines for Two Particles*

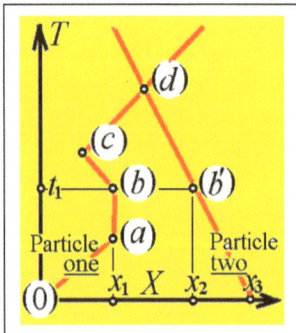

Particle one: *Traveling from*
$(0) \rightarrow (a)$: *constant speed, moving right*

$(a) \rightarrow (b)$: *at rest, speed equals* 0

$(b) \rightarrow (c)$: *constant speed, moving left*

$(c) \rightarrow \infty$: *constant speed, moving right.*

Particle two: *Always moving left at*
constant speed.

(*In Figure* (4.4.5a), *the axes happen to be perpendicular but, as Figure* (4.4.4) *shows, this is not necessary.*) Particle one and Particle two move in one-dimensional space. Except for the points, (a), (b), and (c), where Particle one experiences a force that causes an idealized instantaneous change in direction, there are no forces involved. At point (d), both particles theoretically seem to collide since they share the same $[x, t]$ coordinates. However, we always assume particles that are in motion relative to each other occupy different parallel frames (3.4.2) that are a negligible distance apart.

(4.5) Linearity of Line-of-Sight Functions

Figure (4.5.1) will provide a graphical context for the line-of-sight function G_v, one instance of which is illustrated in Figure (4.3.2). Moreover, note that

- **To choose an arbitrary observer** at an arbitrary time in $\mathcal{F}^A$, is to choose a point on a black, vertical worldline. Similarly, choosing an arbitrary observer at an arbitrary time in $\mathcal{F}^B$ is to choose a point on a red, slanted worldline.

- **Each** (black, vertical) worldline of $\mathcal{F}^A$ intersects one and only one (slanted, red) worldline of $\mathcal{F}^B$.

- **The two points at the intersection** of worldlines — point $[x, t]_A$ on the (vertical, black) worldline in $\mathcal{F}^A$ and point $[x', t']_B$ on the (slanted, red) worldline in $\mathcal{F}^B$ — define the endpoints of the line-of-sight corridor between $\mathcal{F}^A$ and $\mathcal{F}^B$. That is, these points represent the same event in different coordinate systems so that the line-of-sight function G_v linking these two points is expressed by the equation $G_v([x, t]_A) = [x', t']_B$.

For example, Figure (4.5.1) shows the event defined by the intersection of Ashley's and Bernie's worldlines corresponds to the linked points $[2.0, 3.41]_A$ in Ashley's frame $\mathcal{F}^A$ and $[0.0, 3.0]_B$ in Bernie's frame $\mathcal{F}^B$.

(4.5.1) FIGURE: **A Point Where a Line-of-Sight Corridor Links Ashley's Vertical Worldline in $\mathcal{F}^A$ to a Slanted Worldline in $\mathcal{F}^B$.**

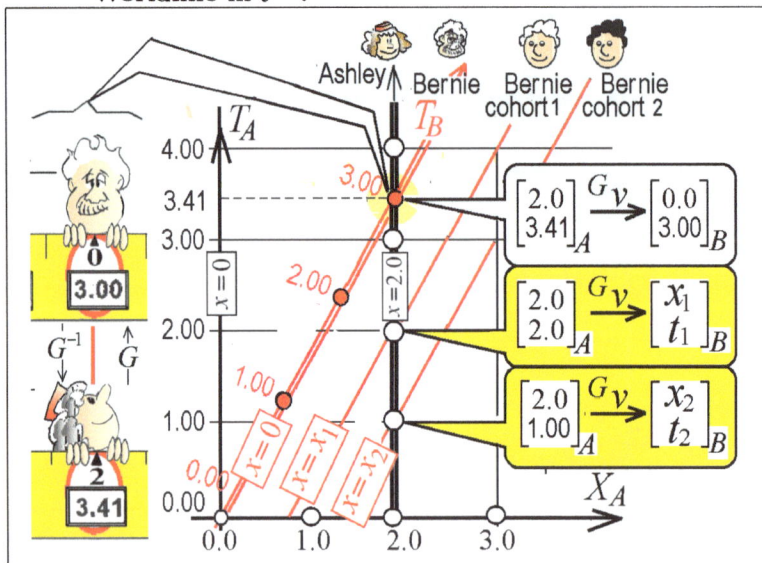

The three colinear points, $[\,2.0,\,3.41\,]_A$, $[\,2.0,\,2.0\,]_A$, and $[\,2.0,\,1.0\,]_A$ in $\mathcal{F}^A$ lie on Ashley's worldline, which is a straight (vertical) line, all points of which are linked by the line-of-sight function G_v to the three colinear points $[\,0,\,3.0\,]_B$, $[\,x_1,\,t_1\,]_B$, and $[\,x_2,\,t_2\,]_B$ which lie on a straight line (not a worldline) in $\mathcal{F}^B$. Intuitively, the straight worldline of Ashley in $\mathcal{F}^A$ casts a "shadow" of itself onto an identical straight line in $\mathcal{F}^B$. The following holds true:

(4.5.2)

CONCLUSION: Any line-of-sight function $G_v : \mathcal{F}^A \to \mathcal{F}^B$ sends straight lines of the X_A-T_A spacetime graph to straight lines of the X_B-T_B spacetime graph.

In fact, G_v may be interpreted as a function that leaves each point of any straight <u>fixed</u>, where only the coordinates (names) are changed from $\mathcal{F}^A$ to $\mathcal{F}^B$ coordinates.

The following figure illustrates a line-of-sight "name-changing" function $G_v : \mathcal{F}^A \to \mathcal{F}^B$ that sends points (vectors) of a straight line L in $\mathcal{F}^A$ to points (vectors) $G_v(L)$, which also form a straight line in

$\mathcal{F}^B$.

(4.5.3) ꜰɪɢᴜʀᴇ: **Line-Preserving Functions**
$$\boldsymbol{G_v : \mathcal{F}^A \to \mathcal{F}^B,\ G_v^{-1}: \mathcal{F}^B \to \mathcal{F}^A}$$

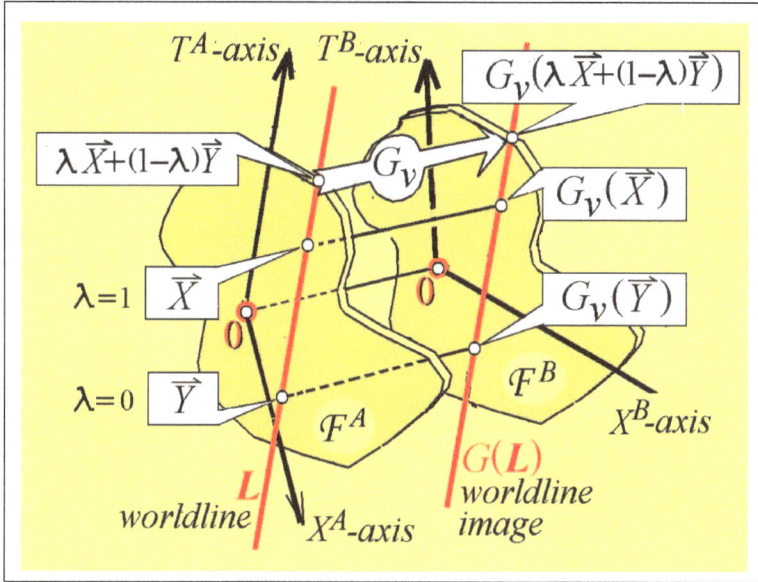

In any vector space, a straight line L containing points $\vec{X} \neq \vec{Y}$ must be the collection of points of the form

$$\lambda\vec{X} + (1-\lambda)\vec{Y} \ \text{ for } -\infty < \lambda < \infty \qquad\qquad Def\ (A.2.14).$$

The line-of-sight function G_v sends the three colinear points $\vec{X}$, $\vec{Y}$, and $\lambda\vec{X} + (1-\lambda)\vec{Y}$ in $\mathcal{F}^A$, respectively, to the three colinear points $G_v(\vec{X})$, $G_v(\vec{Y})$, and $G_v(\lambda\vec{X} + (1-\lambda)\vec{Y})$ in $\mathcal{F}^B$.

(4.5.4) *Any line-of-sight function G_v preserves proportions* since, from (4.5.2), L in $\mathcal{F}^A$ and $G_v(L)$ in $\mathcal{F}^B$ are, point-for-point, the same line but with different coordinates (names).

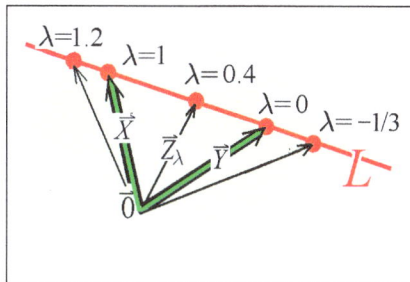

Hence, proportions are preserved in the sense that the mid-point of L is sent to the mid-point of $G_v(L)$, a point $\vec{Z}$ that is 0.4 the distance from $\vec{Y}$ to $\vec{X}$ has an image $G_v(\vec{Z})$ that is 0.4 the distance from $G_v(\vec{Y})$ to $G_v(\vec{X})$, etc. Definition

(4.5.5) provides, in mathematical terms, the equation that character-
izes line-preserving and proportion-preserving functions $G_v : \mathcal{F}^A \to \mathcal{F}^B$.

NOTE: In what follows, we will deal with more general vector spaces
$\mathcal{F}^A$ and $\mathcal{F}^B$ that are not necessarily spacetime vector spaces.
Hence, for the function $G_v : \mathcal{F}^A \to \mathcal{F}^B$, we suppress the subscript v and write
$$G \quad \text{instead of} \quad G_v.$$

(4.5.5)

> **DEFINITION:** *A function G from a vector space $\mathcal{F}^A$ to a vector
> space $\mathcal{F}^B$* PRESERVES PROPORTION *and is* LINE-PRESERVING
> *if and only if for any fixed vectors $\vec{X} \neq \vec{Y}$ in $\mathcal{F}^A$ where $G(\vec{X}) \neq G(\vec{Y})$,*
>
> **(4.5.5a)** $G(\underbrace{\lambda\vec{X} + (1-\lambda)\vec{Y}}_{\text{line in } \mathcal{F}^A}) = \underbrace{\lambda G(\vec{X}) + (1-\lambda)G(\vec{Y})}_{\text{line in } \mathcal{F}^B}$
>
> *for all real λ, $-\infty < \lambda < \infty$.*

To say that a line-of-sight function preserves straight lines (see (4.5.2))
is to describe a visual *geometric property*. We now show that this geometric property is equivalent to an *algebraic property* — more exactly,
the *linear-algebraic property* that is given in the following theorem.

(4.5.6)

> **THEOREM:** *Line-Preserving Functions = Linear Func-
> tions.*   *Given vector spaces $\mathcal{F}^A$ and $\mathcal{F}^B$ and a function
> $G : \mathcal{F}^A \to \mathcal{F}^B$ where $G(\vec{0}) = \vec{0}$. Then*
>
> **(4.5.6a)**   *G is linear in the sense of* (A.4.1)
>
> *if and only if*
>
> **(4.5.6b)**   *G preserves straight lines and proportions.*

PROOF:

G linear	G line-preserving
(4.5.6a) $\Longrightarrow$	**(4.5.6b)**
(*sufficient*)	(*necessary*)

If G is assumed to be linear then, in Definition (A.4.1a), we can replace $\vec{P}_1$ with $\vec{X}$ and $\vec{P}_2$ with $\vec{Y}$ to obtain

(4.5.6c) $G(\lambda_1 \vec{X} + \lambda_2 \vec{Y}) = \lambda_1 G(\vec{X}) + \lambda_2 G(\vec{Y})$

for *all* scalars λ_1 and λ_2. In (4.5.6c), set $\lambda_1 = \lambda$ and replace λ_2 with $(1 - \lambda)$ to arrive at the form (4.5.5a) for a typical line-preserving function.

G *line-preserving*	G *linear*
(4.5.6b) $\implies$	**(4.5.6a)**
(sufficient)	*(necessary)*

We first note that if G preserves lines, then from (4.5.5), for all scalars λ,

$$G(\lambda \vec{P}_1 + (1 - \lambda)\vec{P}_2) = \lambda G(\vec{P}_1) + (1 - \lambda)G(\vec{P}_2).$$

Set $\vec{P}_2 = \vec{0}$ and invoke the hypothesis that $G(\vec{0}) = \vec{0}$ to finally obtain

(4.5.6d) $G(\lambda \vec{P}_1) = \lambda G(\vec{P}_1).$ *scalar λ "factors out" of G*

First, choose some λ not equal to 0 or 1. Then for all vectors $\vec{P}_1$ and $\vec{P}_2$ and for all real scalars λ_1 and λ_2 we write

(4.5.6e) $G(\lambda_1 \vec{P}_1 + \lambda_2 \vec{P}_2) = G\left(\lambda \left[\dfrac{\lambda_1}{\lambda}\right]\vec{P}_1 + (1-\lambda)\left[\dfrac{\lambda_2}{(1-\lambda)}\right]\vec{P}_2\right)$

$$= \lambda G\left(\left[\dfrac{\lambda_1}{\lambda}\right]\vec{P}_1\right) + (1-\lambda)G\left(\left[\dfrac{\lambda_2}{(1-\lambda)}\right]\vec{P}_2\right) \qquad \text{\textit{since } G \textit{ preserves}} \atop \textit{lines (4.5.5a)}$$

$$= \lambda\left[\dfrac{\lambda_1}{\lambda}\right]G(\vec{P}_1) + (1-\lambda)\left[\dfrac{\lambda_2}{(1-\lambda)}\right]G(\vec{P}_2) \qquad \substack{\textit{"factor out"} \\ \left[\frac{\lambda_1}{\lambda}\right], \left[\frac{\lambda_2}{(1-\lambda)}\right] \\ \textit{from } G \textit{ (4.5.6d)}}$$

$$= \lambda_1 G(\vec{P}_1) + \lambda_2 G(\vec{P}_2). \qquad \textit{numerator/denominator cancellation}$$

Equating the very first and last terms in this string of equalities proves that G is linear as required by (A.4.1a). The proof is done. ■

(4.5.7) ‖ NOTE: **Necessary vs. Sufficient.** The proof of Theorem (4.5.6) consists of showing that the two statements, $P = (4.5.6a)$ and $Q = (4.5.6b)$, are logically equivalent in the sense that the truth of one implies the truth of the other. This illustrates the important difference between "*necessary*" and "*sufficient*" since converse implications of theform $P \Rightarrow Q$ and $Q \Rightarrow P$ are, in general, *not* logically equivalent.

(4.6) Exercises

(4.6.1) **Cartesian Coordinate Systems.** In Figure (4.4.5a), what are the X-T coordinates of the three points indicated as a, b, and b'?

Exercise

(4.6.2) **Pictures of Higher Dimensions.** With Cartesian coordinates, we usually draw mutually perpendicular axes X_1, X_2, and X_3 to represent points $[x_1, x_2, x_3]$ in 3-space, as is done in the figure of (17.3.1) and in the left hand panel of Figure (A).

Exercise

Figure (A): Two Representations of Points in 3-Space.

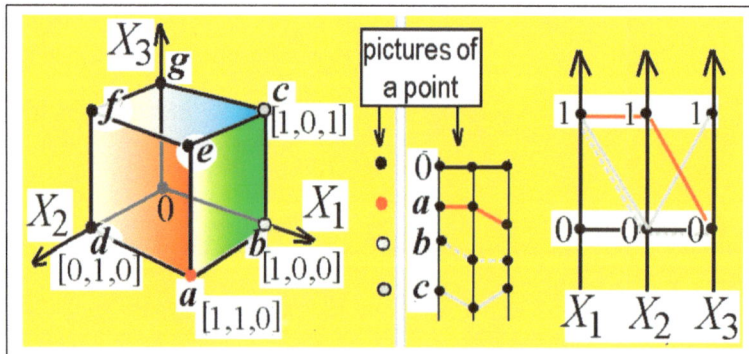

However, when the three axes, X_1, X_2, X_3, are drawn vertically and parallel to each other, then, in Figure (A), the points (dots) in the *left hand* panel are represented as straight lines in the *right hand* panel.

One advantage of the representation of the right hand panel of Figure (A) is that we can easily draw spaces of dimension n greater than three. This is done by drawing n parallel vertical axes.

One disadvantage of the right hand panel of Figure (A) is that familiar shapes in 2-space and 3-space can take on unfamiliar forms. For example, the four corners of the cube, 0, a, b, and c, which are *points* in the left hand panel, look like *lines* in the right hand panel.

. . . .

As another example, the set of points of a sphere does not have a pleasant "ball-like" shape (*see Exercise* (4.6.2c)).

‖ NOTE: *Even though the triples,* $[x_1, x_2, x_3]$, *are represented as jag-*
‖ *ged lines in the right hand panel of Figure* (A), *these "lines" are*
‖ *still called "points."*
. . . .

(4.6.2a) **Question:** Draw the other four points of the cube, d, e, f, and g, in the manner of the right hand panel, i.e., using parallel axes.

(4.6.2b) **Question:** Show that in the right hand panel, using parallel axes to represent 3-space, the set of all points of a cube fill the band between $[0, 0, 0]$ and $[1, 1, 1]$.

(4.6.2c) **Question:** Draw any four points $[x_1, x_2, x_3]$ that lie on the surface of the unit sphere in the manner of both the left hand panel (Cartesian coordinates) and the right hand panel (with parallel axes).

Hint: *A point* $[x_1, x_2, x_3]$ *on the surface of the unit sphere has the property* $x_1^2 + x_2^2 + x_3^2 = 1$.

(4.6.3)
Exercise

Film Strip Model of Motion. It can be argued that the graph in the right hand panel of Figure (4.1.1) is more informative than the "reality" of photographs provided by the film strip (left hand panel) since photographs record only a *finite* set of distances and times, $[x, t]$. On the other hand, the mathematical model of the right hand panel fills in the gaps, showing *infinitely* many points $[x, t]$.

For example, the photographs do not tell us the time t_0 when Bernie was at the point 2.6 on the rightward leg of his journey. But (once we know the equation of a straight line in X-T space), the mathematical model shown in the right hand panel of (4.1.1) gives all $[x, t]$ pairs, including $[2.6, t_0]$.

· ·

(4.6.3a)   **Question:** In Figure (4.1.1), at what time t_0 was Bernie
            at point 2.6 while he was traveling rightward?

(4.6.3b)   **Question:** In Figure (4.1.1), at what time t_1 was Bernie
            at point 2.6 while he was traveling leftward?

(4.6.3c)   **Question:** In Figure (4.1.1), at what distance does Ber-
            nie find himself at $t = 18.7$ seconds?

(4.6.4)   **Synchronizing at $[\mathbf{0,0}]$** Why do we not synchronize observers at
Exercise   points other than $[0,0]$? That is, why do we not have observers in
            two frames synchronize at time $t = 3.0$?

            Hint: *Consider the* linearity *of line-of-sight functions that connect
            observers in separate frames by a line-of-sight corridor (see (4.5.6)).*

(4.6.5)   **Physics and Geometry.**   We have seen geometric portraits of
Exercise   motion that are *not* straight lines (4.2.6)) along with graphs that *are*
            straight lines (Figure (4.1.1)).

(4.6.5a)   **Question:** Show that the physics description of unac-
            celerated motion (4.2.1) is equivalent to the mathemat-
            ical notion of a straight line path $t = ax + b$, in the
            2-dimensional X-T plane where t is time, x is distance,
            and a and b are certain constants.

            Hint: *Physics requires that equal distances are covered
            in equal times (see Definition (4.2.1)). The mathematical
            description of a straight line says the slope of its graph,
            $\Delta t / \Delta x$, on an X-T plane is constant at every point $[x, t]$.*

II. Galilean Transformations of Frames

Chapter 5 *explores the relation between pairs of frames in a linear algebraic context. A well-defined matrix arises in the line-of-sight connection between Galilean frames.*

The over-riding Galilean assumption is that time is universal so that there is one universal clock whose readings are valid for all observers, regardless of their relative motion. Consistent with this property is that the speed of light is infinite.

$$G_v \begin{bmatrix} 1 & -v \\ 0 & 1 \end{bmatrix} * \begin{bmatrix} x_1 \\ t_1 \end{bmatrix}_A = \begin{bmatrix} x_2 \\ t_2 \end{bmatrix}'_B$$

I see speed v — of these $\mathcal{F}^B$ coordinates — when I stand here in $\mathcal{F}^A$.

(5) Galilean Transformations

(5.1) Key Ideas

We assume frame $\mathcal{F}^B$ moves with constant speed v relative to frame $\mathcal{F}^A$.

(5.1.1) From (3.2.1): The set of all spacetime ordered pairs $\{[x,t]\}$ for inertial frame $\mathcal{F}$ form a mathematical <u>vector space</u> (A.2.1).

From (4.5.6): The line-of-sight function, $G_v : \mathcal{F}^A \to \mathcal{F}^B$ sending vector space $\mathcal{F}^A$ to vector space $\mathcal{F}^B$, which takes straight lines to straight lines, where $G_v(\vec{0}) = (\vec{0})$, is necessarily <u>linear</u> (A.4.1b).

From (A.4.6): If the spacetime graph showing overlapping frames $\mathcal{F}^A$ and $\mathcal{F}^B$ is 2-dimensional, then each output vector of the function $G_v : \mathcal{F}^A \to \mathcal{F}^B$ is the 2×1 column $G_v(X)$ which Theorem (A.4.6), pg. 346, tells us can always be computed by multiplying the input vector $\vec{X}$ on the left by a unique 2×2 matrix $\text{MX}(G_v)$. Specifically,

$$
\textbf{(5.1.1a)}
$$
$$
G_v(\vec{X}) = \text{MX}(G_v) * \vec{X} = \underbrace{\left[G_v(\vec{E}_1), \, G_v(\vec{E}_2) \right]}_{2 \times 2 \ \text{MX}(G)} * \underbrace{\vec{X}}_{\substack{2 \times 1 \\ \text{matrix}}}
$$

where the two columns of $\text{MX}(G_v)$ are the 2×1 columns $G_v(\vec{E}_1)$ and $G_v(\vec{E}_2)$. Moreover, $\vec{E}_1 = \begin{bmatrix} 1 \\ 0 \end{bmatrix}$ and $\vec{E}_2 = \begin{bmatrix} 0 \\ 1 \end{bmatrix}$ are the standard basis vectors (A.2.11) of vector space $\mathcal{F}$.

Our goal is to characterize elements of 2×2 $\text{MX}(G_v)$ of (5.1.1a).

(5.1.2) ‖ **NOTE:** **Since the equation** $G_v(\vec{X}) = \text{MX}(G_v) * \vec{X}$ appears in (5.1.1a) we often think of the *matrix* $\text{MX}(G_v)$ as being the same as the *function* G_v. We have already encountered this shortcut in (A.4.8), pg. 347 where it is observed that for any $\vec{X}$, $\text{MX}(G_v) * \vec{X}$ always equals $G_v(\vec{X})$. Hence, we think of the two notations $\text{MX}G_v$ and G_v as equivalent to each other. Symbolically $\text{MX}(G_v) = G_v$.

(5.1.3) ‖ **How Matrices Induce Linear Functions.** As per (A.4.5), display (5.1.1a) states that $\text{MX}(G_v)$ is the 2×2 matrix whose pair of 2×1 columns are $G_v(\vec{E}_1)$ and $G_v(\vec{E}_2)$ where $\vec{E}_1$ and $\vec{E}_2$ are the standard basis vectors of $\boldsymbol{R}^2$ (A.2.11). It is this mechanism of multiplying a column vector $\vec{X}$ on the left by a matrix $\text{MX}(G_v)$ that tells us <u>how</u> any linear function G_v is implemented.

,

(5.2) Galilean Spacetime Diagrams

We construct *mathematical* spacetime graphs to model moving *physical* parallel frames (3.3.1), (3.4.4), (3.4.4) that reflect the Galilean assumption that time is the same for all observers in all frames.

Mathematically, the spacetime graph will show that the link between frames — the line-of-sight-corridors — have equal times in any pair of the linked frames, regardless of relative speeds. Specifically, the linking coordinates at the endpoints of line-of-sight corridors of frames $\mathcal{F}^A$ and $\mathcal{F}^B$, say, have the form $[x, t]_A$ and $[x', t]_B$ with equal time coordinate t.

(5.2.1)

> **DEFINTION:** A GALILEAN GRAPH or DIAGRAM is a spacetime graph of synchronized frames (3.4.5) whose (*horizontal*) X-axes (*and hence, their origins,* $[0, 0]$) overlap. All frames share common (*horizontal*) lines of simultaneous events (*see Figure* (4.4.4)).

This diagram shows points on worldlines of frames $\mathcal{F}^A$ and $\mathcal{F}^B$ at the common time $t = 3.0$.

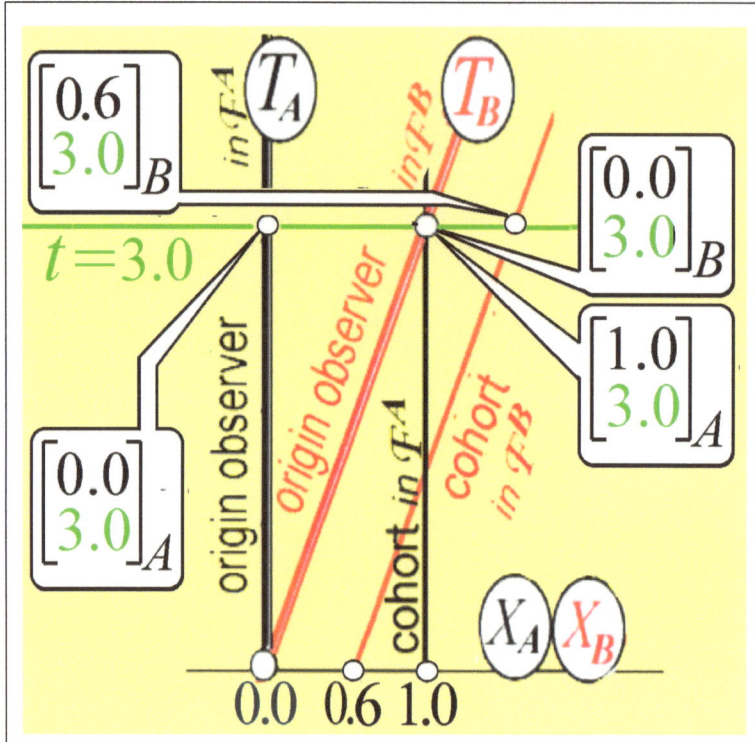

(5.3) The Galilean Matrix

(5.3.1)

> **DEFINITION:** *A* GALILEAN TRANSFORMATION *between inertial frames* $\mathcal{F}^A$ *and* $\mathcal{F}^B$ *where* $\mathcal{F}^B$ *has constant speed* v *relative to* $\mathcal{F}^A$ *is a function* (A.3.2)
>
> $$G_v(\begin{bmatrix} x \\ t \end{bmatrix}_A) = \begin{bmatrix} x' \\ t \end{bmatrix}_B$$
>
> *that links coordinates* $[\,x, t\,]_A$ *and* $[\,x', t\,]_B$ *of frames* $\mathcal{F}^A$ *and* $\mathcal{F}^B$ *respectively, where the time coordinates* t *are always equal regardless of the relative speed* v.

(5.3.2) NOTE: **Theorem (4.5.6)** tells us that G_v is linear in the algebraic sense of (4.5.5) since G_v preserves straight lines (worldlines) and proportions of straight lines.

From Theorem (A.4.2), we can characterize the 2×2 matrix G_v, once we know its values $G_v(\vec{X}_1)$, $G_v(\vec{X}_2)$ on only two basis vectors (A.2.9) $\{\vec{X}_1, \vec{X}_2\}$ of the 2-dimensional X-T spacetime graph.

In fact, if we know the values for $G_v(\vec{X}_1)$ and $G_v(\vec{X}_2)$ and if an arbitrary vector $\vec{Y}$ of $\boldsymbol{R}^2$ has the form $\vec{Y} = \alpha_1 \vec{X}_1 + \alpha_2 \vec{X}_2$, then $G_v(\vec{Y}) = \alpha_1 G_v(\vec{X}_1) + \alpha_2 G_v(\vec{X}_2)$ shows how we produce the value of $G_v(\vec{Y})$ from the known values of $G_v(\vec{X}_1)$ and $G_v(\vec{X}_2)$ alone.

(5.3.3) FIGURE: *The Galilean Transformation $G_v: \mathcal{F}^A \to \mathcal{F}^B$ when $\mathcal{F}^B$ has Speed v Relative to $\mathcal{F}^A$.*

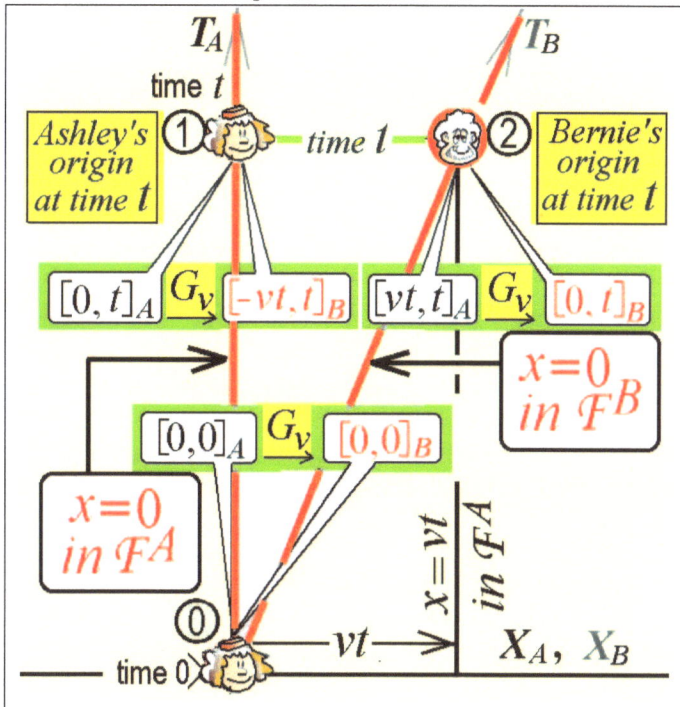

(5.3.4)

> $\textsc{Theorem}$: (*The Galilean Matrix*) *Inertial frame $\mathcal{F}^B$ travels with speed v relative to inertial frame $\mathcal{F}^A$. Then the Galilean matrix $G_v : \mathcal{F}^A \to \mathcal{F}^B$ linking coordinates at opposite ends of line-of-sight corridors between $\mathcal{F}^A$ and $\mathcal{F}^B$ is given by*
>
> **(5.3.4a)** $\begin{bmatrix} 1 & -v \\ 0 & 1 \end{bmatrix} = \textsc{mx}(G_v) = G_v.$

$\textsc{Note}$: *Linear transformations and their matrices are often used interchangeably in the literature. Hence, we write* $\textsc{mx}(G_v) = G_v$ *in (5.3.4a). See (5.1.2) and (A.4.8).*

$\textsc{Proof}$: *The line-of-sight linear function G_v has the following values for points* ⓪, ①, *and* ② *of Figure (5.3.3):*

(5.3.4b)

	$\mathcal{F}^A$		$\mathcal{F}^B$
⓪	$[0,0]_A$		$[0,0]_B$
①	$[0,t]_A$	$\xrightarrow{G_v}$	$[-vt,t]_B$
②	$[vt,t]_A$		$[0,t]_B$

Justifying ⓪ of (5.3.4b):   The origins of $\mathcal{F}^A$ and $\mathcal{F}^B$ are synchronized at $t=0$ (3.4.5). At common time $t=0$, the origins are at opposite ends of a line-of-sight corridor, which is expressed mathematically by the equation $G_v([0,0]_A) = [0,0]_B$. Also, at this time, the distance between the origins is zero.

Justifying ① of (5.3.4b):   Since $\mathcal{F}^A$ is moving leftward at speed $-v$ relative to $\mathcal{F}^B$, $\mathcal{F}^B$ sees the origin of $\mathcal{F}^A$ (the T^A-axis) from $\mathcal{F}^B$ coordinates $[-vt,t]_B$. But at time t, the origin of $\mathcal{F}^A$ has $\mathcal{F}^A$ coordinates $[0,t]_A$. Mathematically, the linking of these coordinates at the ends of a line-of-sight corridor is expressed by the equation $G_v([0,t]_A) = [-vt,t]_B$.

Justifying ② of (5.3.4b):   Since $\mathcal{F}^B$ is moving rightward at speed $+v$ relative to $\mathcal{F}^A$, the observer in $\mathcal{F}^A$ at time t, at coordinates $[vt,t]_A$, looks down the line-of-sight corridor into $\mathcal{F}^B$ and sees the $\mathcal{F}^B$ origin (the T^B-axis) has $\mathcal{F}^B$ coordinates $[0,t]_B$ Mathematically, the linking of these coordinates at the ends of a line-of-sight corridor is expressed by the equation $G_v([vt,t]_A) = [0,t]_B$.

Now (5.3.4b) assures us that we have the values of linear function $G_v : \mathcal{F}^A \to \mathcal{F}^B$ on two basis vectors of $\mathcal{F}^A$ which, as noted in (5.3.2), suffices to characterize the 2×2 matrix G_v. These two basis vectors are $[\,0,\, t\,]_A$ and $[\,vt,\, t\,]_A$. Thus, from (A.4.2), there is a unique quartet of numbers, $\{\,a,\, b,\, c,\, d\,\}$ such that

(5.3.4c) $$\begin{bmatrix} a & b \\ c & d \end{bmatrix} * \begin{bmatrix} 0 \\ t \end{bmatrix}_A = \begin{bmatrix} -vt \\ t \end{bmatrix}_B \quad and \quad \begin{bmatrix} a & b \\ c & d \end{bmatrix} * \begin{bmatrix} vt \\ t \end{bmatrix}_A = \begin{bmatrix} 0 \\ t \end{bmatrix}_B.$$

Matrix multiplication (A.4.3e) applied to (5.3.4c) generates four linear equations with four unknowns which resolve as

$$\begin{aligned} a &= 1 \\ b &= -v \\ c &= 0 \\ d &= 1 \end{aligned}.$$

This proves (5.3.4a) and hence, Theorem (5.3.4). ■

(5.4) Pattern of the Galilean Matrix

(5.4.1) **An Intuitive Reading of Matrix G_v (5.3.4a).** Figure (5.3.3) and Table (5.3.4b) list some linked pairs of coordinates $[x, t]_A$ and $[x', t']_B$, that lie at opposite ends of line-of-sight corridors connecting two frames of reference. However, once we apply the 2×2 Galilean matrix G_v (5.3.4a), the domain (input) vector, $[x, t]_A$, is considered to be stationary and the range (output) vector, $[x', t']_B$ is perceived by an observer in $\mathcal{F}^A$ to be moving with relative speed v.

This viewpoint is embedded in the matrix itself as summarized in the figure following:

(5.4.1a)   FIGURE: *Pattern Recognition in Reading the*
Galilean Matrix

$$G_v \begin{bmatrix} 1 & -v \\ 0 & 1 \end{bmatrix} * \begin{bmatrix} x_1 \\ t_1 \end{bmatrix}_A = \begin{bmatrix} x_2 \\ t_2 \end{bmatrix}'_B$$

I see speed v   *of these $\mathcal{F}^B$ coordinates*   *when I stand here in $\mathcal{F}^A$.*

Matrix $G_v \colon \mathcal{F}^A \to \mathcal{F}^B$ *has 1s on the diagonal and a 0 in the lower left hand corner. Whatever appears in the upper right hand corner is always* the negative *of the speed of $\mathcal{F}^B$ as observed from frame $\mathcal{F}^A$.*

(5.4.2)   NOTE: **Time is invariant** in all synchronized Galilean frames $\mathcal{F}^A$, $\mathcal{F}^B$ as indicated in Figure (5.3.3). This is shown linear algebraically (i.e., from the matrix) of Figure (5.4.1a) where we see that for all synchronized frames, t_1 in $\mathcal{F}^A$ equals t_2 in $\mathcal{F}^B$. Coordinates $[\,x_1, t_1\,]_A$ and $[\,x_2, t_2\,]_B$ are at opposite ends of a line-of-sight corridor and, according to Definition (5.3.1), $t_1 = t_2$ — for Galilean transformations, *time is invariant for all observers.*

(5.5) Addition of Speeds via Matrices

In the next theorem, we present an algebraic property of Galilean matrices.

(5.5.1)

THEOREM: (*An Algebraic Property of Galilean Matrices*)
For a Galilean matrix G and any scalars v and w, we have

(5.5.1a) $G_v * G_w = G_{v+w}.$

PROOF: $G_v * G_w = \begin{bmatrix} 1 & -v \\ 0 & 1 \end{bmatrix} * \begin{bmatrix} 1 & -w \\ 0 & 1 \end{bmatrix}$ *from (5.3.4a)*

$$= \begin{bmatrix} 1 & -(v+w) \\ 0 & 1 \end{bmatrix}$$ *from (A.4.3e) defining the matrix product*

$$= G_{v+w}.$$ *from Figure (5.4.1a), which shows the upper-right corner holds the negative of the speed*

Property (5.5.1a) is now established, proving the theorem. ∎

NOTE *that the first line of the proof uses the symbol G_v as a shorthand for the matrix* MX(G_v) *as is explained in (5.1.2)*

_____

The following theorem confirms our intuition about the way we add speeds of different frames. Specifically, we have:

(5.5.2)

> THEOREM: (***Galilean Addition of Speeds***)   *Given three frames of reference, $\mathcal{F}^A$, $\mathcal{F}^B$, and $\mathcal{F}^C$, in 1-dimensional space that are synchronized at time $t = 0$ and distance $x = 0$. Given*
>
> (5.5.2a) $\mathcal{F}^A \xrightarrow{G_v} \mathcal{F}^B$ *Observers in $\mathcal{F}^A$ see $\mathcal{F}^B$ moving at speed v.*
>
> (5.5.2b) $\mathcal{F}^B \xrightarrow{G_w} \mathcal{F}^C$ *Observers in $\mathcal{F}^B$ see $\mathcal{F}^C$ moving at speed w.*
>
> *Then*
>
> (5.5.2c) $\mathcal{F}^A \xrightarrow{G_{v+w}} \mathcal{F}^C$ *Observers in $\mathcal{F}^A$ see $\mathcal{F}^C$ moving at speed $v + w$.*

PROOF: Hypotheses (5.5.2a) and (5.5.2a) say

(5.5.2d) $G_v([x, t]_A) = [x', t']_B$ $\begin{cases} [x, t]_A \ in \ \mathcal{F}^A \\ [x', t']_B \ in \ \mathcal{F}^B \end{cases}$

and

(5.5.2e) $G_w([x',t']_B) = [x'',t'']_C$ $\begin{cases} [x',t']_B \ in \ \mathcal{F}^B \\ [x'',t'']_C \ in \ \mathcal{F}^C. \end{cases}$

Substituting $[x',t']_B$ of (5.5.2d) into (5.5.2e), we obtain

(5.5.2f) $G_w(G_v([x,t]_A)) = [x'',t'']_C$ $\begin{cases} [x,t]_A \ in \ \mathcal{F}^A \\ [x'',t'']_C \ in \ \mathcal{F}^C. \end{cases}$

Now, $G_w(G_v([x,t]_A)) = G_w \circ G_v([x,t]_A)$ *from* (A.3.6),

$$= G_{v+w}([x,t]_A) \qquad and \ Theorem \ (5.5.1)$$

so that (5.5.2f) rewrites itself as $G_{v+w}([x,t]_A) = [x'',t'']_C$ or

$$G_{v+w}{:}\mathcal{F}^A \to \mathcal{F}^C$$

which confirms (5.5.2c). The proof is done. ∎

. . . .

The following corollary is a mathematical statement of the symmetry principle (1.6.1).

(5.5.3)

> **COROLLARY: (*Confirming the Line of Sight*)**
>
> **(5.5.3a)** $G_v * G_{-v} = I_2,$
>
> *the 2×2 identity matrix.*

PROOF: In (5.5.1), set $v = v$ and $w = -v$ to obtain $G_v * G_{-v} = G_0$. But from (5.3.4a) (or (5.4.1a)), $G_0 = I_2$, which proves the corollary. ∎

(5.5.4) ‖ **NOTE:** *Proposition (3.4.6e) tells us that, in general, $G_v^{-1} = G_{-v}$, which means that $G_v * G_{-v} = I_2$. Corollary (5.5.3) confirms this result in another way through the use of calculations with the specific matrices G_v and G_{-v}.*

(5.6) Addition of Speeds via Areas

(5.6.1) NOTE: *To be dimensionally consistent, as per Chapter (??), pg. ??, we* **define** *the area in a Minkowski X-T space-time graph as follows: Given horizontal Δx and vertical Δt, then*

(5.6.1a) $\mathbf{Area}(\Delta x, \Delta t) \stackrel{def}{=} \Delta x(distance) \times \Delta t(time) \times 1\underbrace{\dfrac{1}{time^2}}_{\substack{dimension \\ factor}}$

The reader may reasonably ask why we need to multiply by the factor $1/time^2$ in (5.6.1a). The reason is that the goal of this section is to provide a graphical way of adding *speeds* by identifying them with appropriate *areas* in a Minkowski diagram. The only purpose of the factor $1/time^2$, then, is to give the areas appropriate units. Consequently, Definition (5.6.1a) ensures that $\mathbf{Area}(\Delta x, \Delta t)$ has units $\dfrac{distance}{time}$. For simplicity, however, the symbols $(1/time^2)$, will henceforth be omitted.

The following theorem shows how constant speeds can also be determined from areas of triangles in the spacetime graph.

(5.6.2)
Trig
THEOREM: (***Relative Speed in Terms of Areas***) *Given synchronized frames $\mathcal{F}^A$ and $\mathcal{F}^B$ with respective time axes T' and T''. Let the line$(A'A'')$ intersect time axes T' and T'' at time $t = \sqrt{2}$. Then frame $\mathcal{F}^B$ has speed v relative to frame $\mathcal{F}^A$ if and only if*

(5.6.2a) $Area(\triangle OA'A'') = v.$

PROOF: (*Our fundamental strategy is to use the fact that the length of a triangle's* <u>base</u> *at any time t is the distance between the origins of the two frames at time t.*)

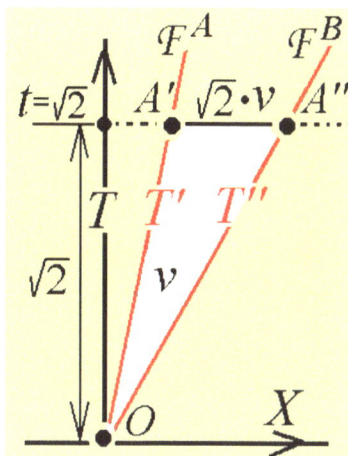

Accordingly, with the notation of (A.2.13a), denote the length of line

$\overline{A'A''}$ by $\|\overline{A'A''}\|$, which, according to the figure of (5.6.2), is the length of the base of $\triangle OA'A''$. The length of the base, in turn, corresponds to the distance between the origins of synchronized frames, $\mathcal{F}^A$ and $\mathcal{F}^B$, after time $t = \sqrt{2}$. But the relative speed between the frames is v if and only if this separation distance, $\|\overline{A'A''}\|$, has the property

(5.6.2b) $\|\overline{A'A''}\| = speed \cdot time = (v \cdot \sqrt{2}).$

Now $Area(\triangle OA'A'') = \dfrac{1}{2} \cdot base \cdot altitude$ $\qquad$ *plane geometry*

$$= \frac{1}{2} \cdot \|\overline{A'A''}\| \cdot \sqrt{2} \qquad \begin{array}{l} \textit{from figure of (5.6.2),} \\ \|\overline{A'A''}\| = base, \\ \sqrt{2} = altitude \end{array}$$

$$= \frac{1}{2} \cdot (v \cdot \sqrt{2}) \cdot \sqrt{2} \qquad \textit{from (5.6.2b)}$$

$$= v,$$

which proves (5.6.2a) and completes the proof. ∎

Theorem (5.6.2) provides a geometric way to prove Theorem (5.5.2), as follows:

(5.6.3) RESTATEMENT OF THEOREM **(5.5.2)**
Given:
(5.6.3a)   *Frame $\mathcal{F}^B$ has speed v relative to $\mathcal{F}^A$, and*

(5.6.3b)   *frame $\mathcal{F}^C$ has speed w relative to $\mathcal{F}^B$.*

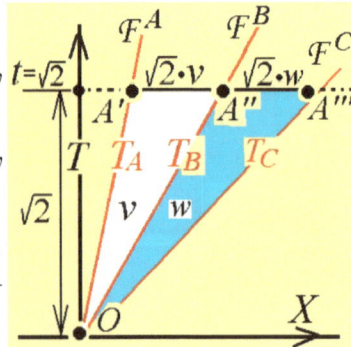

Then............................
(5.6.3c)   *frame $\mathcal{F}^C$ has speed $v + w$ relative to $\mathcal{F}^A$.*

PROOF: From Theorem (5.6.2), frame $\mathcal{F}^B$ has speed v as seen from frame $\mathcal{F}^A$ if and only if $Area(\triangle OA'A'') = v$. Similarly, frame $\mathcal{F}^C$ has speed w as seen from frame $\mathcal{F}^B$ if and only if $Area(\triangle OA''A''') = w$. (*See diagram above.*) Now the two triangles together form this single triangle, $\triangle OA'A'''$, which has area $v + w$. From Theorem (5.6.2), this area represents the speed of $\mathcal{F}^C$ relative to $\mathcal{F}^A$. Having established (5.6.3c), the alternate, geometric proof of Theorem (5.5.2) is done. ∎

III. The Speed of Light is Constant

Chapter 6 *shows why a pair of events occurring simultaneously at two separate locations in one frame can never occur simultaneously from the point of view of another moving frame.*

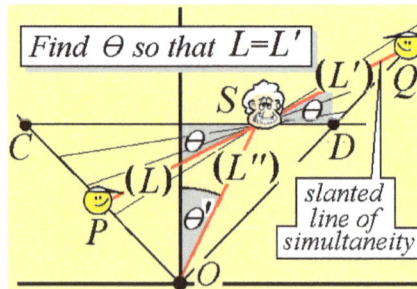

Find θ so that L=L'

(6) Constant c in Spacetime

(6.1) Minkowski Spacetime Diagrams

In the definition following, we extend the notion of a spacetime diagram (4.1.2) to a Minkowski diagram by adding the time-paths (worldlines) of leftward traveling photons ←☺ and rightward traveling photons ☺→ (e.g., Figure (6.1.2)).

(6.1.1)

> **DEFINITION:** *The* MINKOWSKI DIAGRAM
> - *is a spacetime graph of all points (x,t) for one or more synchronized (3.4.5a) frames of reference moving with constant velocity, each of which is represented in spacetime by its own X and T axes.*
> - *At time $t = 0$ two photons are emitted from point $x = 0$ — one traveling leftward and the other traveling rightward. Relative to any inertial frame, $\mathcal{F}$, the world lines of these synchronized photons form straight lines emanating from the origin.*

(6.1.2) **FIGURE:** *Minkowski Diagram for 2 Observers and 2 Photons*

Diagram (6.1.2) shows perpendicular X_A and T_A axes for Ashley's frame. There are four synchronized worldlines (*see Def.* (3.4.5a)) namely,

• The two worldlines of the synchronized leftward and rightward traveling photons, respectively.

• Ashley's vertical worldline in frame $\mathcal{F}^A$ which coincides with T^A, the $\mathcal{F}^A$ time axis,

• Bernie's slanted worldline in frame $\mathcal{F}^B$ which coincides with T^B, the $\mathcal{F}^B$ time axis that forms angle θ with the vertical T_A. All frames must have their T-axes (the worldline of $x = 0$) overlap at opposite ends of a line-of-sight corridor at common time $t = 0$. This is the mutual synchronization requirement of (3.4.5) and (3.4.3).

(6.1.3) | NOTE: *As shown in Figure* (4.4.4), *lines of simultaneous events within a frame $\mathcal{F}^A$ are drawn parallel to the X-axis of that frame. In Galilean/Newtonian relativity, one master (universal) clock serves all observers in all frames — simultaneous events at time t_0 for all frames $\mathcal{F}^A$, $\mathcal{F}^B$, $\mathcal{F}^C$, ... , can always be represented as straight lines of simultaneous $t = t_0$ events that are pairwise parallel (see Figure* (5.3.3)).

However, in Section (6.3), *we will see that if the speed of light c is the same in all frames, then simultaneous events at time t_0 in various frames, can no longer be represented as straight lines that are mutually parallel.*

(6.2) Constant c and Simultaneity

(6.2.1) Synchronized Observers are Always Equidistant from Synchronized Photon Pairs.

The Setup: Since a Minkowski diagram will require two synchronized photons, there will always be (at least) three actors who are simultaneously synchronized at $x = 0$ and time $l = 0$ (3.4.5a), namely,

(1) an observer, who is traveling at speed $\pm v$ (which may be zero),

(2) a leftward moving photon ←☺— traveling at speed $-c$,

(3) a rightward moving photon —☺→ traveling at speed $+c$.

Display (6.2.2) is a reformulation of (1.5.2b), which says the speed of light is the same for all observers:

(6.2.2)

> **Constant Speed of Light and Equidistance.** *An observer at constant speed arrives at source point $x = 0$ and time $t = 0$ from which leftward and rightward traveling synchronized (2.2.5) photons are emitted.*
>
> *To say the speed of light (of the two photons) is absolute (the same for all observers), is to say that regardless of relative speeds of the observer and the source point $x = 0$, the observer is, at all times, SIMULTANEOUSLY EQUIDISTANT from each of the two photons. That is, at any time, the distances from the observer to each to the two photons will be the same.*

Ashley's Simultaneous Events. In Figure (6.1.2), a horizontal line that captures Ashley's position on her T_A-axis also intersects the slanted worldlines at a' and a'' where, for <u>any</u> time t_0, the two photons are at equal distances from Ashley. Hence, for any time t_0, Ashley's lines of simultaneity in Figure (6.1.2) are <u>horizontal</u> and <u>parallel</u> to her X_A-axis.

We shall now see why Bernie's line of simultaneity (*the line on which each point shows the same time*) can never be horizontal.

(6.2.3) *Bernie's Simultaneous Events.*

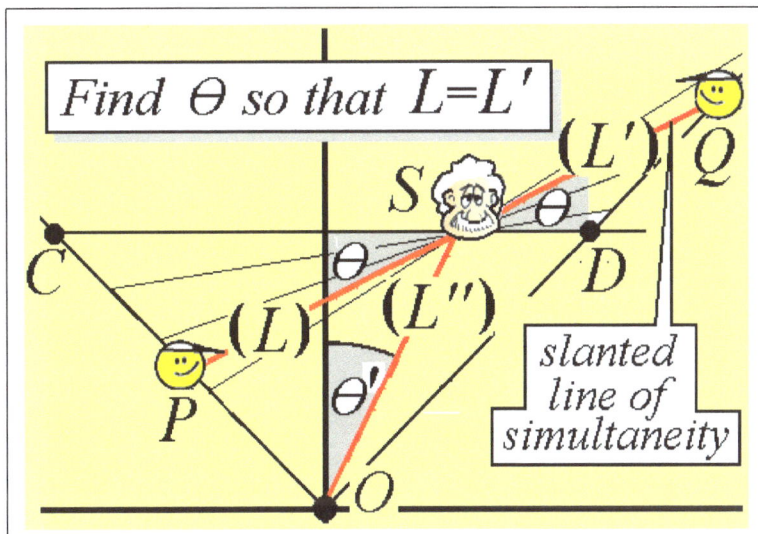

This figure shows Bernie at the pivot point S on a swiveling line $\overline{PSQ}$ of simultaneity. Each point on $\overline{PSQ}$ has the same time coordinate in Bernie's frame. The swiveled line forms an angle θ with the horizontal. For all angles θ, the lengths L and L' on line $\overline{PSQ}$ are the respective distances between Bernie and the left and right photons. (*This follows from* (4.4.1a) *and* (4.4.3) *since lines of simultaneity like* $\overline{PSQ}$ *represent (a portion of) the X-axis at a particular time.*)

Hence, *The only way* Bernie at speed v is always at the midpoint between the two photons is when line $\overline{PSQ}$ is swiveled to the exact angle $\theta \neq 0$ at which $L = L'$. This appropriately-swiveled line becomes Bernie's ***line of simultaneity which cannot be horizontal.*** This easily follows from Figure (6.2.3) where a horizontal line $\overline{PSQ}$ ($\theta = 0$) shows $L > L'$.

(6.3) How Constant c Destroys Simultaneity

More accurately, we should label this section, *"How Constant c Destroys Universal Simultaneity."* With the Minkowski graph in hand, we have a visual model that explains how two observers who are in motion with respect to each other can not see the same events at the same time according to their previously synchronized and identical clocks.

(**6.3.1**) **To fix ideas,** the left photon is red and the right photon is blue.

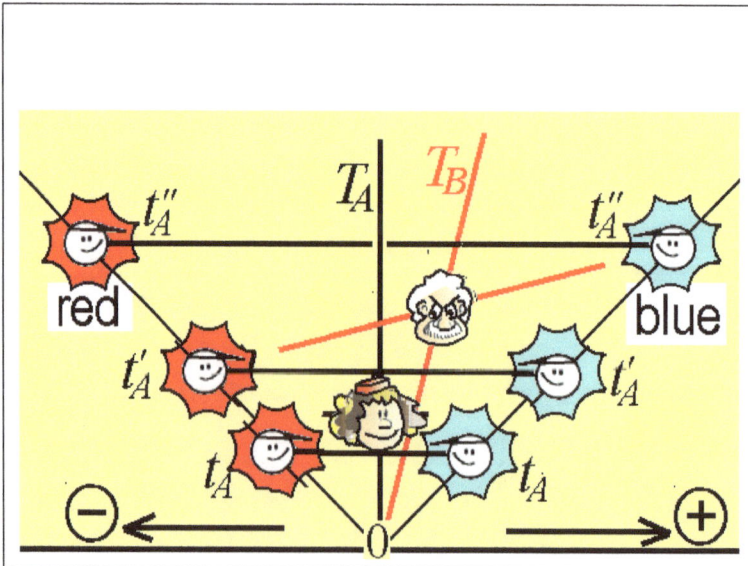

Then at time t_0 Ashley sees the red and blue photons along a *horizontal* line of simultaneity. That is, the three clocks at the positions of the two photons and of Ashley each read time t_A in $\mathcal{F}^A$. The very pair of red/blue photons that Ashley perceives *simultaneously* along her horizontal line of constant $\mathcal{F}^A$ time t_A, are perceived sequentially by Bernie along two *separate slanted lines* of constant $\mathcal{F}^B$ times, t_B and $t_B + \Delta$.

The blue photon will be observed first by Bernie at time t_B where the clocks at Bernie's position and the blue photon's position both read t_B in frame $\mathcal{F}^B$. Similarly, the red photon will be observed by Bernie later at $\mathcal{F}^B$ time $t_B + \Delta$ when the two clocks, one at Bernie's position and the other at the red photon's position, both read $t_B + \Delta$.

This is how the constant value of c for all observers destroys common simultaneity for any two observers who are in relative motion with respect to each other.

Summary

This chapter presents a prime example of how the fusion of mathematics, cartoons, and reality brings us to an unexpected feature of reality and a new level of understanding. We first encountered the fundamental property of special relativity — that the speed of light is constant for all observers (1.5.2b). With the rephrasing of (6.2.2), we then established the cartoon of Figure (6.2.3) and applied mathematical analysis to lines of simultaneity that correspond to reciprocal slopes. This analysis further led us to the illustration of (6.3.1) that produced a new observation, namely,

(6.3.2) *If two observers are in motion relative to each other, then two events that are simultaneous for one can never be simultaneous for the other.*

Property (6.3.2) is not likely to be revealed through our five senses alone.

(6.4) Exercise

(6.4.1)
Exercise
The Swiveled Line Theorem.
(*see animation at* `www.Special-Relativity-Illustrated.com`)
Assume the speed of light $c = 1$ so that the worldlines of the two
photons of Figure (6.2.3) are at angles $\pm 45°$,i.e., the lines have slopes
± 1. Then show that:

- If v is the speed of Bernie's frame $\mathcal{F}^B$ relative to Ashley's frame
  $\mathcal{F}^A$, then $v = \tan(\theta')$.

- In order to swivel line $\overline{PSQ}$ to the exact angle θ so that distances
  $L = L'$, then necessarily $\theta = \theta'$ and all three lengths L, L', and
  L'' are equal to each other.

IV. Lorentz Transformations of Frames

Chapter 7 *explores the relation between pairs of frames in a linear algebraic context (Compare with the description of Chapter 5.) A well-defined matrix arises in the line-of-sight connection between Lorentzian, relativistic frames. The overriding relativistic assumption is that time is not universal — it varies with each observer. Consistent with this property is that c, the speed of light, is finite and does not vary with either the relative speeds of the source or of any observer.*

$$L_v$$

$$\frac{1}{\sqrt{1-(v/c)^2}}\begin{bmatrix}1 & -v \\ -(v/c^2) & 1\end{bmatrix} * \begin{bmatrix}x_1 \\ t_1\end{bmatrix}_A = \begin{bmatrix}x_2 \\ t_2\end{bmatrix}_B$$

I see speed v — *of these coordinates* — *when I stand here.*

Chapter 8 *studies the track of origins of moving frames that trace out a well-defined hyperbola in the Minkowski X-T diagram. This hyperbola in relativistic physics corresponds to a straight line in classical Galilean/Newtonian physics.*

(7) Lorentz Transformations

The Setup for Figure (7.1.2)

(7.1) The Lorentz Matrix

(7.1.1) **PREVIEW:** In the Minkowski diagram (7.1.2) following, the world-lines are plotted for four players:

- Ashley at the origin in inertial frame $\mathcal{F}^A$ whose worldline is vertical,

- Bernie at the origin in synchronized frame $\mathcal{F}^B$ (3.4.5a) moving with relative speed v. Bernie's worldline coincides with T_B, the time axis of $\mathcal{F}^B$.

- Leftward and a rightward traveling photons whose worldlines contain the line segments from $\boxed{1}$ to $\boxed{3}$ and from $\boxed{2}$ to $\boxed{4}$, respectively.

(7.1.2) FIGURE: *Lorentz Linking of Synchronized Frames*

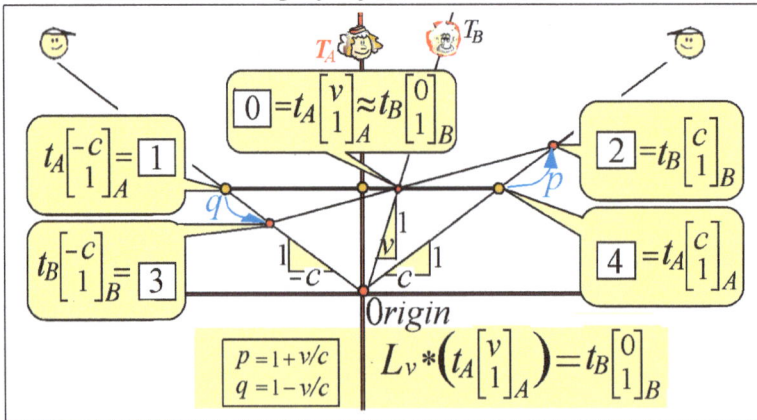

The values $p = (1 + v/c)$ and $q = (1 - v/c)$ in Figure (7.1.2) will be established in (7.1.3f).

Line-of-Sight Coordinates of Point $\boxed{0}$**, Fig. (7.1.2).** Ashley and Bernie, who are fixed at their respective origins $x = 0$, are synchronized at time $t = 0$ in each frame. Therefore, after time t_A in Ashley's frame $\mathcal{F}^A$, Bernie travels vt_A units to Ashley's right in frame $\mathcal{F}^A$. At this time t_A, Bernie in $\mathcal{F}^B$ is seen by an observer in $\mathcal{F}^A$ from his/her $\mathcal{F}^A$ coordinates $[\, vt_A, t_A \,]_A = t_A[\, v, 1 \,]_A$.

At the other end of the line-of-sight corridor in $\mathcal{F}^B$, Bernie is positioned in $\mathcal{F}^B$ at coordinates $[0, t_B]_B$ for some $\mathcal{F}^B$ time t_B. (*Bernie always stands at his own origin* $x = 0$.)

The times t_A and t_B at the opposite ends of the line-of-sight corridor are not assumed to be equal. We can say, however, that the line-of-sight linkage of coordinates at Point $\boxed{0}$ of Figure (7.1.2),

$$t_A \begin{bmatrix} v \\ 1 \end{bmatrix}_A \quad with \quad t_B \begin{bmatrix} 0 \\ 1 \end{bmatrix}_B$$

is expressed by the Lorentz transformation equation

(7.1.2a) $\qquad t_A L_v * \begin{bmatrix} v \\ 1 \end{bmatrix}_A = t_B \begin{bmatrix} 0 \\ 1 \end{bmatrix}_B .$

Coordinates Reflecting Constant Speed of Light c. All four traveling actors, namely, (i) Ashley, (ii) Bernie, (iii) the leftward traveling photon, and (iv) the rightward traveling photon, find themselves at point $x = 0$ at the common synchronization time, $t = 0$.

It follows that at time t_A in $\mathcal{F}^A$, both left and right photons are at a distance $\pm t_A c$ from the $\mathcal{F}^A$ origin where Ashley is always located. Figure (7.1.2) illustrates the $\mathcal{F}^A$ coordinates at time t_A of the two photons as

(7.1.2b) $\qquad Point \boxed{1} = t_A \begin{bmatrix} -c \\ 1 \end{bmatrix}_A , \quad and \quad Point \boxed{4} = t_A \begin{bmatrix} c \\ 1 \end{bmatrix}_A .$

Similarly, at arbitrary time t_B in $\mathcal{F}^B$, both left and right photons are $\pm t_B c$ units from the $\mathcal{F}^B$ origin where Bernie is always located. Figure (7.1.2) shows the $\mathcal{F}^B$ coordinates at time t_B of the two photons, namely,

(7.1.2c) $\qquad Point \boxed{3} = t_B \begin{bmatrix} -c \\ 1 \end{bmatrix}_B , \quad and \quad Point \boxed{2} = t_B \begin{bmatrix} c \\ 1 \end{bmatrix}_B .$

(7.1.2d)

. . . .

NOTE: *As shown in Figures* (6.2.3) *and* (7.1.2), *the line of constant time t_B in frame $\mathcal{F}^B$ cannot be horizontal if we require equal distances from the origin to the photons at equal times.*

The coordinates of Figure (7.1.2) are the basis for the following theorem that establishes the Lorentz line-of-sight transformation.

(7.1.3)

THEOREM: (***The Lorentz Matrix***) *An inertial frame $\mathcal{F}^B$ has speed v relative to a synchronized frame $\mathcal{F}^A$. The line-of-sight Lorentz linear transformation $L_v : \mathcal{F}^A \to \mathcal{F}^B$ has the matrix form*

(7.1.3a)
$$L_v = \gamma_v \begin{bmatrix} 1 & -v \\ -v/c^2 & 1 \end{bmatrix}$$

where the LORENTZ FACTOR *γ_v (and* SHRINKAGE FACTOR *σ_v) are defined by*

(7.1.3b)
$$\gamma_v \stackrel{def}{=} \frac{1}{\sqrt{1 - (v/c)^2}} = \frac{1}{\sigma_v} \qquad (see \ (1.8.2)).$$

PROOF: The strategy is to find the values of the linear transformation $L_v : \mathcal{F}^A \to \mathcal{F}^B$ on just the two vectors $\{[-c, 1]_A, [c, 1]_A\}$ which happen to form a basis of the two-dimensional spacetime graph (7.1.2). With these two vector values in hand, the full Lorentz 2×2 matrix is determined as per (A.4.2).

- **Eigenvector-Eigenvalues of the Lorentz Transformation:** (*see Figure* (7.1.2).) Multiply Point $\boxed{1}$ of $\mathcal{F}^A$ by a (shrinking) factor $q < 1$ so that it "slides" down its worldline toward the origin to meet and align with Point $\boxed{3}$ of $\mathcal{F}^B$. Thus, we have the pair of overlapping coordinates

(7.1.3c)
$$\underbrace{q\, t_A[-c, 1]_A}_{q \times \boxed{1}} \ \leftrightarrow \ \underbrace{t_B[-c, 1]_B}_{\boxed{3}}$$

that are at opposite ends of a line-of-sight corridor that runs between frames $\mathcal{F}^A$ and $\mathcal{F}^B$. Similarly, multiply Point $\boxed{4}$ by an (expanding) factor $p > 1$ so that it "slides" up its worldline away from the origin to align with Point $\boxed{2}$. That is, the $\mathcal{F}^A$ and $\mathcal{F}^B$ coordinates

$$(\textbf{7.1.3d}) \qquad \underbrace{p\,t_A[c,1]_A}_{p\times\boxed{4}} \quad \leftrightarrow \quad \underbrace{t_B[c,1]_B}_{\boxed{2}}$$

are at opposite ends of a line-of-sight corridor that runs between $\mathcal{F}^A$ and $\mathcal{F}^B$. But to be paired at opposite ends of a line-of-sight corridor is to be linked by the Lorentz transformation L_v as follows:

$(\textbf{7.1.3e})$

$$L_v\left(\begin{bmatrix} -c \\ 1 \end{bmatrix}_A\right) = \frac{t_B/t_A}{q} \begin{bmatrix} -c \\ 1 \end{bmatrix}_B \quad and \quad L_v\left(\begin{bmatrix} c \\ 1 \end{bmatrix}_A\right) = \frac{t_B/t_A}{p} \begin{bmatrix} c \\ 1 \end{bmatrix}_B$$

where $[-c,1]_A$ and $[c,1]_A$ are eigenvectors of L_v with respective eigenvalues $(t_B/t_A)/q$ and $(t_B/t_A)/p$ (*see Definition* (A.5.1)).

• **Values of numbers (scalars)** p **and** q **in (7.1.3e).** The $\mathcal{F}^A$ coordinates of midpoint $\boxed{0}$ are $t_A[v,1]_A$ which is the average of the endpoints, $\boxed{2}$ and $\boxed{3}$, in $\mathcal{F}^A$ coordinates, which we read from the left hand sides of (7.1.3c) and (7.1.3d). The average $(\boxed{3}+\boxed{2})/2=\boxed{0}$ in $\mathcal{F}^A$ coordinates implies

$$\underbrace{q t_A \begin{bmatrix} -c \\ 1 \end{bmatrix}_A}_{\boxed{3}} + \underbrace{p t_A \begin{bmatrix} c \\ 1 \end{bmatrix}_A}_{\boxed{2}} = \underbrace{2 t_A \begin{bmatrix} v \\ 1 \end{bmatrix}_A}_{2\boxed{0}} \implies \begin{array}{l} (p-q) = 2v/c \\ (p+q) = 2 \end{array}$$

which gives us the values

$$(\textbf{7.1.3f}) \qquad \boxed{\begin{array}{l} p = (1 + v/c) \\ q = (1 - v/c) \end{array}} \quad \begin{array}{l} which \\ in\ turn \\ implies \end{array} \quad \boxed{\begin{array}{l} \dfrac{1}{p} - \dfrac{1}{q} = \dfrac{-2(v/c)}{1 - (v/c)^2} \\[2mm] \dfrac{1}{p} + \dfrac{1}{q} = \dfrac{2}{1 - (v/c)^2} \end{array}} \ .$$

• **Using the Decomposition Theorem (A.5.4):** Since (7.1.3e) provides a pair of eigenvectors and eigenvalues for L_v we use the decomposition theorem (A.5.4), pg. 348 to write

$$L_v = S * D * S^{-1}$$

$$= \begin{bmatrix} -c & c \\ 1 & 1 \end{bmatrix} * \begin{bmatrix} (t_A/t_B)/q & 0 \\ 0 & (t_A/t_B)/p \end{bmatrix} * \frac{-1}{2c} \begin{bmatrix} 1 & -c \\ -1 & -c \end{bmatrix}$$

$$= \frac{(t_A/t_B)}{2} \begin{bmatrix} \left(\frac{1}{p} + \frac{1}{q}\right) & c\left(\frac{1}{p} - \frac{1}{q}\right) \\ \frac{1}{c}\left(\frac{1}{p} - \frac{1}{q}\right) & \left(\frac{1}{p} + \frac{1}{q}\right) \end{bmatrix}$$

(7.1.3g) $$= \frac{t_A/t_B}{(1 - (v/c)^2)} \begin{bmatrix} 1 & -v \\ -v/c^2 & 1 \end{bmatrix}$$ *from* (7.1.3f).

• **Symmetry and Inverses.** L_v is the line-of-sight function where frame $\mathcal{F}^B$ has speed v relative to frame $\mathcal{F}^A$. By symmetry (1.6.1), if we view the line-of-sight corridor from the opposite direction, then L_{-v} is the line-of-sight function where frame $\mathcal{F}^A$ has speed $-v$ relative to frame $\mathcal{F}^B$. That is, L_{-v} is the inverse of L_v (*see also Proposition* (3.4.6e)), which is to say

$$\underbrace{\frac{(t_B/t_A)}{1 - (v/c)^2} \begin{bmatrix} 1 & v \\ v/c^2 & 1 \end{bmatrix}}_{\substack{L_{-v} \\ \textit{replace } v \textit{ with } -v \textit{ in} \\ L_v \textit{ of (7.1.3g)}}} = \underbrace{\frac{1}{(t_B/t_A)} \begin{bmatrix} 1 & v \\ v/c^2 & 1 \end{bmatrix}}_{\substack{L_v^{-1} \\ \textit{apply (A.4.4d) to } L_v \\ \textit{in (7.1.3g)}}}.$$

Equating the scalar multiples of the (equal) matrices above yields the time dilation formula

(7.1.3h) $(t_B/t_A) = \sqrt{1 - (v/c)^2}.$ { *note:* $t_B \leq t_A$ }

Substituting the value of (t_B/t_A) from (7.1.3h) into (7.1.3g) finally yields (7.1.3a) which ends the proof. ∎

(7.2) Pattern of the Lorentz Matrix

(7.2.1) **An Intuitive Reading of Matrix L_v (7.1.3a).** Once we apply the 2×2 Lorentz matrix L_v (7.1.3a), the domain frame, $\mathcal{F}^A$, containing the (input) vector, $[x, t]_A$, is considered to be stationary and the range frame, $\mathcal{F}^B$, containing the (output) vector, $[x', t']_B$, is seen to be moving with speed v relative to $\mathcal{F}^A$.

All this information is embedded in the Lorentz matrix itself as summarized in Figure (7.2.2), a counterpart to the pattern (5.4.1a) of the Galilean matrix.

(7.2.2) FIGURE: *Pattern of the Lorentz Matrix* (7.1.3a)

Lorentz matrix $L_v : \mathcal{F}^A \to \mathcal{F}^B$ is the product of the number (scalar) $\gamma_v = 1/\sqrt{1-(v/c)^2}$ and the 2×2 matrix with 1s on the diagonal and scalars $-v$ and $-v/c^2$ in the northeast and southwest corners respectively. The speed v appears twice in the 2×2 matrix and once in the radical term, $\sqrt{1-(v/c)^2}$. The matrix corners always contain the <u>negative</u> of the speed of $\mathcal{F}^B$ as observed from frame $\mathcal{F}^A$.

(7.3) The Lorentz Sum of Speeds

We define a special addition between any pair of numbers, v and w.

(7.3.1)

> Definition: *For numbers v and w, the symbol $v \oplus w$ denotes their Relativistic Sum or Lorentz Sum and is defined to be*
>
> **(7.3.1a)** $\qquad v \oplus w \overset{def}{=} \dfrac{v + w}{1 + (vw/c^2)}$
>
> *where c is the speed of light, and the numerator $(v + w)$ and denominator $(1 + (vw/c^2))$ are the ordinary arithmetic sums.*

(7.3.2) The odd-looking sum **(7.3.1a)** may seem to be unmotivated. However, as we shall see in Theorem (7.3.3), the sum $v \oplus w$ is exactly what is needed to describe relative speeds between moving frames. Also, we have already noted in (1.13.3), pg. 39, that it is the this very sum $v \oplus w$ of (7.3.1a) that resolves the Pea-Shooter Paradox.

(7.3.3)

> Theorem: (*An Algebraic Property of Lorentz Matrices*)
> *For Lorentz matrices L_v and L_w where v and w are arbitrary speeds (or scalars), we have*
>
> **(7.3.3a)** $\qquad \boxed{L_v * L_w = L_{v \oplus w}}$
>
> *where $v \oplus w$ is the Lorentz sum (7.3.1).*

Proof: From (5.3.4a), we have

$$L_v * L_w = \frac{1}{\sqrt{1 - v^2/c^2}} \begin{bmatrix} 1 & -v \\ -v/c^2 & 1 \end{bmatrix} * \frac{1}{\sqrt{1 - w^2/c^2}} \begin{bmatrix} 1 & -w \\ -w/c^2 & 1 \end{bmatrix}$$

which, after matrix multiplication (A.4.3e) that entails some disagreeable algebra, allows us to continue the equality with

$$= \frac{1}{\sqrt{1 - \left[\frac{(v+w)/c}{1+(vw/c^2)}\right]^2}} \begin{bmatrix} 1 & -\left[\frac{v+w}{1+(vw/c^2)}\right] \\ -\frac{1}{c^2}\left[\frac{v+w}{1+(vw/c^2)}\right] & 1 \end{bmatrix}.$$

$$= \frac{1}{\sqrt{1-[(v \oplus w)/c]^2}} \begin{bmatrix} 1 & -(v \oplus w) \\ -(v \oplus w)/c^2 & 1 \end{bmatrix} \quad \begin{array}{l} \textit{from (7.3.1)} \\ \textit{defining} \\ (v \oplus w) \end{array}$$

$$= L_{v \oplus w}. \qquad \begin{array}{l} \textit{from Figure (7.2.2), which shows} \\ \textit{the pattern for the three entries} \\ \textit{that contain the speed } (v \oplus w) \end{array}$$

Property (7.3.3a) is now established, proving the theorem. ∎

(7.4) Addition of Speeds via Matrices

The following theorem, which justifies the form for $v \oplus w$ based on the Lorentz matrix transformation, is the relativistic counterpart to Theorem (5.5.2).

(7.4.1)

> **THEOREM:** (***Lorentzian Addition of Speeds***) *Given three frames of reference, $\mathcal{F}^A$, $\mathcal{F}^B$, and $\mathcal{F}^C$, in 1-dimensional space that are synchronized at time $t = 0$ and distance $x = 0$. Given*
>
> **(7.4.1a)** $\quad \mathcal{F}^A \xrightarrow{L_v} \mathcal{F}^B.$ *Observers in $\mathcal{F}^A$ see $\mathcal{F}^B$ moving at speed v.*
>
> **(7.4.1b)** $\quad \mathcal{F}^B \xrightarrow{L_w} \mathcal{F}^C.$ *Observers in $\mathcal{F}^B$ see $\mathcal{F}^C$ moving at speed w.*
>
> *Then* ....
>
> **(7.4.1c)** $\quad \mathcal{F}^A \xrightarrow{L_{v \oplus w}} \mathcal{F}^C.$ *Observers in $\mathcal{F}^A$ see $\mathcal{F}^C$ moving at speed $v \oplus w$ (7.3.1a).*

PROOF: Hypotheses (7.4.1a) and (7.4.1a) say

(7.4.1d) $\quad L_v([x,t]_A) = [x',t']_B \qquad \left\{ \begin{array}{l} [x,t]_A \textit{ in } \mathcal{F}^A \\ [x',t']_B \textit{ in } \mathcal{F}^B \end{array} \right.$

and

(7.4.1e) $\quad L_w([x',t']_B) = [x'',t'']_C \qquad \left\{ \begin{array}{l} [x',t']_B \textit{ in } \mathcal{F}^B \\ [x'',t'']_C \textit{ in } \mathcal{F}^C. \end{array} \right.$

Substituting $[x',t']_B$ of (7.4.1d) into (7.4.1e), we obtain

(7.4.1f) $\quad L_w(L_v([x,t]_A)) = [x'',t'']_C. \qquad \left\{ \begin{array}{l} [x,t]_A \textit{ in } \mathcal{F}^A \\ [x'',t'']_C \textit{ in } \mathcal{F}^C. \end{array} \right.$

Now, $\qquad L_w(L_v([x,t]_A)) = L_w \circ L_v([x,t]_A) \qquad \textit{from (A.3.6)}$

$$= L_{v \oplus w}([x,t]_A).$$ *from (A.3.6) and Theorem (7.3.3)*

so that (7.4.1f) rewrites itself as $L_{v\oplus w}([x,t]_A) = [x'',t'']_C$ or $L_{v\oplus w}: \mathcal{F}^A \to \mathcal{F}^C$ which confirms (7.4.1c). The proof is done. ∎

(7.4.2) **The sum** $v \oplus w$ **does not have properties** we might expect of ordinary, familiar addition. For example, it is $\boxed{\textbf{\textit{not}}}$ true that for *all* u, v, and for *any* scalar α,

(7.4.2a) $(\alpha u) \oplus v = u \oplus (\alpha v) = \alpha(u \oplus v).$ *Exercise (7.6.1)*

We do have, however, the following structure:

(7.4.3)

> **PROPOSITION:** *There are only three instances for which the Lorentz sum, $v \oplus w$ (7.3.1a), and the ordinary arithmetic sum, $v + w$, agree, namely,*
>
> $$u \oplus v = u + v \quad \textit{if and only if} \quad \begin{cases} u = 0, & or \\ v = 0, & or \\ u + v = 0. \end{cases}$$

PROOF: Exercise (7.6.3) ∎

(7.4.4) **NOTE:** *If u and v are speeds, then the Lorentz sum $u \oplus v$ of (7.3.1a) shows that speeds of material objects can never exceed c. That is, $-c < u, v < c$. The straightforward proof is left as Exercise (7.6.2).*

(7.5) Addition of Speeds via Areas

In Theorem (7.5.6) following, we present the relativistic counterpart of Theorem (5.6.2) that measures speeds of one frame relative to another by the addition of areas.

Results in this section depend on properties of the particular hyperbola

- $y^2 - x^2 = 1$ that is symmetric about the Y-axis where

- the hyperbola asymptotes are the $\pm 45^o$ lines as measured from the X-axis.

The $\pm 45^o$ slopes of the asymptotes force the speed of light to be $c = 1$ (see Def. (1.13.5a), pg. 39). In the special case where $c = 1$, we have from (B.3.5a) that relative speeds between frames have the form $v = \tan(\theta) = \tanh(r_{A,B})$. This is the first time the speed c is associated with the trigonometric quantity, $\tan(\theta)$ (*see* (B.1.1c), *pg. 350 for definition of the hyperbolic function* $\tanh(\cdot)$.)

(7.5.1) **Rescaling the X-axis:** To pass from the special case $c = 1$ to more general and arbitrary units for c, we rescale the original X-axis (*where $c = 1$*) by replacing each x with $c\,x$ (*where the value of c can now be arbitrary*). The rescaling now allows for *non-unity* units for c such as $c = 2.997\,924\,58 \times 10^8$ meters/sec and $c = 186,000$ miles/sec.

We interpret the special-case Figure (B.3.4), pg. 355 as a Minkowski spacetime diagram where the T_A and T_B axes represent the world-lines of the origins of synchronized inertial frames $\mathcal{F}^A$ and $\mathcal{F}^B$, respectively. The two $\pm 45^o$ asymptotes will represent the worldlines of leftward and rightward traveling photons that are simultaneously released at time $t = 0$ from the synchronized origins.

(7.5.2) FIGURE: *worldlines of Two Photons (Leftward and Rightward) and Three Origins of Synchronized Inertial Frames*

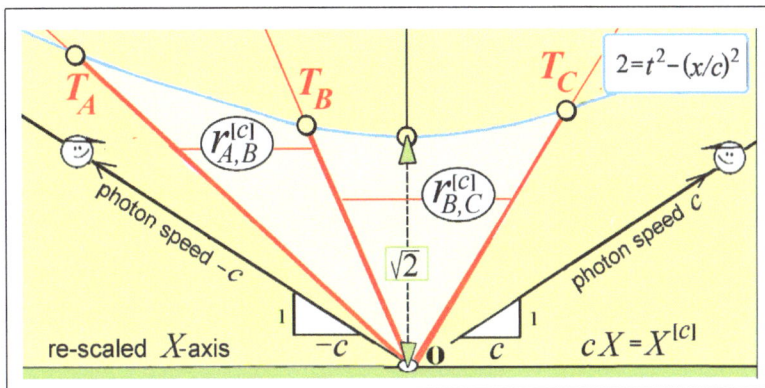

(7.5.3) NOTATION: *Unscaled and Scaled Quantities.* As noted in (7.5.1), the X-axis $X^{[c]}$ of the Minkowski diagram of (7.5.2) is constructed by replacing each x of the original X-axis (*where* $c = 1$) with the value cx. Then

(7.5.3a)

Figure (7.5.2)	Unscaled $c=1$		Scaled $c>0$
Axes	x	$\rightarrow$	$cx = x^{[c]}$
Speed of one frame relative to another	v	$\rightarrow$	$v^{[c]}$
Triangular area bounded by hyperbola, T_A *and* T_B-*axes*	$r_{A,B}$	$\rightarrow$	$r_{A,B}^{[c]}$

(7.5.4) | | **NOTE:** In the notation of (7.5.3) above, $v^{[1]}$ and $r_{A,B}^{[1]}$ coincide with the speed v and area $r_{A,B}$ of the unscaled graph, i.e., $v^{[1]} = v$ and $r_{A,B}^{[1]} = r_{A,B}$. We abandon this somewhat awkward superscript notation $v^{[c]}$ and $r_{A,B}^{[c]}$ when the units for c are clearly established and we no longer need to compare diagrams with differently scaled X-axes.

(7.5.5)
Calculus

PROPOSITION: *In Figure* (B.3.4), *replace the unscaled axis* X *with the scaled axis* $X^{[c]}$ *in which each* $x \in X$ *is replaced with* $cx \in X^{[c]}$. *Then*

(7.5.5a) *Hyperbola* $2 = t^2 - x^2$ *is transformed (rescaled) into the hyperbola* $2 = t^2 - (x^{[c]}/c)^2$. *Moreover,*

(7.5.5b) $\boxed{v^{[c]} = cv \quad \text{and} \quad r_{A,B}^{[c]} = c\,r_{A,B}}$

where $v^{[c]}$ and $r_{A,B}^{[c]}$ are defined in (7.5.3a).

PROOF: From the first line of Table (7.5.3a), we replace $x^{[c]}$ with cx in the scaled equation $2 = t^2 - (x^{[c]}/c)^2$ to obtain the unscaled equation $2 = t^2 - x^2$, which proves (7.5.5a).

Speed of $\mathcal{F}^B$ relative to $\mathcal{F}^A$ is $v^{[c]} = (x_1^{[c]} - x_0^{[c]})/(t_1 - t_0)$ where $[x_1^{[c]}, t_1]_A$ and $[x_0^{[c]}, t_0]_A$ are line-of-sight coordinates in $\mathcal{F}^A$ from which a particular point of $\mathcal{F}^B$ is observed at $\mathcal{F}^A$ times t_1 and t_0.

$$v^{[c]} = \frac{(x_1^{[c]} - x_0^{[c]})}{(t_1 - t_0)} = \frac{(c\,x_1 - c\,x_0)}{(t_1 - t_0)} = c\frac{(x_1 - x_0)}{(t_1 - t_0)} = c\,v,$$

which proves part of (7.5.5b). Now suppose the areas in Figure (7.5.2) are defined by a double integral over region R. In particular, since $x^{[c]} = c\,x$,

$$r_{A,B}^{[c]} = \iint_R dx^{[c]}\ dt = \iint_R c\ dx\ dt = c\iint_R dx\ dt = c\ r_{A,B}$$

Statements (7.5.5a) and (7.5.5b) are fully established, which ends the proof. ∎

(7.5.6) THEOREM: ***Sums of Areas as Speeds.*** *Given the scaled space-time graph (7.5.2) with general units for c where the areas $r_{A,B}^{[c]}$ and $r_{B,C}^{[c]}$ denote the areas (under the hyperbola $2 = t^2 - (x/c)^2$) between the T_A-T_B pair of axes and the T_B-T_C pair of axes, respectively.*

If $v^{[c]}$ denotes the speed of frame $\mathcal{F}^B$ relative to frame $\mathcal{F}^A$, and $w^{[c]}$ is the speed of frame $\mathcal{F}^C$ relative to frame $\mathcal{F}^B$, then

(7.5.6a) $\quad v^{[c]} = c \tanh\left(\dfrac{1}{c}\,r_{A,B}^{[c]}\right), \quad w^{[c]} = c \tanh\left(\dfrac{1}{c}\,r_{B,C}^{[c]}\right).$

The speed of frame $\mathcal{F}^C$ relative to frame $\mathcal{F}^A$ is

(7.5.6b) $\quad v^{[c]} \oplus w^{[c]} = \dfrac{v^{[c]} + w^{[c]}}{1 + (v^{[c]}w^{[c]})/c^2}.$

PROOF: **Proving (7.5.6a).**

$$
\begin{aligned}
v^{[c]} &= c\,v & &\text{\textit{from} (7.5.5b)}\\
&= c \tanh(r_{A,B}) & v &= \tanh(r_{A,B})\ \text{\textit{from} (B.3.5a)}\\
&= c \tanh(\tfrac{1}{c}r_{A,B}^{[c]}) & r_{A,B}^{[c]}/c &= r_{A,B}\ \text{\textit{from} (7.5.5b).}
\end{aligned}
$$

A similar argument, using $r_{B,C}$ instead of $r_{A,B}$, establishes (7.5.6a).

. . . .

Proving (7.5.6b).　　Using (7.5.6a), the speed of frame $\mathcal{F}^C$ relative to frame $\mathcal{F}^A$ is determined by the "double" area $r_{A,C}^{[c]}$ of Figure (7.5.2). To echo the notation of (1.13.3), we denote this speed of $\mathcal{F}^C$ relative to $\mathcal{F}^A$ by $(v^{[c]} \oplus w^{[c]})^{[c]}$. Then

$$v^{[c]} \oplus w^{[c]} = c\,\tanh(r_{A,C}^{[c]}/c) \qquad\qquad \textit{from (7.5.6a)}$$

$$= c\,\tanh(r_{A,C}) \qquad\qquad r_{A,C}^{[c]}/c = r_{A,C}\ \textit{from (7.5.5b)}$$

$$= c\,\tanh(r_{A,B} + r_{B,C}) \qquad\qquad \textit{addition of areas}$$

$$= c\,\frac{\tanh(r_{A,B}) + \tanh(r_{B,C})}{1 + \tanh(r_{A,B})\tanh(r_{B,C})} \qquad\qquad \begin{array}{l}\textit{property (B.1.3e)}\\\textit{of }\tanh(x+y)\end{array}$$

$$= c\,\frac{v + w}{1 + vw} \qquad\qquad \textit{from (B.3.5a), pg. 355}$$

$$= \frac{c\left(\dfrac{v^{[c]}}{c} + \dfrac{w^{[c]}}{c}\right)}{1 + (v^{[c]}/c)(w^{[c]}/c)} \qquad\qquad \textit{from (7.5.5b)}$$

$$= \frac{v^{[c]} + v^{[c]}}{1 + (v^{[c]}v^{[c]})/c^2}\ . \qquad\qquad \textit{cancel c in numerator}$$

Having established (7.5.6a) and (7.5.6b), the proof is done. ∎

(7.6)　Exercises

(7.6.1)
Exercise　　**Properties of $u \oplus v$.** Show examples that prove (7.4.2a).

(7.6.2) **Property of $u \oplus v$.** Show that for all numbers (speeds) u and v,
Exercise we have $-c < u \oplus w < c$.

(7.6.3) **Three cases of $u \oplus v$.** Prove Proposition (7.4.3)
Exercise

(8) The Hyperbola of Time-Stamped Origins

(8.0.4)

One of the most revolutionary results in modern science is Einstein's use of hyperbolic geometry to connect space and time. The explanation will require that we go through a series of theorems (no royal road). Nevertheless, we invite the reader to join us along this challenging yet rewarding path.

(8.1) Invariance of Minkowski Length

As noted on page 83, non-relativistic Galilean physics assumes that time is universal, which implies there is one universal clock whose readings are valid for all observers, regardless of their relative motion.

This universal clock physical property is interpreted mathematically by saying that time is invariant under the Galilean line-of-sight transformation (5.4.2), pg. 90, $G_v : [x, t]_A \rightarrow [x', t]_B$. That is to say, the <u>time</u> entries of the input coordinate pair $[x, t]_A$ and the output coordinate pair $[x', t]_B$ are always equal.

Similarly, we may ask what quantity might be invariant under the Lorentz transformation $L_v : [x, t]_A \rightarrow [x', t']_B$.

A first glance might not reveal an obvious invariant quantity for L_v. However, Theorem (8.1.2) will show a special "length" that remains intact for the linked coordinates $[x, t]_A$ and $[x', t']_B$. We define this "length" now:

(8.1.1)

DEFINITION: *For any event* $[x, t]$ *for any frame* $\mathcal{F}$ *in X-T spacetime, the* <u>MINKOWSKI LENGTH</u> *is defined to be the quantity*

(8.1.1a) $M([x, t]) \overset{def}{=} c^2 t^2 - x^2 \leftarrow \{\text{units} = distance^2\}$

where c is the speed of light in a vacuum.

Invariance of Minkowski Lengths Between Pairs of Frames.
Note that (8.1.1) defines $M([x,t])$ for a point (event) $[x,t]$ in *any* inertial frame $\mathcal{F}$. The following theorem shows that if two points $[x,t]_A$ and $[x',t']_B$ are in different frames $\mathcal{F}^A$ and $\mathcal{F}^B$ and lie on opposite ends of a line-of-sight corridor, then they share the same Minkowski length.

(8.1.2)

> THEOREM: (***Lorentz Invariance***) *Given the Lorentz transformation $L_v:\mathcal{F}^A \to \mathcal{F}^B$ (7.1.3a) If $L_v([x,t]_A) = [x',t']_B$, then the Minkowski length M of (8.1.1) is invariant for both line-of-sight points $[x,t]_A$ and $[x',t']_B$ in that*
>
> **(8.1.2a)**
>
> $$L_v([x,t]_A) = [x',t']_B \quad implies \quad M([x,t]_A) = M([x',t']_B).$$

PROOF: (*As usual, we write vectors of $\mathbf{R}^2$ either as horizontal row vectors or as vertical column vectors, depending on the context.*) Consider the following chain of equalities:

$$L_v\left(\begin{bmatrix} x \\ t \end{bmatrix}_A\right) = \underbrace{\gamma_v \begin{bmatrix} 1 & -v \\ -v/c^2 & 1 \end{bmatrix} * \begin{bmatrix} x \\ t \end{bmatrix}_A}_{L_v \text{ from (7.1.3a)}} = \underbrace{\begin{bmatrix} \gamma_v(x-vt) \\ \gamma_v(-vx/c^2+t) \end{bmatrix}_B}_{\substack{\text{from} \\ \text{(A.4.3e)}}} = \underbrace{\begin{bmatrix} x' \\ t' \end{bmatrix}_B}_{\substack{\text{from} \\ \text{(8.1.2a)}}}.$$

where $\gamma_v = 1/\sqrt{1 - (v/c)^2}$ (7.1.3b). It is the final (rightmost) equality that tells us

(8.1.2b) $x' = \gamma_v(x - vt)$ and $t' = \gamma_v\left(\dfrac{-vx}{c^2} + t\right).$

With the specific form (8.1.2b), algebraic manipulation shows that $(ct')^2 - (x')^2 = (ct)^2 - (x)^2$, which is to say, the Minkowski lengths, $M([x',t']_B)$ and $M([x,t]_A)$, as defined in (8.1.1), are equal. This confirms statement (8.1.2a) and ends the proof. ■

(8.2) The Time-Stamped Origins Theorem

One way to view the time axis T of any frame $\mathcal{F}$ is to view each of its coordinates $[0, t]$ as the origin $x = 0$ with a time-stamp t. More formally, we state

(8.2.1)

> DEFINITION: *For time t in frame $\mathcal{F}$ we say that $[0, t]$ is the* ORIGIN OF $\mathcal{F}$ WITH TIME-STAMP t.

NOTE: The set of all time-stamped origins of a <u>single frame $\mathcal{F}$</u> is T-axis of $\mathcal{F}$.

(8.2.2)

> NOTATION: (***Moving Frames and Coordinates***) *Given a fixed frame $\mathcal{F}_0$ with coordinates denoted $[x, t]_0$. Then any frame moving with constant speed v relative to frame $\mathcal{F}_0$ is denoted by $\mathcal{F}_v$ and its coordinates are written in the form $[x', t']_v$.*

Using the terminology of (8.2.1), we have a useful hyperbolic relation among a set of t_0-stamped origins for infinitely many frames (all origins are stamped at fixed time t_0).

(8.2.3)

> THEOREM: (***The Time-Stamped Origins Theorem***)
>
> *The graph of frame $\mathcal{F}_0$ is drawn with perpendicular X_0 and T_0 axes. For each speed v, $\mathcal{F}_v$ denotes the synchronized frame with speed v relative to $\mathcal{F}_0$.*
>
> *Suppose the time-stamped origin $[0, t_0]_v$ in $\mathcal{F}_v$ corresponds to $[x, t]_0$ in frame $\mathcal{F}_0$. That is, $L_v([0, t_0]_0) = [x, t]_v$.*
>
> *Then* ....
>
> *for any fixed time, $t_0 > 0$, and all speeds, $-c < v < c$, the set of all time- stamped origins, $[0, t_0]_v \in \mathcal{F}_v$, forms the hyperbola*
>
> (8.2.3a) $t_0^2 = t^2 - (x/c)^2,$
>
> *where x and t are coordinates in any of the frames $\mathcal{F}_v$.*

PROOF: Let the time-stamped origin, $[0, t_0]_v \in \mathcal{F}_v$ and coordinates $[x, t]_0$ lie at opposite ends of a Lorentz transformation line-of-sight corridor. That is, for $[x, t]_0$ in $\mathcal{F}_0$ and $[0, t_0]$ in $\mathcal{F}_v$, we have $L_v([x, t]_0) = [0, t_0]_v$. Then, from (8.1.2a), $[0, t_0]$ and $[0, t_0]_v$ have the same Minkowski length (8.1.1a), or

$$M([x, t]_0) = c^2 t^2 - x^2 \text{ and } M([0, t_0]_v) = c^2 t_0^2$$

for variables x and t in $\mathcal{F}_v$. Having established the equality $(ct_0)^2 = (ct)^2 - x^2$ (*equivalently,* $t_0^2 = t^2 - (x/c)^2$), (8.2.3a) is proved. ∎

(8.3) Interpreting the Time-Stamped Origins Theorem

In Section (7.5) we showed that the hyperbola $t^2 - (x/c)^2 = k^2$ was useful in providing a geometric interpretation of the formula $v \oplus w$, the relative addition of speeds v and w. The following theorem provides a physical interpretation of this hyperbola.

(8.3.1) FIGURE: *Galilean vs. Relativistic Time Axis Scales*

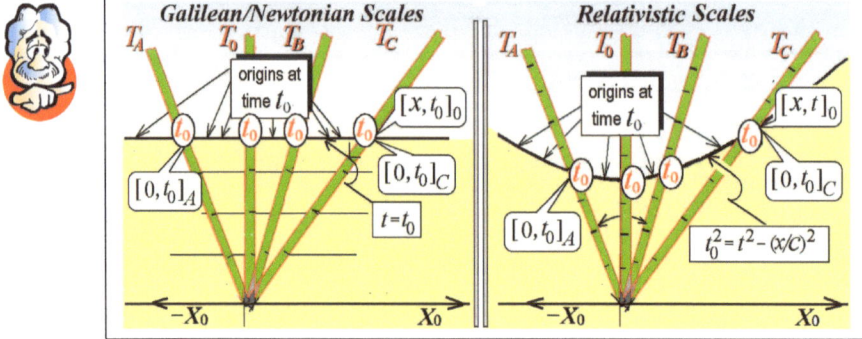

In the Galilean and relativistic spacetime diagrams above, each frame with relative speed v has its time axis forming a certain angle with the vertical time axis.

In the **Galilean case** (left panel of Fig. (8.3.1)), the various origins $[0, t_0]_0$, $[0, t_0]_A$, $[0, t_0]_B$, and $[0, t_0]_C$, at time t_0 lie on the horizontal line $[x, t_0]$, irrespective of relative speeds between frames $\mathcal{F}^A$, $\mathcal{F}^B$, $\mathcal{F}^C$, etc., where x and t_0 are coordinates in frame $\mathcal{F}_0$ with perpendicular X_0 and T_0 axes.

In the **relativistic case** (right panel of Fig. (8.3.1)), the various origins $[0, t_0]_0$, $[0, t_0]_A$, $[0, t_0]_B$, $[0, t_0]_C$, at time t_0 lie on the hyperbola $t_0^2 = t^2 - (x/c)^2$ where x and t are coordinates in frame $\mathcal{F}_0$ with perpendicular X_0 and T_0 axes (8.2.3a).

For both the Galilean (left panel) and relativistic case (right panel) of Figure (8.3.1), each time axis can be viewed as a ruler where the distance from 0 to t_0 varies when it is measured in the $\mathcal{F}_0$ frame having perpendicular X_0 and T_0 axes.

(8.4) Tangent Lines of Simultaneity

When the spacetime graph for frame $\mathcal{F}$ has perpendicular axes X and T, then the synchronized frame $\mathcal{F}'$, moving with speed v with respect to $\mathcal{F}$ has its time axis, T', forming a certain angle with the vertical T axis.

The following theorem describes the slopes and positions of lines of simultaneity (*at fixed time* t') which we already know represent the X-axis at time t' of the moving frame $\mathcal{F}'$ (*see* (4.4.3), *pg.* 73).

(8.4.1)
Calculus

T<small>HEOREM:</small> (*Tangent Lines = Simultaneous Events*)
Given a spacetime graph of frame $\mathcal{F}$ with perpendicular X and T axes, and a synchronized frame $\mathcal{F}'$, moving with relative speed v, with distance axis X' and time axis T'.

Then ....

(8.4.1a)   *$1/v$ is the slope of time axis T', and*

(8.4.1b)   *v/c^2 is the slope of any and all straight lines of simultaneous events in frame $\mathcal{F}'$. The X'-axis is the set of $t' = 0$ events in $\mathcal{F}'$.*

In particular, the line of simultaneous events at any time t' in $\mathcal{F}'$, expressed in $\mathcal{F}'$ coordinates x, t, is the tangent line to the hyperbola

(8.4.1c)   *$(t')^2 = t^2 - (x/c)^2$.*

For all times t' in $\mathcal{F}'$, tangent line (8.4.1c) has slope v/c^2 and intersects the vertical T-axis of $\mathcal{F}$ at the $\mathcal{F}$ coordinates $[0, t']$.

P<small>ROOF:</small>
Proof of (8.4.1a).   The speed of frame $\mathcal{F}'$ as seen from frame $\mathcal{F}$ is $v = \Delta x/\Delta t$ while slope of lines in the X-T plane is given by $\Delta t/\Delta x$. This proves (8.4.1a).

Proof of (8.4.1b). Points of constant (simultaneous) time t' in $\mathcal{F}'$ have form $[x', t']$ where x' varies and time t' is fixed. The Lorentz line-of-sight transformation $L_v : [x, t] \to [x', t']$ (7.1.3a) tells us that

$$\underbrace{\gamma_v \begin{bmatrix} 1 & -v \\ -v/c^2 & 1 \end{bmatrix}}_{L_v} \begin{bmatrix} x \\ t \end{bmatrix} = \begin{bmatrix} \gamma_v(x - vt) = x' \\ \gamma_v(t - vx/c^2) = t' \end{bmatrix} \quad \text{in frame } \mathcal{F}'$$

where, from (7.1.3b), $\gamma_v = 1/\sqrt{1 - (v/c)^2}$.

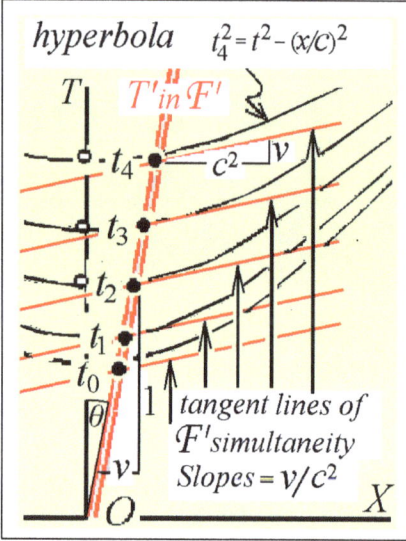

Relative to the frame for the X-T graph, there are infinitely many frames each moving with speed v, where $-c < v < c$. If we fix a time, t_i, then the plot of each of the origins (with age t_i) of the infinitely many frames, yields the hyperbola $t_i^2 = t^2 - (x/c)^2$. This graph shows hyperbolas of origins of ages $t_0 < t_1 < t_2 < t_3 < t_4$. For one frame $\mathcal{F}'$, we plot its set of origins at all times (its time axis T'). The five tangent lines shown at the intersections of T', one for each t_i, is the line of simultaneous events that occur at time t_i in frame $\mathcal{F}'$.

Now the second entry, $\gamma_v(t - vx/c^2) = t'$, is constant. Solving for t, we produce the following straight-line form $t = mx + b$ in $\mathcal{F}$ variables x and t:

$$t = \left(\frac{v}{c^2}\right)x + \frac{t'}{\gamma_v}.$$

This line in the X-T plane has slope (v/c^2), which proves (8.4.1b).

Proof of (8.4.1c). Solve for dt/dx via implicit differentiation of the hyperbola $(t')^2 = t^2 - (x/c)^2$ to obtain $0 = 2t - 2(x/c^2)(dx/dt)$. This equation implies the slope (derivative) dt/dx at any $[x, t]$ in $\mathcal{F}$ is

$$\frac{dt}{dx} = \frac{(x/t)}{c^2} = \left(\frac{v}{c^2}\right) \qquad (when\ x = vt\,)$$

where $[x, t]$ lies on the T' axis in graph of frame $\mathcal{F}$ only when $x = vt$. Substitution of $x = vt$ into the previous equation shows that the slope $dt/dx = (v/c^2)$ which is exactly what is required by (8.4.1b). This justifies (8.4.1c) which ends the proof. ∎

(8.5) Exercises

(8.5.1)
Exercise

Use of Lorentz transformation: The diagram of (12.2.2) shows four coordinates that are listed in the table to the right — the two $\mathcal{F}^A$ coordinates for P and Q' and the two $\mathcal{F}^B$ coordinates of P' and Q.

Point	Frame	
	$\mathcal{F}^A$	$\mathcal{F}^B$
P	$[\,0, t_0\,]_A$	*
Q'	$[\,d_0, t_0\,]_A$	*
P'	*	$[\,-d_0, t_0\,]_B$
Q	*	$[\,0, t_0\,]_B$

Fill in the four missing coordinates in the table (*marked* *) that lie at opposite ends of the line-of-sight corridors of the given coordinates.

Hint: *Use the Lorentz transformation* $L_v : \mathcal{F}^A \to \mathcal{F}^B$ (*equivalently,* $L_{-v} : \mathcal{F}^B \to \mathcal{F}^A$).

(8.5.2)
Exercise

(1) From the coordinates of the table (8.5.1) show that $P' = \sigma_v P$ and $Q' = \sigma_v Q$ where $\sigma_v = \sqrt{1 - (v/c)^2}$ (1.8.2).

(2) Show how the results of (1) prove that moving clocks run slower than stationary clocks by a factor of σ_v.

Hint: *Recall that measuring moving clocks requires two readings of the same moving clock at different times* (**??**).

V.   Graphic Resolution of the Paradoxes

Chapter 9 The Accommodating Universe. *An intergalactic traveler apparently travels faster than light.*

Chapter 10 The Comparison Paradoxes. *Any two observers see each other's moving clock run slower than his own clock.*

Chapter 11 The Twin Paradox. *A (stationary) Earthbound twin sees her (moving) spaceship twin age more slowly than Earth residents. Yet, the (stationary) spaceship twin does not see the (moving) Earthbound twin age more slowly than spaceship residents.*

Chapter 12 The Train-and-Tunnel Paradox. *A pair of lightning bolts strike simultaneously when viewed by observers on the ground but they are viewed as sequential lightning bolts when viewed by passengers on the train.*

Chapter 13 The Pea-Shooter Paradox. *A ball on a moving train has speed 0.8c relative to the train while the train, as seen from the embankment, has speed 0.9c. From the train embankment, the speed of the ball is measured to be 0.988c instead of the "expected" speed of 0.8c + 0.9c = 1.7c.*

Chapter 14 Bug-Rivet Paradox. *The collision of a rivet with a metal plate shows that the rivet cannot remain rigid as it accelerates out of an inertial frame. Moreover, if rigidity is assumed, then the sequence of cause and effect can be reversed — cause can result <u>after</u> its effect.*

(9) The Accommodating Universe Paradox

(9.1) Preview

(9.1.1) **Summary of the Accommodating Universe, Section (1.9).** If a skateboarder in frame $\mathcal{F}^S$ races on a straight-line track in $\mathcal{F}^E$ towards the finish line, then the faster he runs, the shorter the track becomes from his point of view in frame $\mathcal{F}^S$ — the Universe is *accommodating* in the sense that the faster he goes, the shorter the track becomes from his point of view and, hence, the closer the finish line is to the runner.

The paradox lies in the statement that if the length of the track can be made arbitrarily small as viewed from fame $\mathcal{F}^S$, then the skateboarder can complete the run in an arbitrarily short amount of time — one year, say. If the at-rest track length in frame $\mathcal{F}^E$ is 100 light years, then a photon requires 100 years to reach the finish line. This implies that the photon may require *more* time than the runner does to run on the same track.

The specious reasoning springs from using clocks from separate frames to time the skateboarder and the photon. The Minkowski diagram (9.2.1) following offers a graphical analysis.

(9.2) Setup for the Minkowski Diagram

Figure (9.2.1) shows four worldlines:

- The line of a photon with slope $1/c$ that indicates the photon is traveling rightward at speed c.

- In frame $\mathcal{F}^E$, the vertical line of Earth which rests at $x = 0$ and the vertical line of Star XYZ which is at a distance d_E from Earth and rests at $x = d_E$.

- In frame $\mathcal{F}^S$, the <u>line of the skateboarder</u> has slope $1/v$. This means the skateboarder travels with speed v relative to Earth's frame $\mathcal{F}^E$.

(9.2.1) FIGURE: *The Accommodating Universe*

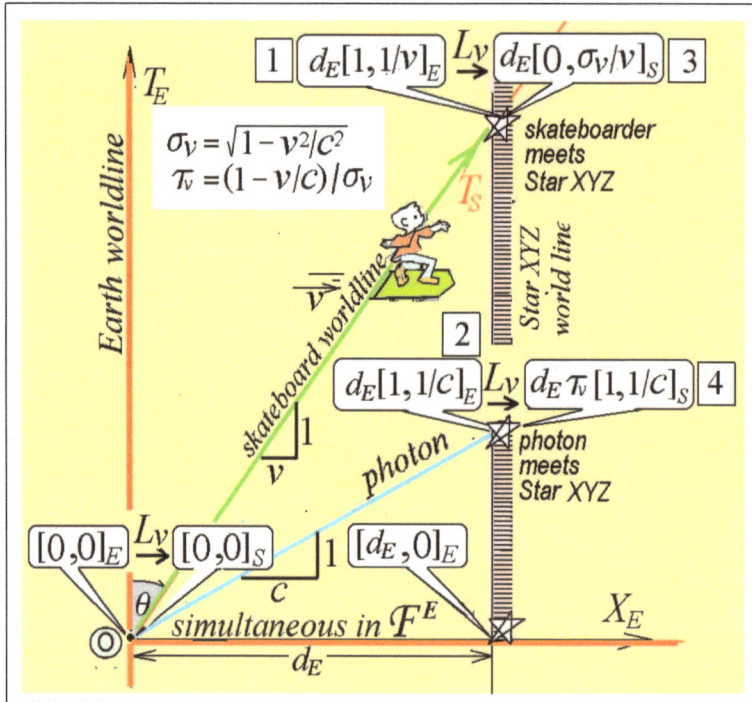

$\mathcal{F}^E$ **Coordinates.** Relative to the Earth's frame, $\mathcal{F}^E$, the skateboarder moves rightward with speed v and a photon travels with speed c over a distance d_E to Star XYZ. The distance for both journeys is d_E so the times in $\mathcal{F}^E$ are d_E/v and d_E/c, respectively. Figure (9.2.1) lists these $\mathcal{F}^E$ coordinates as $\boxed{1}$ and $\boxed{2}$ at Star XYZ.

$\mathcal{F}^S$ **Coordinates.** The Lorentz transformation, $L_v([x,t]_E) = [x',t']_S$, converts coordinates $[x,t]_E$ in $\mathcal{F}^E$ to line-of-sight coordinates $[x',t']_S$ in $\mathcal{F}^S$. These values of Figure (9.2.1) are summarized as follows:

(9.2.1a)

At Star XYZ	$\mathcal{F}^E$ coordinates			$\mathcal{F}^S$ coordinates	
skateboarder	$\boxed{1}$	$d_E \begin{bmatrix} 1 \\ 1/v \end{bmatrix}_E$	$\boxed{3}$	$d_E \begin{bmatrix} 0 \\ \sigma_v/v \end{bmatrix}_S$	
photon	$\boxed{2}$	$d_E \begin{bmatrix} 1 \\ 1/c \end{bmatrix}_E \cdot$	$\boxed{4}$	$d_E \tau_v \begin{bmatrix} 1 \\ 1/c \end{bmatrix}_S \cdot$	

$$L_v \longrightarrow$$

where

(9.2.1b) $L_v = \dfrac{1}{\sigma_v} \begin{bmatrix} 1 & -v \\ -v/c^2 & 1 \end{bmatrix}$, $\sigma_v = \sqrt{1 - (v/c)^2}$ *from (7.1.3b).*

(9.3) Resolving the Accommodating Universe

(9.3.1) **With unsound reasoning,** we concluded in (1.9.5), that the skateboarder is traveling "faster than light" because his journey takes only 1% of the time that light requires to traverse the same path from Earth to StarXYZ. We neglected to mention that we were clocking a race with two separate participants (*photon and skateboarder*) using clocks from two separate frames (*Earth frame $\mathcal{F}^E$ and skateboarder frame $\mathcal{F}^S$*).

The Minkowski diagram (9.2.1) shows the worldline of the photon below that of the skateboarder. This means the photon collision with Star XYZ occurs before (*below*) the skateboarder collision with Star XYZ.

Analytically, we may compare the travel times that appear as the second coordinate entries of the column vectors in Table (9.2.1a). If we choose these times <u>either</u> from the $\mathcal{F}^E$ column (*vectors* $\boxed{1}$, $\boxed{2}$) <u>or</u> from the $\mathcal{F}^S$ column (*vectors* $\boxed{3}$, $\boxed{4}$), we will see that in both cases,

(9.3.1a) $$\frac{skateboarder\ time}{photon\ time} > 1.$$

(9.3.2) **Stephen Hawking** implies that the Accommodating Universe Paradox (9.2.1) will help us ensure our long-term survival. In a 13 June 2006 press conference in Hong Kong, he declared that humankind must colonize outer space.

How can a crew of humans complete voyages to planets that may lie millions of light years from Earth? The answer lies in being able to travel fast enough according to the Accommodating Universe Paradox, which says that at sufficiently high speeds, both *distance* and *travel time* from Earth to the desired planet can be as small as desired. In this way, a trip that requires a few million years (*as measured on Earth*) could be achieved within a a human lifetime (*as measured on the spaceship*).

There are at least two major problems.
- First, after we determine the appropriate planet of a remote solar system, there is the challenge of how to generate the enormous amount of energy required for this high-speed trip. [1]

- Second, even if the trip takes only a few spaceship months, it will take millions of Earth years to complete.

  So if the travelers forget something and go back, they will have aged a year or two, but the civilization that sent them will have long vanished. The memory of the colonizing effort will probably be gone. The spaceship travelers will be strangers in a strange land. (*To see why this is so, consult Figure (9.2.1), or the resolution of the Twin Paradox at (1.11.3), (1.11.4), pg. 31, and (12.3), pg. 158.*)

---

[1]The question of how the human race might best survive in the long run was raised by Hawking on the Internet when the people at Yahoo! invited Hawking, as one of ten celebrities, to submit a question online at `answer.yahoo.com`. This was part of their "Ask the Planet" campaign. Hundreds of replies were submitted although Hawking claimed to have "known the answer" in advance.

(9.4) Exercises

(9.4.1)
Exercise
Computing $\mathcal{F}^E$ Time Ratio Using times from vectors $\boxed{1}$ and $\boxed{2}$ in $\mathcal{F}^E$ coordinates of Table (9.2.1a), show that the ratio of (9.3.1a) is greater than 1.

(9.4.2)
Exercise
Computing $\mathcal{F}^S$ Time Ratio Using times from vectors $\boxed{3}$ and $\boxed{4}$ in $\mathcal{F}^S$ coordinates of Table (9.2.1a), show that the ratio of (9.3.1a) is greater than 2.

---

(10) The Length-Time Comparison Paradoxes

(10.1) An Overview of the Paradoxes

(10.1.1) We have two inertial reference frames, $\mathcal{F}^A$ and $\mathcal{F}^B$, with respective observers A and B that are at rest. Observer A sees Observer B move with speed v and B sees A move with speed $-v$. Both A and B have identical clocks and rulers. Then the first results, (1.7.2a) and (1.7.2b), along with the illustrations of (1.10.4a) and (1.10.4b) lead to the following observations:

• **Both moving clocks run slow.** Each observer (*on the platform, or in the railway car*) sees the other's moving clock run slower than his/her own clock.

• **Both moving rulers shrink.** Each observer sees the other's moving ruler as shorter than his/her own ruler.

(10.1.2) FIGURE: *Synchronized Parallel Rods*

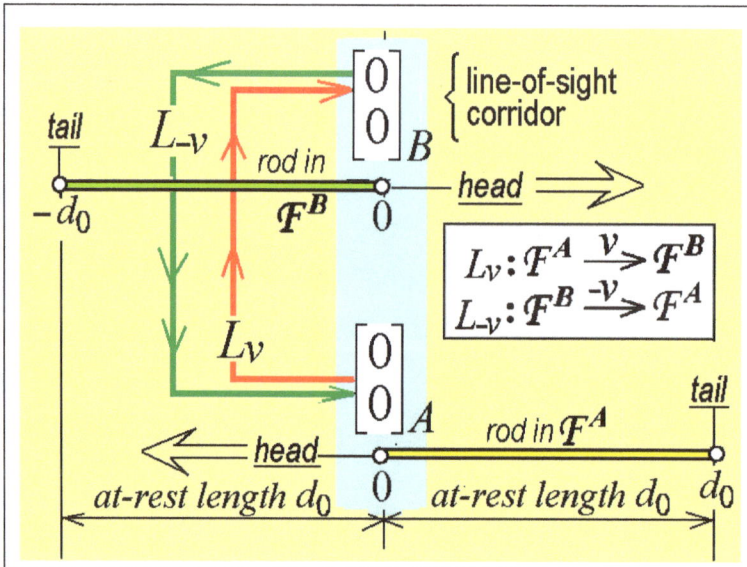

$\mathbf{T}$*he rod (ruler) in inertial frame $\mathcal{F}^B$ has positive rightward speed v relative to frame $\mathcal{F}^A$. The rod in inertial frame $\mathcal{F}^A$ has negative leftward speed $-v$*

relative to frame $\mathcal{F}^B$. The frames $\mathcal{F}^A$ and $\mathcal{F}^B$ are synchronized in the sense of (3.4.5)*, pg.* 61.

(10.1.3) Question: How can both bullet statements of (10.1.1) be true?

Answer: They can not. However, if we replace "Each observer" with "A pair of observers in each frame," then, as we shall see in Theorem (10.2.1), there is no inconsistency. For example, from (stationary) frame $\mathcal{F}^A$, we declare the length of a moving rod that lives in (moving) frame $\mathcal{F}^B$, to be the $\mathcal{F}^A$ distance between two (stationary) observers in frame $\mathcal{F}^A$ who simultaneously stand at the positions of the head and the tail of the moving rod. (*Also see Figures* (2.4.2) *and* (2.4.3)*, pg.* 51.)

Since frames $\mathcal{F}^A$ and $\mathcal{F}^B$ are synchronized (3.4.5), pg. 61, we enlist the aid of two other cohort members, one in each frame so that we can measure

(10.1.4) **(a)** the ticking rate of the moving clock,

(b) the speed of the rod in the other moving frame (*see* (2.4.2)*, pg.* 51)*,* and

(c) the length of the rod in the other moving frame (*see* (2.4.3)).

Specifically, choose any two arbitrary times $t_A > 0$ in $\mathcal{F}^A$ and $t_B > 0$ in $\mathcal{F}^B$. Then, in each frame, choose the (stationary) cohorts with the following coordinates:

$$\begin{bmatrix} x_A - d_A \\ t_A \end{bmatrix}_A, \quad \begin{bmatrix} x_A \\ t_A \end{bmatrix}_A \text{ in } \mathcal{F}^A \quad \text{and} \quad \begin{bmatrix} x_B - d_B \\ t_B \end{bmatrix}_B, \quad \begin{bmatrix} x_B \\ t_B \end{bmatrix}_B \text{ in } \mathcal{F}^B$$

who, through their respective line-of-sight corridors, observe the end-points of the moving rods — the end-points are marked 0 and $\pm d_0$. The two cohorts are depicted in Figure (2.4.3). Moreover, the as yet unknown distances between the two cohorts are d_A in $\mathcal{F}^A$ and d_B in frame $\mathcal{F}^B$. (*See Figures* (10.1.5) *and* (10.1.10) *where d_A is the length of $\mathcal{F}^B$'s rod as measured in $\mathcal{F}^A$, and similarly, d_B is the length of $\mathcal{F}^A$'s rod as measured in $\mathcal{F}^B$.*)

(10.1.5) FIGURE: *Line-of-sight Coordinates in $\mathcal{F}^B$ Where Frame $\mathcal{F}^A$ is Fixed, Relative Speed of Frame $\mathcal{F}^B = v$.*

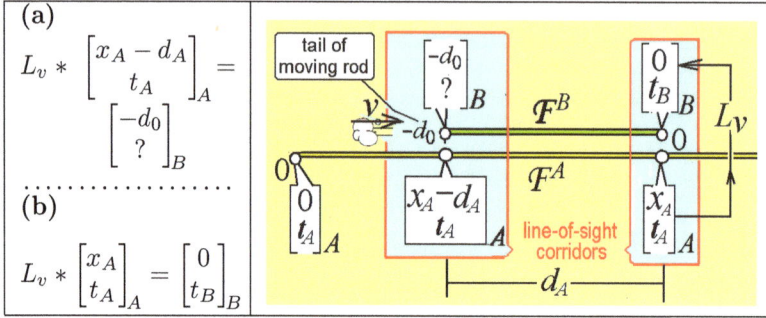

(10.1.6) ‖ NOTE: The column vector $\mathrm{col}[-d_0, \,?\,]_B$ appears in Figure (10.1.5) with the symbol "?" since the time at the tail of the rod in $\mathcal{F}^B$ (i.e., at $x = -d_0$) does not need to be calculated.

(10.1.7) **Calculating Quantities of (10.1.4) as Measured in $\mathcal{F}^A$.** These results follow from the vector equations in the left hand panel of Figure (10.1.5). Specifically, equation (10.1.5)(b) takes the form

$$\underbrace{\frac{1}{\sigma_v}\begin{bmatrix} 1 & -v \\ -v/c^2 & 1 \end{bmatrix}}_{\text{Lorentz matrix } L_v} * \begin{bmatrix} x_A \\ t_A \end{bmatrix}_A = \begin{bmatrix} 0 \\ t_B \end{bmatrix}_B$$

which gives us the two equations

(10.1.7a) $x_A - vt_A = 0$ *which implies* $\begin{cases} x_A = vt_A \\ \boxed{v = x_A/t_A} \end{cases}$

  (i) $t_A - \dfrac{v}{c^2} x_A = \sigma_v t_B$.

Substituting $x_A = vt_A$ from (10.1.7a) into equation (i) above, we obtain

(10.1.7b) $t_A \underbrace{\left(1 - \dfrac{v^2}{c^2}\right)}_{\sigma_v^2} = \sigma_v t_B$ *which implies* $\boxed{\sigma_v t_A = t_B}$.

Finally, equation (10.1.5)(a) takes the form

$$\underbrace{\frac{1}{\sigma_v} \begin{bmatrix} 1 & -v \\ -v/c^2 & 1 \end{bmatrix}}_{\text{Lorentz matrix } L_v} * \begin{bmatrix} x_A - d_A \\ t_A \end{bmatrix}_A = \begin{bmatrix} -d_0 \\ * \end{bmatrix}_B$$

which, after equating the top entry of the left and right hand sides, gives us the single equation

$$x_A - d_A - vt_A = -\sigma_v d_0.$$

Substitute $x_A = vt_A$ from (10.1.7a) in the equation above to obtain

(10.1.7c) $\boxed{d_A = \sigma_v d_0}$.

(10.1.8) <u>When frame $\mathcal{F}^A$ is fixed</u>, **the boxed equations** of (10.1.7a), (10.1.7b), and (10.1.7c) answer the respective questions of (10.1.4)(a), (10.1.4)(b), and (10.1.4)(c). That is, from frame $\mathcal{F}^A$, the speed of frame $\mathcal{F}^B$ is measured to be v and the clock in $\mathcal{F}^B$ ticks at a rate that is only σ_v times the rate of all $\mathcal{F}^A$ clocks ($\sigma_v t_A = t_B$) and the length d_A of the moving rod as measured in stationary $\mathcal{F}^A$ is σ_v times the at-rest length d_0 ($d_A = \sigma_v d_0$).

This is no optical illusion. From frame $\mathcal{F}^A$, the length of the rod in (*moving*) frame $\mathcal{F}^B$ is measured to be $\sigma_v d_0$ which is shorter than its at-rest length d_0.

(10.1.9) Nᴏᴛᴇ: The values of times t_A and t_B in Figure (10.1.5) and the analysis that follows are not necessarily equal to the values of t_A and t_B in Figure (10.1.10) and the analysis that will follow.

(10.1.10) Fɪɢᴜʀᴇ: *Line-of-sight Coordinates in $\mathcal{F}^A$ Where Frame $\mathcal{F}^B$ is Fixed, Relative Speed of Frame $\mathcal{F}^A = -v$.*

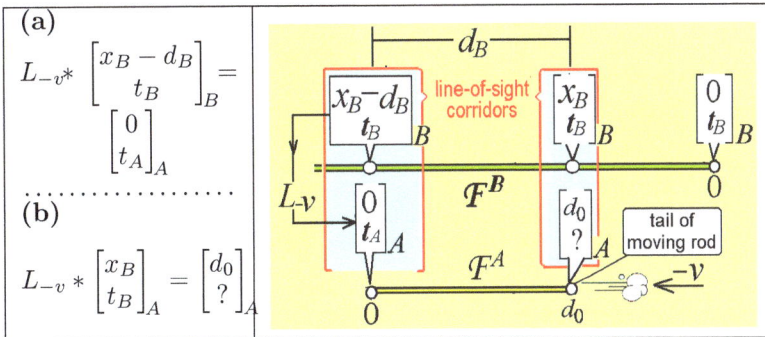

(10.1.11) NOTE: As in Note (10.1.6), the column vector col[d_0, ?] in Figure (10.1.10) contains the symbol "?" since the time at the tail of the rod in $\mathcal{F}^A$ (i.e., at $x = d_0$) does not need to be calculated.

(10.1.12) Calculating Quantities of (10.1.4) When $\mathcal{F}^B$ is Fixed. These results follow from the vector equations in the left hand panel of Figure (10.1.10). Specifically, equation (10.1.5)(a) takes the form

$$\underbrace{\frac{1}{\sigma_v} \begin{bmatrix} 1 & v \\ v/c^2 & 1 \end{bmatrix}}_{\text{Lorentz matrix } L_{-v}} * \begin{bmatrix} x_B - d_B \\ t_B \end{bmatrix}_A = \begin{bmatrix} 0 \\ t_A \end{bmatrix}_A$$

which gives us the two equations

(10.1.12a)

$$x_B - d_B + v t_B = 0 \qquad \text{which implies} \begin{cases} (x_B - d_B) = -v t_B \\ \boxed{-v = (x_B - d_B)/t_B} \end{cases}$$

(ii) $\quad t_B + \dfrac{v}{c^2}(x_B - d_B) = \sigma_v t_A.$

Substitute $(x_B - d_B) = -v t_B$ from (10.1.12a) into equation (**ii**) above to obtain

(10.1.12b) $\quad t_B \underbrace{(1 - \dfrac{v^2}{c^2})}_{\sigma_v^2} = \sigma_v t_A \quad \text{which implies} \quad \boxed{\sigma_v t_B = t_A}.$

Finally, Equation (10.1.10)(b) takes form

$$\underbrace{\frac{1}{\sigma_v} \begin{bmatrix} 1 & v \\ v/c^2 & 1 \end{bmatrix}}_{\text{Lorentz matrix } L_{-v}} * \begin{bmatrix} x_B \\ t_B \end{bmatrix}_B = \begin{bmatrix} d_0 \\ * \end{bmatrix}_A$$

which, after equating the top entry of the left and right hand sides, gives us the single equation

$$x_B + v t_B = \sigma_v d_0.$$

Substitute $x_B = d_B - v t_B$ from (10.1.12a) in the equation above to obtain

(10.1.12c) $\qquad \boxed{d_B = \sigma_v d_0}.$

(10.1.13) With frame $\mathcal{F}^B$ fixed, the boxed equations of (10.1.12a), (10.1.12b), and (10.1.12c) answer the respective questions of (10.1.4)(a), (10.1.4)(b), and (10.1.4)(c). That is, the speed of frame $\mathcal{F}^A$ is measured to be $-v$ and the clock in moving frame $\mathcal{F}^A$ ticks at a rate that is only σ_v times the rate of all $\mathcal{F}^B$ clocks ($\sigma_v t_B = t_A$) and the length d_B of the moving rod as measured in stationary $\mathcal{F}^B$ is σ_v times the at-rest length d_0 ($d_B = \sigma_v d_0$).

Again, similar to (10.1.8), this is no optical illusion. From frame $\mathcal{F}^B$, the length of the rod in (*moving*) frame $\mathcal{F}^A$ is measured to be $\sigma_v d_0$ which is shorter than its at-rest length d_0.

(10.2) Resolving the Mutual Length-Time Paradoxes

The groundwork has now been laid for the following

(10.2.1)

> THEOREM: ***Mutual Length-Shrinking and Time-Slowing:***
> - *Frame $\mathcal{F}^B$ moves with observed speed v relative to frame $\mathcal{F}^A$ (frame $\mathcal{F}^A$ has speed $-v$ relative to frame $\mathcal{F}^B$).*
>
> - *Inertial frames $\mathcal{F}^A$ and $\mathcal{F}^B$ have rods with at-rest length d_0.*
>
> - $\sigma_v = \sqrt{1 - (v/c)^2}$.
>
> *Then*
>
> **(10.2.1a)** *Observers in each frame measure the ticking rate of moving clocks in the other frame to be $\sigma_v \leq 1$ times the ticking rate of clocks in the observer's frame.*
>
> **(10.2.1b)** *Observers in each frame measure the moving ruler in the other frame to have length $\sigma_v d_0 \leq d_0$. It is important to note that only lengths along the direction of motion are affected. Lengths that are perpendicular to the direction of motion remain the same.*

PROOF of (10.2.1a)
From (10.1.7b): If an observer jumps into frame $\mathcal{F}^A$ so that frame $\mathcal{F}^B$ is seen to have relative speed v, then the moving clocks in $\mathcal{F}^B$ tick at a rate that is σ_v times the rate of all synchronized stationary

clocks in $\mathcal{F}^B$.

Frpm (10.1.12b): Conversely, if an observer jumps into frame $\mathcal{F}^B$ instead, so that frame $\mathcal{F}^A$ has relative speed $-v$, then the moving clocks in $\mathcal{F}^A$ tick at a rate that is σ_v times the rate of all synchronized stationary clocks in $\mathcal{F}^A$.

This proves hypothesis (10.2.1a).

PROOF of (10.2.1b)
From (10.1.7c): If an observer jumps into frame $\mathcal{F}^A$ so that frame $\mathcal{F}^B$ is seen to have relative speed v, then a moving rod with at-rest length d_0 in $\mathcal{F}^B$ is measured from stationary frame $\mathcal{F}^A$ to have length $\sigma_v d_0$.

Frpm (10.1.12c): Conversely, if an observer jumps into frame $\mathcal{F}^B$ so that frame $\mathcal{F}^A$ is seen to have relative speed $-v$, then a moving rod with at-rest length d_0 in $\mathcal{F}^A$ is measured from stationary frame $\mathcal{F}^B$ to have length $\sigma_v d_0$.

This proves hypothesis (10.2.1b) and the theorem is done. ■

(10.3) Summary

(10.3.1) **Time and Distance Comparison Paradoxes Are Resolved.**
The misleading questions of Section (10.1) suggest that there is a *single* observer measuring the length of a moving rod, or the ticking rate of a moving clock. In this way, the illusion of a contradiction is created. To measure lengths of moving rods, two observers in each (stationary) frame are necessary. Figures (2.4.3), pg. 51, shows that there is no inconsistency. To measure speeds of moving frames (*although one observer may suffice*), Figure (2.4.2) shows that two observers at a known distance between them may also be used to measure the speed of a moving object.

(10.4) Exercises

(10.4.1)
Exercise

Modified Proof of Theorem (10.2.1). Theorem (10.2.1) uses three cohort observers to measure the slower ticking rates of moving clocks and the shorter length of moving rods. Prove this theorem using only two cohort observers in each frame.

> **Hint:** *In the current proof, there are three cohorts in each frame. The origin observer in the stationary frame (who views the head of the moving rod) is joined by two cohorts who, at a later arbitrary time, view the tail and the head of the moving rod. To reduce the number of cohorts, let the origin observer take on the role of the cohort that views the tail of the moving rod. (The price of reducing the work force is that the later observations of the rod's tail and head can not be made at an arbitrary time.)*

(10.4.2)
Exercise

Mutual Shrinkage of Moving Rods — Slowing of Moving Clocks: Conditions (10.2.1a) and (10.2.1b) were proved using the physical ruler-clock model of an inertial frame. In this exercise, prove these statements ((10.2.1a) *and* (10.2.1b)) using the Lorentz transformation and the space time diagrams (10.4.2a) below.

(10.4.2a) **Space Time Diagrams for the Mutual Shrinkage Theorem (10.2.1)**

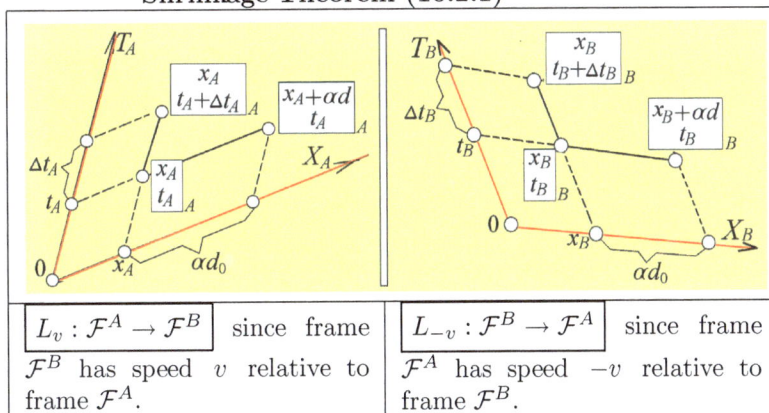

$L_v : \mathcal{F}^A \to \mathcal{F}^B$	since frame	$L_{-v} : \mathcal{F}^B \to \mathcal{F}^A$	since frame
$\mathcal{F}^B$ has speed v relative to frame $\mathcal{F}^A$.		$\mathcal{F}^A$ has speed $-v$ relative to frame $\mathcal{F}^B$.	

Hint: *As an example, we show that a rod with at-rest length d_0 in moving frame $\mathcal{F}^A$ is measured in stationary frame $\mathcal{F}^B$ to have length $\sigma_v d_0$. As indicated in the second panel of Figure (10.4.2a), frame $\mathcal{F}^B$*

is stationary and sees frame $\mathcal{F}^A$ moving with speed $-v$. The points of interest are:

(10.4.2b)
$$\vec{P} = \begin{bmatrix} x_B \\ t_B \end{bmatrix}_B \qquad \text{starting point in } \mathcal{F}^B$$

(10.4.2c)
$$\vec{P} + \begin{bmatrix} \alpha d_0 \\ 0 \end{bmatrix}_B = \begin{bmatrix} x_B + \alpha d_0 \\ t_B \end{bmatrix}_B \qquad \text{point that is } \alpha d_0 \text{ units from } \vec{P} \text{ at } \mathcal{F}^B \text{ time } t_B$$

$$\vec{P} + \begin{bmatrix} 0 \\ \Delta t_B \end{bmatrix}_B = \begin{bmatrix} x_B \\ t_B + \Delta t_B \end{bmatrix}_B \qquad \text{point that is } \Delta t_B \text{ units later than } \vec{P} \text{ in } \mathcal{F}^B.$$

The $\mathcal{F}^A$ line-of-sight coordinates are, respectively, $L_{-v} * \vec{P}$ and $L_{-v} * \left(\vec{P} + \begin{bmatrix} \alpha d_0 \\ 0 \end{bmatrix}_B \right)$ (see (7.1.3)). The vector difference is

(10.4.2d)
$$L_{-v}*\vec{P} - L_{-v}*\left(\vec{P} + \begin{bmatrix} \alpha d_0 \\ 0 \end{bmatrix}_B \right) = L_{-v}*\left(\begin{bmatrix} \alpha d_0 \\ 0 \end{bmatrix}_B \right) = \frac{1}{\sigma_v} \begin{bmatrix} \alpha d_0 \\ \alpha d_0 v/c^2 \end{bmatrix}_A.$$

Interpretation: *Now the topmost entry of the final vector of (10.4.2d) is $\alpha d_0/\sigma_v = d_0$ since d_0 is the at-rest distance in $\mathcal{F}^A$ between the head and tail of the moving rod. Consequently, $\alpha = \sigma_v$. Replacing α with σ_v in Equation (10.4.2c), we see that the distance between the line-of-sight corridors in frame $\mathcal{F}^B$ is $\alpha d_0 = \sigma_v d_0$. That is, the head-to-tail distance of the moving rod in $\mathcal{F}^A$ is measured in stationary $\mathcal{F}^B$ to be only $\sigma_v d_0$.*

(11) The Twin Paradox

(11.1) An Overview of the Paradox

A graphical analysis of Figure (11.3.1) completely resolves the twin paradox (*Section* (1.11)) which we restate as follows:

(11.1.1) • At time 0 years, one twin leaves Earth traveling at an outbound speed of $0.8c$. After 25 years by the Earth clock, the traveling twin instantly boards a returning spaceship which travels in the opposite direction, at an inbound speed of $0.8c$. After 25 more years, or at time 50 years by the Earth clock, the traveling twin returns to Earth to greet his 50 year-old twin while he, himself, is only 30 years old.

(11.1.2) • **The Apparent Contradiction.** There is a *symmetry*. During the journey, each twin "sees" the (moving) clock of the other twin run slower than his own (stationary) clock (*see* (1.7.2a)). Thus, the spaceship twin may conclude that the earthbound twin with the slower (moving) earthbound clock will be younger. Likewise, the earthbound twin may conclude that the spaceship twin with the slower (moving) spaceship clock will be younger when they meet.

That each twin is simultaneously younger than the other is clearly not possible: Hence, the paradox.

(11.2) A Simplifying Assumption

(11.2.1) Special Relativity *requires frames to have constant speeds relative to each other.* Hence, when the traveling twin switches from the outgoing spaceship (*at speed* v) to the inbound spaceship (*at speed* $-v$), we assume an instantaneous transfer that requires zero transfer time which we denote $\Delta t_{tfr} = 0$.

However, in (11.5), we apply **Einstein's Theory of General Relativity** to analyze the non-zero time of transfer $\Delta t_{\text{tfr}} > 0$. When $\Delta t_{\text{tfr}} \to 0$, then we see that *both* special relativity and general relativity predict that the transferring twin measures a 32-year forward jump into the future. of the Earth clock's readings as he passes from the outbound spaceship to the inbound spaceship. (*Computations for the general relativity confirmation will be found in* (11.5).)

A*n instantaneous change of frame and speed approximates the continuous speed change in the diagram.*

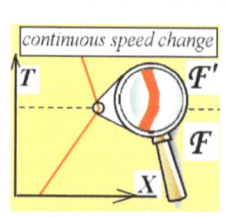

(**Note:** Instantaneous speed change was also assumed in Figure (1.11.2c) of the twin paradox.)

(11.3) Setup for the Minkowski Diagram

(11.3.1) FIGURE: *Three Frames of Reference for the Twin Paradox*

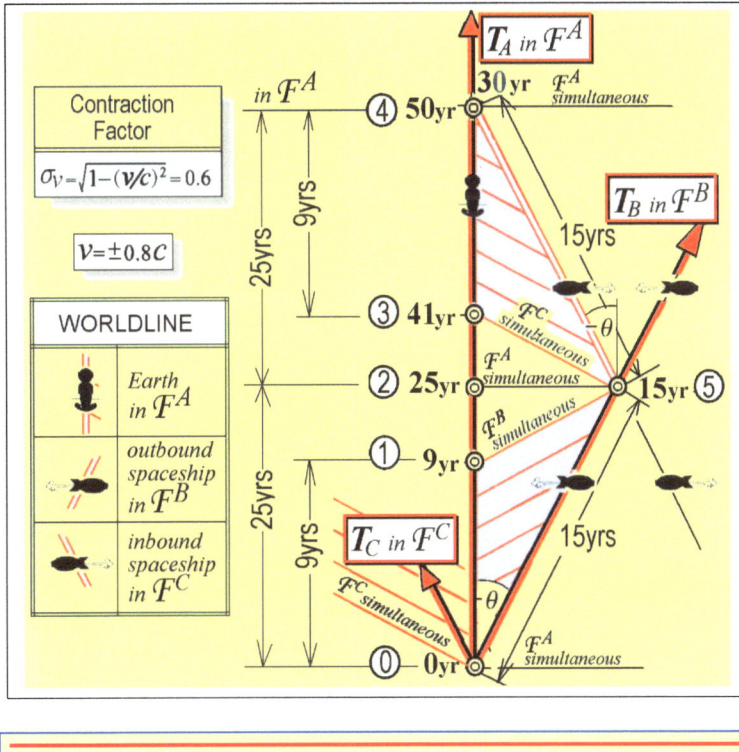

(11.4) Resolving the Twin Paradox

(11.4.1) To fix ideas, we retain the numerical values of Section (1.11) where the relative speed between Earth and the spaceship is $\pm 0.8c$ so that the Lorentz contraction factor (1.8.2) is

(11.4.1a) $\quad$ *when* $c = 1, \quad \sigma_v = \sqrt{1 - (0.8c)^2} = 0.6$.

With σ_v of (11.4.1a) we have:

(11.4.1b)

> *Whenever a stationary clock reads time* t, *and is simultaneously compared (through a line-of-sight corridor) to a synchronized clock moving with speed* v *that reads time* t', *then*
>
> $$v = 0.8\,c \quad implies \quad t' = (\sigma_v \cdot t) = 0.6 \cdot t.$$

(11.4.2) <u>**Traveling with the Earthbound Twin in $\mathcal{F}^A$:**</u>

Worldline of earthbound twin in $\mathcal{F}^A$:
   The vertical T_A axis containing points ⓪-④.
Simultaneous events in $\mathcal{F}^A$:
   Along horizontal lines (*see Theorem* (8.4.1)).

(11.4.2a)   **Earth in $\mathcal{F}^A$ Sees Outbound Times in $\mathcal{F}^B$**

STATIONARY $\mathcal{F}^A$ TIME	OUTBOUND SPACESHIP TIME ($0.6 \times \mathcal{F}^A$ Time)
⓪ $t_A = 0$ yrs. →	⓪ $t_B = 0$ yrs. in $\mathcal{F}^B$ (*synchronization*)
② $t_A = 25$ yrs. →	⑤ $t_B = 15$ yrs. in $\mathcal{F}^B$
② $t_A = 25$ yrs. →	⑤ $t_C = 15$ yrs. in $\mathcal{F}^C$ ($\mathcal{F}^C$ *clock reset*)
④ $t_A = 50$ yrs. →	④ $t_C = 30$ yrs. in $\mathcal{F}^C$

⓪→⓪: The Earthbound twin and the spaceship twin at $x = 0$ are synchronized at time $t_A = 0$ in $\mathcal{F}^A$ and $t_B = 0$ in $\mathcal{F}^B$. At this moment, the spaceship twin accelerates instantly to constant relative speed $v = 0.8\,c$.

②→⑤: At Earth time $t_A = 25$ yrs. in $\mathcal{F}^A$, a cohort of the Earthbound twin (*who is also in $\mathcal{F}^A$*) peers inside the spaceship in $\mathcal{F}^B$ that is traveling at speed $v = 0.8\,c$ and reads $t_B = 0.6 \times 25 = 15$ yrs. on its onboard slow-running clock (*see* (11.4.1b)).

②→⑤ (*continued*): An infinitesimal instant after $t_A = 25$ years on Earth in $\mathcal{F}^A$, the spaceship twin switches spaceships, changing from the outbound frame $\mathcal{F}^B$ to an inbound frame $\mathcal{F}^C$ which has relative speed $v = -0.8\,c$. The clock aboard the incoming spaceship in $\mathcal{F}^C$ must be re-set to read its new coordinated time, $t_C = t_B = 15$ yrs.

④→④ The clock on the inbound spaceship has been running slow at 0.6 times the rate of the Earth clock. Thus, at Earth time $t_A = 50$ yrs., the spaceship clock reads $t_B = 0.6 \times t_A = 0.6 \times 50 = 30$ yrs. (*see* (11.4.1b)). *The spaceship twin has returned 20 years younger than the Earthbound twin.*

(11.4.3) <u>**Traveling with the Spaceship Twin in $\mathcal{F}^B$:**</u>

Worldline of outbound twin in $\mathcal{F}^B$:
   The slanted $\nearrow$ T_B axis which forms angle θ with the vertical axis
   that also contains the points ⓪ and ⑤.

Simultaneous events in $\mathcal{F}^B$:
   All slanted lines $\nearrow$ parallel to line ①-⑤, that is, all lines with
   slope $\tan(\theta)$ (*see Theorem* (8.4.1)).

(11.4.3a) *Spaceship in $\mathcal{F}^B$ Sees Earth Clock Times in $\mathcal{F}^A$*

STATIONARY $\mathcal{F}^B$ TIME	MOVING EARTH TIME ($0.6 \times \mathcal{F}^B$ *Time*)
⓪ $t_B = 0$ yrs. $\rightarrow$	⓪ $t_A = 0$ yrs. in $\mathcal{F}^A$(*synchronization*)
⑤ $t_B = 15$ yrs. $\rightarrow$	① $t_A = 9$ yrs. in $\mathcal{F}^A$
$\underbrace{}$ *15 yrs. in $\mathcal{F}^B$*	$\underbrace{}$ *9 yrs. in $\mathcal{F}^A$*

⓪→⓪: The Earthbound twin and the spaceship twin at $x = 0$ are
synchronized at time $t_A = 0$ in $\mathcal{F}^A$ and $t_B = 0$ in $\mathcal{F}^B$. At this moment, the spaceship twin takes off and accelerates instantly so that
the Earth's relative (constant) speed is measured to be $v = -0.8\,c$.

⑤→①:At spaceship time $t_B = 15$ yrs., the (cohort of the) spaceship
twin reads time $t_A = 0.6 \times 15 = 9$ yrs. (11.4.1b) on the slowly-
running Earth clock in $\mathcal{F}^A$ that has observed speed $v = -0.8\,c$.

(11.4.3b) Spaceship Twin Accelerates into a New Frame:

The spaceship twin, traveling at speed $v = 0.8\,c$ as measured from
Earth in $\mathcal{F}^A$, decides at time $t_B = 15$ yrs. in frame $\mathcal{F}^B$, to instantaneously change to an incoming spaceship in frame $\mathcal{F}^C$ that has speed
$v = -0.8\,c$ relative to Earth.

It is now clear why the frames for each the twins are not equivalent
or symmetric. While the Earth twin spends 50 Earth years in only
one inertial frame, $\mathcal{F}^A$, the traveling spaceship twin lives in several
frames:
- an inertial outbound frame $\mathcal{F}^B$,
- an inertial inbound frame $\mathcal{F}^C$, and, for an extremely brief time,
- *non-inertial* frames during the "instant" transition from $\mathcal{F}^B$ to $\mathcal{F}^C$
  when strong forces eventually change the worldline of the spaceship
  twin from line ⓪-⑤ in inertial frame $\mathcal{F}^B$ to the line ⑤-④ in inertial
  frame $\mathcal{F}^C$ (*see Figure* (11.3.1)).

As the time of the transition becomes shorter and shorter, these forces felt by the traveling twin become arbitrarily large and the intermediate reference frames deviate more and more from being inertial.

As (11.3.1) indicates, the transition from inertial frame $\mathcal{F}^B$ to inertial frame $\mathcal{F}^C$ severely changes the slope of the lines of simultaneity from positive $v = \tan(\theta)$ to negative $-v = -\tan(\theta)$. All lines of simultaneity in $\mathcal{F}^C$ are parallel to line ⑤-③.

(11.4.3c)   A Jump into the Future:

Special relativity: As Figure (11.3.1) confirms, when spaceship time $t_B = 15$ yrs. in frame $\mathcal{F}^B$, a cohort (*also in frame $\mathcal{F}^B$*) simultaneously reads the moving Earth in $\mathcal{F}^A$ time as $t_A = 0.6 \times t_B = 0.6 \times 15 = 9$ yrs. (*see* (11.4.1b)). An infinitesimal moment later, when the traveling twin is in frame $\mathcal{F}^C$, another cohort (*in frame $\mathcal{F}^C$*) simultaneously reads the moving Earth time in $\mathcal{F}^A$ as $t_A = 41$ yrs., an advance of 32 years!

General relativity: Section (11.5) has computational details.

(11.4.4) Traveling with the Spaceship Twin in $\mathcal{F}^C$:

Worldline of inbound twin in $\mathcal{F}^C$:
   The slanted ⬂ line that contains the points ⑤ and ④.

Simultaneous events in $\mathcal{F}^C$:
   All slanted lines parallel to line ⑤-③, that is, all lines with slope equal to $-\tan(\theta)$ (*see Theorem* (8.4.1)).

(11.4.4a) Spaceship in $\mathcal{F}^C$ Sees Earth Clock Times in $\mathcal{F}^A$

STATIONARY $\mathcal{F}^C$ TIME	MOVING EARTH TIME
⑤ $t_C = 15$ yrs. $\rightarrow$	③ $t_A = 41$ yrs. in $\mathcal{F}^A$
④ $t_C = 30$ yrs. $\rightarrow$	④ $t_A = 50$ yrs. in $\mathcal{F}^A$
15 yrs. pass in $\mathcal{F}^C$	*9 yrs. pass in $\mathcal{F}^A$*

⑤→③: The stationary clock in the newly-entered spaceship is reset to read $t_C = 15$ yrs. Simultaneously, the (cohort of the) Earth-bound twin in $\mathcal{F}^A$ sees the moving Earth clock time to be $t_A = 41$ yrs.

(11.4.4b)Why Earth time is $t_A = 41$ yrs. when $t_C = 15$ yrs.
Symmetry in Figure (11.3.1) and Table (11.4.2a) tell us that until
the spaceship returns to Earth, the spaceship clock will run for 15
more years in $\mathcal{F}^C$ and the Earth clock will run for $0.6 \times 15 = 9$ more
years in $\mathcal{F}^A$. Since the Earth clock will read 50 yrs. at the end of
the journey, we calculate backwards 9 years to see that Earth time
reads $t_A = 50 - 9 = 41$ years at the start of the returning journey.

④→④:The inbound spaceship journey ends, during which time, as
discussed above, 15 Earth years and 9 spaceship years pass. The
Earth clocks reads 50 yrs. and the inbound spaceship clock reads
30 yrs.

This concludes the analysis of the Twin Paradox using the tools of
special relativity.

(11.5) General Relativity Confirmation

General relativity predicts and quantifies the 32-year jump
mentioned in (11.4.3c). We analyze the non-linear portion of the
diagram of (11.2.1) where acceleration of the traveling twin disallows
the special relativity rules of inertial frames. Here is the overview.

(11.5.1)
Calculus
One of the central tenets of general relativity (GR) is that
within a sufficiently small region, the effects of acceleration are indistinguishable from those produced by an appropriate gravitational
force. Forces that are time-independent and relativity weak (*as is
the case with the Twin Paradox*), can be described as follows:

For $\vec{r} = [\,x,\,y,\,z\,]$ in three-dimensional space $\boldsymbol{R}^3$, there is a real-valued
function $\Phi(x, y, z)$ called the GRAVITATIONAL POTENTIAL. When
this potential function Φ is differentiable (16.3.4), we construct the
vector-valued GRADIENT, **grad** of Φ, given by

(11.5.1a) $\mathbf{grad}(\Phi(\underbrace{x,\,y,\,z}_{\vec{r}\,\in\,\boldsymbol{R}^3})) = \left[\dfrac{\partial\Phi(x,y,z)}{\partial x},\ \dfrac{\partial\Phi(x,y,z)}{\partial y},\ \dfrac{\partial\Phi(x,y,z)}{\partial z}\right].$

EXAMPLE: If $\Phi(x, y, z) = x^2 y + z + \sin(z),$
then $\mathbf{grad}(\Phi(x, y, z)) = [\ 2xy,\ \ x^2,\ \ 1 + \cos(z)\ ].$

EXAMPLE: If point $\vec{r} = x$ lies on the one-dimensional X-axis, then the gravitational potential function $\Phi(x)$ is a function of the single variable x and, according to (11.5.1a),

(11.5.1b) $\mathbf{grad}(\Phi(x)) = \Phi'(x)$, the derivative of Φ at point x.

. . . .

The gravitational force, on a mass m at point $\vec{r}$ is the three-dimensional vector

(11.5.1c) $\quad \vec{F}(\vec{r}) \ = \ -\mathbf{grad}(\Phi(\vec{r})) \cdot m.$

Finally, Newton's classic relation between force and acceleration on mass m at a point r' is given by

(11.5.1d) $\qquad \vec{F}(\vec{r}) = m \cdot \vec{a}(\vec{r}).$

(11.5.2)
Calculus

Another result from GR is that clocks slow down in a region where gravitational effects are strongest. (*Equivalently, clocks speed up as they pass to weaker regions of gravitational effects. It is this interpretation that is used in (23.5.3), pg. 322, in the discussion of clocks aboard GPS satellites*).

Quantitatively, if an observer at position $\vec{r}_1$ where the gravitational potential is $\Phi(\vec{r}_1)$ finds that an event occurs over a time interval $\Delta t(\vec{r}_1)$, and another observer at a different position $\vec{r}_2$ where the gravitational potential is $\Phi(\vec{r}_2)$ finds the same event occurs over a time interval $\Delta t(\vec{r}_2)$, then

(11.5.2a) $\qquad \Delta t(\vec{r}_1) = \left(1 + \dfrac{\Phi(\vec{r}_2) - \Phi(\vec{r}_1)}{c^2}\right) \Delta t(\vec{r}_2).$

At point $\vec{r}$, a final relation between force $\vec{F}(\vec{r})$ on a mass m due to acceleration $\vec{a}(\vec{r})$ is given by Newton's classic formula (11.5.1d).

‖ NOTE: Although these equations can be derived in a straightfor-
‖ ward manner, doing so would take us too far afield.

The time has come to see how these equations justify the 32-year jump (11.4.3c) in the Twin Paradox.

(11.5.3)
Calculus

FIGURE: *Transition of Traveling Twin After 25 Earth Years*

T*he one-dimensional X-axis contains Earth, considered as a point at $x = x_E$, the straight-line paths of the outbound and inbound spaceships and the jump point $x = x_S$ where the traveling twin changes from outbound Spaceship B to inbound Spaceship C. As the twin jumps from one spaceship to the other at $x = x_S$, the change in time on the spaceship clocks is $\Delta t(x_S)$ while the time change on Earth clocks is $\Delta t(x_E)$. (These time changes are related in (11.5.2a).) See Figure (1.3.2), pg. 13.*

(11.5.4)
Calculus

NOTATION: **As shown in Figure** (11.5.3), all vectors belong to the 1-dimensional X-axis. Hence, we eliminate vector notation for force, velocity, acceleration, and position. Hence, we write v instead of $\vec{v}$ for velocity vectors, a instead of $\vec{a}$ for acceleration, and we write x instead of $\vec{r}$ for position vectors.

Our goal is to find $\Delta t(x_E)$ the change in Earth time during the small, almost instantaneous, time interval $\Delta t(x_S) \to 0$ of the jump of the traveling twin from outbound Spaceship B to inbound Spaceship C.

Suppose the jump from the outbound to the inbound spaceship occurs at an arbitrary point x over spaceship time $\Delta t(x)$. During this (short) transfer time, the speed of the traveling twin changes from constant $+v$ (*outbound*) to constant $-v$ (*inbound*), which is a total change of $\Delta v = -2\,v$. Since the *acceleration a* is the change in velocity per unit time, we have

(11.5.4a) $-2\,v/\Delta t(x) = a$ ⠀⠀⠀⠀⠀⠀⠀⠀⠀⠀ *v constant ⇒ a constant*

$\qquad\qquad\qquad = F(x)/m$ ⠀⠀⠀⠀⠀⠀⠀⠀ *solve for a in* (11.5.1d)

$\qquad\qquad\qquad = -\mathbf{grad}(\Phi(x))$ ⠀ *solve for F/m in* (11.5.1c)

$\qquad\qquad\qquad = -\Phi'(x).$ ⠀⠀⠀⠀⠀⠀⠀⠀ *from* (11.5.1b)

Since $-2v/\Delta t(x) = a$ is constant for all points x, we set the jump point in the first term of (11.5.4a) at $x = x_S$. We then compute the definite integral of the (equal) first and last terms of Equation (11.5.4a), taking the limits from $x = x_E$ to $x = x_S$, to obtain

$$2\int_{x_E}^{x_S} (v/\Delta t(x_S))\; dx = \int_{x_E}^{x_S} \Phi'(x)\; dx,$$

which computes to be

(11.5.4b) ⠀⠀⠀⠀⠀⠀ $2(v/\Delta t(x_S))(x_S - x_E) - \Phi(x_S) - \Phi(x_E).$

Finally, in Equation (11.5.2a), set $\vec{r}_1 = x_E$ and $\vec{r}_2 = x_S$. Now replace the term $\Phi(x_S) - \Phi(x_E)$ of (11.5.2a) with the left hand side of (11.5.4b) above so that (11.5.2a) takes the form

$$\Delta t(x_E) = \left(1 + \frac{2v(x_S - x_E)}{\Delta t(x_S)\; c^2}\right)\Delta t(x_S).$$

Since the 25-year journey of the spaceship at speed $v = 0.8c$ covers a distance $(x_s - x_E) = v \cdot 25$ years, the equation above takes the form

(11.5.5)
Calculus

$$\Delta t(x_E) = \Delta t(x_S) + 50\left(\frac{v}{c}\right)^2 \text{ years}$$

$$= \Delta t(x_S) + 32 \text{ years} \qquad \text{\textit{since } } v = 0.8c$$

$$\longrightarrow 32 \text{ years, as } \Delta t(x_S) \longrightarrow 0.$$

(11.6) Exercises

(11.6.1)
Exercise

Actual clock time of inbound spaceship. We know that the outbound spaceship in $\mathcal{F}^B$ traveling at constant speed $v = 0.8c$ relative to Earth will have an onboard clock time of 15 years when it arrives at point ⑤ of Figure (11.3.1). At this moment, the Earth clock (simultaneously) reads 25 years.

Question: What is the clock time on board the inbound spaceship in $\mathcal{F}^C$ when it arrives at the same point ⑤? (*This is the clock that will be reset to read "15 yrs." by the twin who transfers from the outbound spaceship.*)

Hint:
Background. *The clocks of origin observers at $x = 0$, $t = 0$ for the three frames, $\mathcal{F}^A$ (Earth), $\mathcal{F}^B$ (outbound spaceship), and $\mathcal{F}^C$ (inbound spaceship), are synchronized as per Definition (3.4.5a). The earthbound twin and the departing outbound spaceship are Origin Observers, so their clocks are synchronized. But the inbound spaceship is a cohort, not the Origin Observer in $\mathcal{F}^A$.*

Analysis. *At point ⑤, the $\mathcal{F}^A$ $[x, t]$ coordinates are $[v \times 25yrs., 25yrs.]$ Use the Lorentz transformation, L_{-v}, to link these coordinates to $[x'', t'']''$, the distance-time coordinates in $\mathcal{F}^C$. That is,*

$$[v \times 25, 25] \overset{L_{-v}}{=} [x'', t'']'',$$

in which case, $t'' = (1 + v^2)25/\sigma_v$.

(11.6.2)
Exercise

Generalizing Exercise (11.6.1). In Exercise (11.6.1), replace speed $0.8c$ with an arbitrary speed $v_0 < c$. Let the two spaceships meet when the earthbound twin is t_0 years old. Then, in Figure (11.3.1), the simultaneous line-of-sight times at point ⑤ for the three frames are:
In $\mathcal{F}^A$ on Earth, $t = t_0$.
In $\mathcal{F}^B$ on the outbound spaceship, $t = \sigma_v t_0$.
In $\mathcal{F}^C$ on the inbound spaceship, $t = (1 + v_0^2)t_0\sigma_v$.

(11.6.3) **Congruent Symmetric Triangles.** Show that in Figure (11.3.1),
Exercise   triangles ⑤-③-④ and ⑤-①-⓪ are congruent.

Hint: ....

(1) *As Figure (11.3.1) indicates, ∠ ①-⓪-⑤ = ∠③-④-⑤ = θ.*

(2) *From Theorem (8.4.1), lines of simultaneity, line ⑤-③, and line
    ⑤-① form angle θ with the horizontal. Hence, ∠①-⑤-⓪ =
    ∠③-⑤-④ = (90° − 2θ).*

(3) *From the Angle-Side-Angle theorem, △①-⑤-⓪ and △③-⑤-④
    are congruent to each other since they share a common side length
    (length[⑤-④] = length[⑤-⓪] = 15), and two common angles as
    described in (2) above.*

(4) *Congruency tells us that length(⓪-①) = length(④-③) = 9. If
    ③ is 9 units from the 50-year mark ④ in $\mathcal{F}^A$ time, then ③ must
    be at the 41-year mark in $\mathcal{F}^A$.*

(11.6.4) **Slow-Fast-Slow Analysis.** Assume at age 0, one twin instantly
Exercise   travels into space at constant speed v and after t_0 years on the outbound leg, as measured by Earth's clock, travels the inbound leg for
t_0 more years before returning to Earth. Generalize Figure (11.3.1) to
show that for any speed $v = \tan\theta$, the traveling twin, after a round
trip requiring $2 \cdot t_0$ years according to an Earth clock, will always
return at age $2 \cdot \sigma_v \times t_0$ years. *(see (1.8.2) for the definition of σ_v.)*

Hint: *Analytically, use the Lorentz transformation to link line-of-
sight times at points ⓪ through ⑤ between appropriate pairs of frames.
Alternately, use trigonometry on the triangles in the diagram to obtain lengths that are interpreted as times on the clocks of frames $\mathcal{F}^A$,
$\mathcal{F}^B$, and $\mathcal{F}^C$.*

(12) The Train-Tunnel Paradox

(12.1) An Overview of the Paradox

(12.1.1) FIGURE: Static Train and Tunnel are of Equal Length.

- The common at-rest length of the tunnel in frame $\mathcal{F}^A$ and the train in frame $\mathcal{F}^B$ is d_0.

- In $\mathcal{F}^A$, the origin $x = 0$ is at the left entrance of the tunnel — $x = d_0$ marks the right entrance.

- In $\mathcal{F}^B$, $x = 0$ is at the right end (the front) of the train — $x = -d_0$ marks the left end (the rear) of the train.

We will use the fact that objects moving with relative speed $\pm v$ shrink in the direction of motion by a factor of σ_v (1.8.2), pg. 20.

(12.1.2) FIGURE: **Train in Motion Shrinks, Tunnel is Stationary.**

• The frame $\mathcal{F}^B$ of the train has speed v relative to the stationary frame $\mathcal{F}^A$ of the tunnel.

• As measured in the tunnel's frame $\mathcal{F}^A$, the train has reduced length $\sigma_v d_0$. When the (short) train is midway through the tunnel, two bolts of lightning simultaneously strike the ends of the tunnel. Since the train is well inside the tunnel, as observed from $\mathcal{F}^A$, the train *is spared* any damage from the simultaneous lightning strikes.

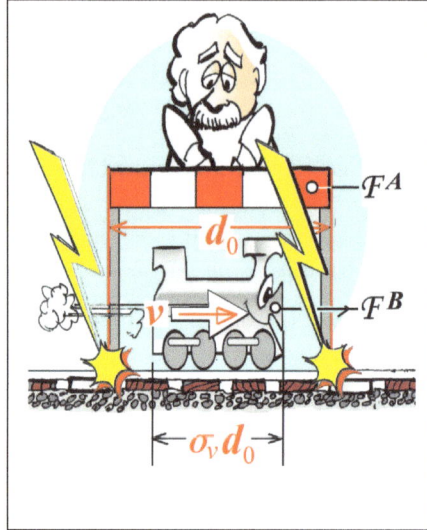

(12.1.3) FIGURE: **Tunnel in Motion Shrinks, Train is Stationary.**

• The frame $\mathcal{F}^A$ of the tunnel has speed $-v$ relative to the stationary frame $\mathcal{F}^B$ of train.

• As measured in the train's frame $\mathcal{F}^B$, the tunnel has reduced length $\sigma_v d_0$. When the midpoints of the moving tunnel and stationary train are aligned (overlap), the short tunnel leaves the ends of the train exposed outside the tunnel entrances. If the two bolts of lightning simultaneously strike the ends of the tunnel, then the (too-long) train *is not spared* electrocution by the simultaneous lightning strikes.

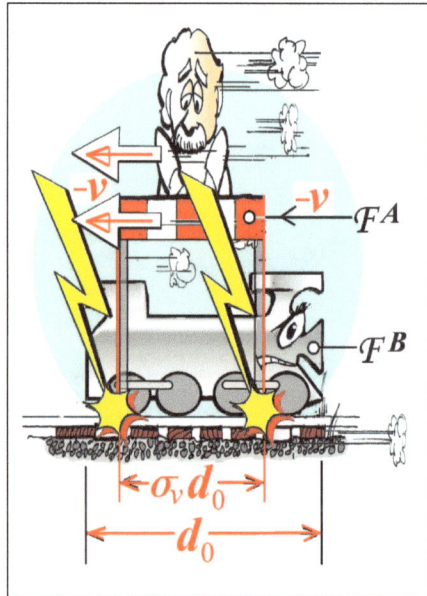

(12.1.4)

> **The Paradox:** The concluding statements of (12.1.2) and (12.1.3) contradict each other. Is the train electrocuted or is it not?
>
> **The Resolution:** The lightning strikes that are simultaneously viewed from the tunnel's frame $\mathcal{F}^A$ (12.1.2) are not simultaneous when viewed from the train's frame $\mathcal{F}^B$. In the final sentence of (12.1.3), there is an invalid use of the adjective "simultaneous.".

(12.2) A Distance Lemma

(12.2.1) **The Minkowski diagram** (12.2.2) following, shows two lengths d_A and d_B which are, apparently, unequal — the horizontal line $\overline{PQ'}$ (*measured in $\mathcal{F}^A$ with length d_A*) *seems to be* shorter the slanted line $\overline{P'Q}$ (*measured in $\mathcal{F}^B$ with length d_B*). Lemma (12.2.2) shows us that in spite of their apparent inequality on the graph, the actual at-rest distances (*or lengths*) measured within their respective frames are, in fact, equal to each other, that is, $d_A = d_B$.

(12.2.2) THEOREM: *Equal Distances* d_0 *for Frames $\mathcal{F}^A$ and $\mathcal{F}^B$*

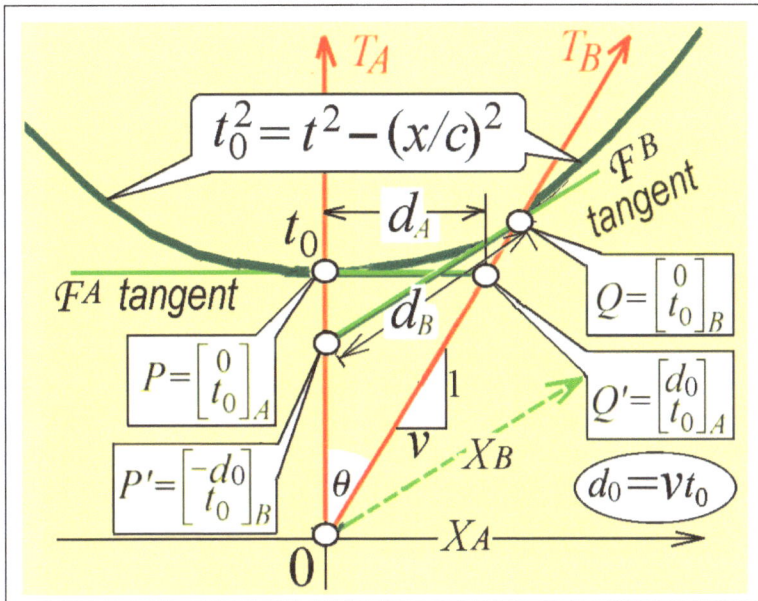

Given: *The Minkowski diagram for the inertial frame* $\mathcal{F}^A$ *has orthogonal spacetime axes* X_A *and* T_A. *The synchronized frame* $\mathcal{F}^B$ *has speed* v *relative to* $\mathcal{F}^A$ *so its time axis* T_B *has slope* $1/v$ *and forms an angle* θ *with the vertical* T_A*-axis where* $\tan(\theta) = v$.

Choose   *arbitrary time* $t_0 > 0$ *which then defines distance* $d_0 = vt_0$.

Construct

(12.2.2a)   *The hyperbola* $t_0^2 = t^2 - (x/c)^2$.

(12.2.2b)   *Points* P *and* Q *on the hyperbola* (12.2.2a) *at the intersections of the* T_A*-axis and the* T_B*-axis of* $\mathcal{F}^A$ *and* $\mathcal{F}^B$, *respectively.*

(12.2.2c)   *Tangent lines* $\text{Tangent}(P)$ *and* $\text{Tangent}(Q)$ *to the hyperbola* (12.2.2a) *at the respective points* P *and* Q.

(12.2.2d)   *Point* P' *at the intersection of line* $\text{Tangent}(P)$ *and the* T_A*-axis,*

                                    and

Point Q' *at the intersection of line* $\text{Tangent}(Q)$ *and the* T_B*-axis.*

THEN ··

(12.2.2e) $\|\overline{PQ'}\|_A = d_0$ *is the length* (A.2.18) *of the line segment* $\overline{PQ'}$ *as measured in* $\mathcal{F}^A$ *coordinates,*

(12.2.2f) $\|\overline{P'Q}\|_B = d_0$ *is the length* (A.2.18) *of the line segment* $\overline{P'Q}$ *is as measured in* $\mathcal{F}^B$ *coordinates,*

(12.2.2g)   *The* X_A*-axis and the* X_B*-axis in the Minkowski diagram* (12.2.2) *are parallel to the line segments* $\overline{PQ'}$ *and* $\overline{P'Q}$, *respectively.*

PROOF:

The importance of hyperbola (12.2.2a).   The Time-Stamped Origin Theorem (8.2.3) tells us that all the points on the hyperbola $[x, t]_A$ in $\mathcal{F}^A$ (where $t_0^2 = t^2 - (x/c)^2$) lie at the $\mathcal{F}^A$ end of a line-of-sight corridor. At the opposite end of each of these line-of-sight corridors lies the time-stamped origin, $[0, t_0]_v$ of frame $\mathcal{F}^v$, which has relative speed v with respect to $\mathcal{F}^A$ where $-\infty < v < \infty$.

Coordinates of points P and Q (12.2.2b).   In our case, the two points P and Q (12.2.2b) whose T-axes intersect the hyperbola of origins (12.2.2a) have home frame coordinates

(12.2.2h) $\qquad P = \begin{bmatrix} 0 \\ t_0 \end{bmatrix}_A$ in $\mathcal{F}^A$ and $Q = \begin{bmatrix} 0 \\ t_0 \end{bmatrix}_B$ in $\mathcal{F}^B$.

The importance of time t_0. Synchronization (3.4.5) of frames $\mathcal{F}^A$ and $\mathcal{F}^B$ tells us that at time $t = 0$ in each frame, their respective origins overlap — or are zero distance apart — in the sense that $[0,0]_A$ and $[0,0]_B$ are at opposite ends of a line-of-sight corridor (*see figure of Theorem* (12.2.2)).

The symmetry principle (1.6.1) says that as time increases, the origins speed away from each other at the same rate in each frame — at speed $\pm v$.

> **(12.2.2i)** *Hence, at common time t_0 in frames $\mathcal{F}^A$ and $\mathcal{F}^B$, the origins (i.e., the T_A and T_B axes on the Minkowski diagram) are separated by the usual Euclidean distance $d_0 = vt_0$.*

Therefore, coordinates of points P' and Q' (12.2.2b) have the form

(12.2.2j) $\qquad P' = \begin{bmatrix} -d_0 \\ t_0 \end{bmatrix}_A$ in $\mathcal{F}^A$ and $Q' = \begin{bmatrix} d_0 \\ t_0 \end{bmatrix}_B$ in $\mathcal{F}^B$.

Using the definition (A.2.18a), pg. 340, for the usual vector lengths in frames $\mathcal{F}^A$ and $\mathcal{F}^B$ respectively, we calculate the Euclidean distance between the $\mathcal{F}^A$ origin and the $\mathcal{F}^B$ origin at time t_0 in $\mathcal{F}^A$ as follows:

$$\|\overline{PQ'}\| = \|P - Q'\| = \left\| \begin{bmatrix} 0 \\ t_0 \end{bmatrix}_A - \begin{bmatrix} d_0 \\ t_0 \end{bmatrix}_A \right\|$$
$$= \sqrt{(0 - d_0)^2 + (t_0 - t_0)^2} = d_0.$$

Similarly, the calculation of the distance between the $\mathcal{F}^A$ origin and the $\mathcal{F}^B$ origin as observed at time t_0 in $\mathcal{F}^B$ is

$$\|\overline{P'Q}\| = \|P' - Q\| = \left\| \begin{bmatrix} -d_0 \\ t_0 \end{bmatrix}_B - \begin{bmatrix} 0 \\ t_0 \end{bmatrix}_B \right\|$$
$$= \sqrt{(-d_0)^2 + (t_0 - t_0)^2} = d_0.$$

The previous two strings of equalities prove (12.2.2e) and (12.2.2f).

Finally, (12.2.2g) holds true since, according to (4.4.1a), pg. 73, (4.4.3), pg. 73, all lines of events with fixed time t_0 — i.e., lines of simultaneous events — are, in fact, (a subset of) their respective X axes at time t_0. The proof is done. ■

(12.3) The Train-Tunnel Minkowski Diagram

(12.3.1) The single Minkowski diagram that illustrates the relative motion of Figures (12.1.2) and (12.1.3), will be based on the Minkowski diagram of Figure (12.2.2), pg. 155 (*repeated on the right*).

This diagram shows the worldlines of a pair of origins at relative speed $\pm v$ after they separate from their line-of-sight overlap (synchronization) at a common coordinate $[0, 0]$. At time t_0 in either frame, the separation of the origins is measured to be $d_0 = v\, t_0$. This allows us to identify the horizontal line segment $\overline{PQ'}$ (of length d_0) with the tunnel and the slanted line segment $\overline{P'Q}$ (of length d_0) with the train.

(12.3.2) ‖ NOTE: *Why Are There Two d_0 Lengths in Figure* (12.2.2)?
‖ Similar to the figure of (8.2.3), pg. 120, where the scale of the *time*
‖ axes change from frame to frame, in Figure (12.3.1), the scale of
‖ *length d_0* changes from frame $\mathcal{F}^A$ to $\mathcal{F}^B$. This is why the slanted
‖ d_0 in $\mathcal{F}^B$ appears longer than the horizontal d_0 in $\mathcal{F}^A$.

The Minkowski diagram (12.3.3) following is based on Figure (12.3.1). For example, the worldlines of the two origins from Figure (12.3.1) — the left entrance to the tunnel in $\mathcal{F}^A$ and the right front of the train in $\mathcal{F}^B$ — are transferred to Figure (12.3.3) as the vertical line (1) and the slanted line (3). The construction of Figure (12.3.3) continues as follows:

For each of the infinitely many points $0 < x \le d_0$ of the <u>tunnel</u> floor in $\mathcal{F}^A$, add its vertical worldline, creating a vertical (green) band of all possible spacetime coordinates for the tunnel. The left and right boundaries of this vertical band are indicated by (1) and (2), respectively.

For the infinitely many points $-d_0 \le x < 0$ of the train in $\mathcal{F}^B$, add its vertical worldline, creating a slanted (gray) band of all possible spacetime coordinates for the train. The left and right boundaries of this slanted band are indicated by (4) and (3), respectively.

(12.3.3) Figure: *The Minkowski Diagram for the Train-in-Tunnel Paradox:*

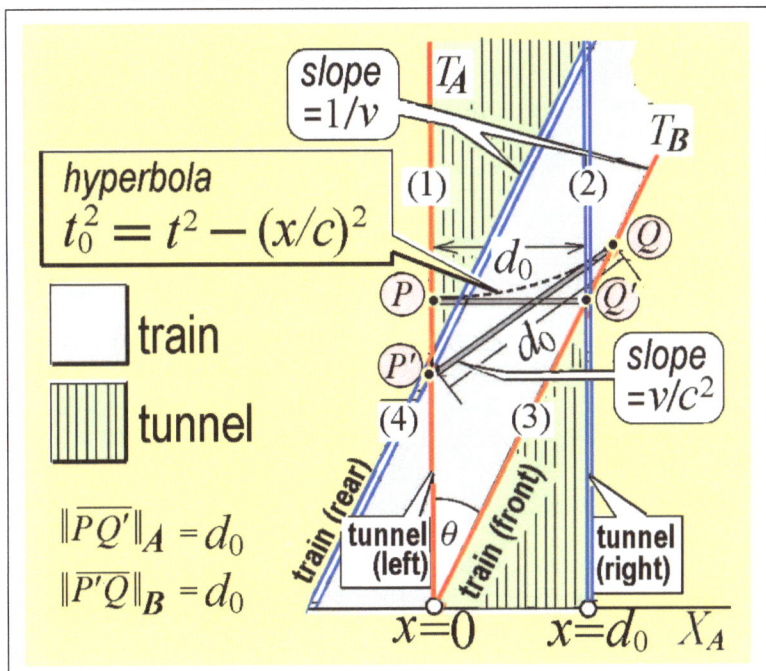

NOTE: Consistent with Theorem (8.4.1), the horizontal lines of simultaneity in frame $\mathcal{F}^A$ also have slope of the form (v'/c^2). In our case, the relative speed v' of $\mathcal{F}^A$ with respect to itself is zero.

(12.4) Explaining Mutual Contraction

The Train-in-Tunnel Paradox addresses an apparent *time anomaly* for the pair of lightning strikes. There is also a question of an apparent *distance anomaly*, namely, when the train is shorter than the tunnel (Figure (12.1.2)), and, at the same time, the tunnel is shorter than the train (12.1.3).

(12.4.1) Question: How can both comparisons be true at the same time?

(12.4.2) **Figure (12.3.3) shows:**

Simultaneity in the tunnel's frame $\mathcal{F}^A$ is always along a single line segment like $\overline{PQ'}$ that is horizontal in Figure (12.3.3). Hence, from frame $\mathcal{F}^A$, the train in $\mathcal{F}^B$ must be viewed along a horizontal line that intersects the train's slanted band of spacetime coordinates. Notice that the part of the horizontal line $\overline{PQ'}$ that intersects the slanted band (*between worldlines (4) and (3)*) is shorter than the width of the tunnel (*horizontal distance between worldlines (1) and (2)*). That is,

- from the frame of the tunnel $\mathcal{F}^A$, the moving train in $\mathcal{F}^B$ is shorter than the width of the tunnel.

Simultaneity in the train's frame $\mathcal{F}^B$ is always along a single line segment like $\overline{PQ'}$ that has slope (v/c^2) in Figure (12.3.3). Hence, the tunnel (*which is in $\mathcal{F}^A$*) must be viewed along the slanted line $\overline{P'Q}$ which is the length of the train in its home frame $\mathcal{F}^B$. From this point of view (*from $\mathcal{F}^B$*) the train extends beyond the green band — the train is longer than the tunnel.

- from the frame of the train $\mathcal{F}^B$, the moving tunnel in $\mathcal{F}^A$ is shorter than the length of the train.

(12.5) Resolving the Train-Tunnel Paradox

(12.5.1) **FIGURE: Two Simultaneous Lighting Strikes at Time t_1 in $\mathcal{F}^A$**

The vertical shaded band between the labels, **tunnel (left)** and **tunnel (right)**, represents all points of the tunnel in frame $\mathcal{F}^A$ at all times.

The slanted shaded band between the labels, **train (rear)** and **train (front)**, represents all points of the train in frame $\mathcal{F}^B$ at all times.

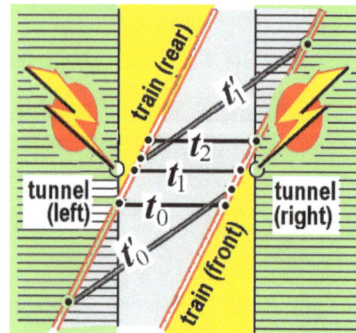

Simultaneity in $\mathcal{F}^A$ of the stationary tunnel is represented by events on any single horizontal line. Three such lines in $\mathcal{F}^A$ are shown for times $t_0 < t_1 < t_2$. The *simultaneous* lightning strikes occur at

$\mathcal{F}^A$ time t_1.

> This agrees with Figure (12.1.2).

Simultaneity in $\mathcal{F}^B$ of the stationary train is represented by events on any single slanted line with slope (v/c^2). Two such lines in $\mathcal{F}^B$ are shown for times $t_0' < t_1'$. The two **sequential** lightning strikes (*that occur simultaneously at time* t_1 *in* $\mathcal{F}^A$) occur in $\mathcal{F}^B$ first at time t_0' then later at time t_1'.

> This does <u>not</u> agree with Figure (12.1.3).

which falsely shows the two lightning strikes occuring at the same time in the train's frame $\mathcal{F}^B$.

(12.5.2) In Figure (12.1.3), the lightning strikes are <u>sequential</u>. As seen in Figure (12.5.1):

- At time t_0' in $\mathcal{F}^B$, all points of the train lie on the slanted t_0' line. The front end of the train has entered the tunnel while its rear end is still outside the tunnel. Straight ahead on the slanted t_0' line, the first lightning strike occurs at the tunnel's right entrance.

- Later, at time t_1' in $\mathcal{F}^B$, all points of the train lie on the slanted t_1' line. The front of the train has exited the right side of the tunnel while its rear end has passed by the left entrance of the tunnel. Behind the train on the slanted t_1' line, the second lightning strike occurs at the tunnel's left entrance.

(12.6) Exercises

(12.6.1)
Lin Alg
The Moving Train Shrinks by a Factor of σ_v: In Figure (12.3.3), show that in $\mathcal{F}^A$ coordinates, the length of the intersection of horizontal line segment $\overline{PQ'}$ with the slanted band is $\sigma_v d_0$.

Hint: *The $\mathcal{F}^A$ coordinates for $\overline{PQ'}$ have the form $[x, t_0]_A$, $0 \le x \le d_0$. The $\mathcal{F}^B$ coordinates for worldline (4) (the left entrance of the tunnel) have the form $[-d_0, t]_B$ where $-\infty < t < \infty$.*

Use the Lorentz transform $L_v : \mathcal{F}^A \to \mathcal{F}^B$ (7.1.3) to find the intersection point $[\hat{x}, t_0]_A$ of $\overline{PQ'}$ and worldline (4) and show that $\hat{x} = (1 - \sigma_v)d_0$, which shows that the intersection of $\overline{PQ'}$ with the slanted band of points on the train has length $\sigma_v d_0$.

The calculations proceed as follows:
The equation $L_v([\hat{x}, t_0]_A) = [-d_0, t]_B$ produces two equations. Eliminate the variable t and solve for $\hat{x} = (1 - \sigma_v)d_0$.

(12.6.2)
Lin Alg

The Moving Tunnel Shrinks by a Factor of σ_v: In Figure (12.3.3), show that in $\mathcal{F}^B$ coordinates, the length of the intersection of slanted $\overline{P'Q}$ with the vertical band is $\sigma_v d_0$.

Hint: *See the hint in Exercise (12.6.1).*

(12.6.3)
Lin Alg

A Moving Origin Clock on Worldline (3) Slows by a Factor of σ_v: Show that in Figure (12.3.3), the origin clock (3.4.3) whose worldline coincides with worldline (3), moving at speed v relative to frame $\mathcal{F}^A$ slows down by a factor of σ_v.

Hint: *Let us use the notation Clock_3 to denote the origin clock of $\mathcal{F}^B$ whose coordinates coincide with worldline (3). Then the origin clocks of $\mathcal{F}^A$ and $\mathcal{F}^B$ (Clock_3) are synchronized at common time $t = 0$. At point $Q' = [d_0, t_0]_A$, clocks in $\mathcal{F}^A$ read time $t = t_0$. Use the Lorentz transformation to read the origin clock Clock_3 at the other end — the $\mathcal{F}^B$ end — of the line-of-sight corridor. That reading will be $\sigma_v t_0$.*

(12.6.4)
Lin Alg

A Moving Origin Clock on Worldline (1) Slows by a Factor of σ_v: Show that in Figure (12.3.3), the origin clock (3.4.3) whose worldline coincides with worldline (1), moving at speed v relative to frame $\mathcal{F}^B$ slows down by a factor of σ_v.

Hint: *See the hint in Exercise (12.6.3).*

(13) The Pea-Shooter Paradox

(13.1) An Overview of the Paradox

Restatement of the pea-shooter paradox (1.13): You observe an airplane traveling away from you at speed w while the plane launches a spaceship (*in the same direction of the plane*) whose instantaneous speed, as measured by the pilot, is v. Your intuition and experience should tell you that you would measure the spaceship speed to be $v + w$. In fact, the actual spaceship speed is the theoretically accurate Lorentz sum, $v \oplus w$ (7.3.1), which, due to the low speeds involved, is practically indistinguishable from the usual arithmetic sum $v + w$ for speeds v and w.

Our powers of perception are inadequate to distinguish between Lorentzian addition $v \oplus w$ (*which is valid for all speeds*) and the imperceptibly larger quantity $v + w$ we usually experience at *small speeds*. When the measured sum of v plus w for *high* values of v and w disagrees with the familiar formula $v + w$ used for *low* values of v and w, we tend to throw up our hands and announce a *paradox*.

(13.1.1) A pea in frame $\mathcal{F}^C$ has speed $v = 0.8c$ as measured in frame $\mathcal{F}^B$ by the peashooter, Ashley. Ashley (and her platform) have speed $w = 0.9c$ as measured in frame $\mathcal{F}^A$ by Bernie. Yet, Bernie does not see the pea moving with the speed we would expect from our experience.

I SEE THE PEA FLEEING AT 0.8 x C.

I SEE ASHLEY FLEEING AT 0.9 x C, BUT THE PEA IS FLEEING AT 0.988 x C.

$\mathcal{F}^C$

$v_0 = 0.8 \times C$

$\mathcal{F}^B$

$v_1 = 0.9 \times C$

$\mathcal{F}^A$

Figure (1.13.2b),   page 37

Galilean/Newtonian addition of speeds tells us that

(13.1.1a) $v + w = 0.8c + 0.9c = 1.7c.$ *classical Galilean addition*

Formula (13.1.1a) was justified in (5.5.2) using <u>matrices</u>, and in (5.6.3), using <u>triangles</u> whose areas correspond to speeds.

(13.1.2) **Pea-Shooter Paradox Already Resolved.** Due to the relativistic constraint that no material object can have a speed greater than the speed of light, $c = 1$, a rule different from (13.1.1a) must apply — this general rule turns out to be the Lorentzian addition

$$\textbf{(13.1.2a)} \quad v \oplus w = \frac{v+w}{1+vw}, \qquad \{-1 < v, w < 1 \; implies \; v \oplus w < 1\}$$

that has *already* been defined in (7.3.1). Formula (13.1.2a) for relativistic addition was justified in (7.4.1) using <u>matrices</u>.

Similar to the Galilean case (5.6.3), we established the rule for Lorentzian addition in (13.1.2a) using <u>triangles</u> whose areas are closely related to relative speeds.

(13.2) The Fizeau Experiment: Adding Speeds

The experiment of Fizeau offered early verification of (13.1.2a), the relativistic formula for the addition of speeds. Here are the details:

The Democratization of Light

(13.2.1) **Wave Speed *Is Not* Affected by Speed of the Wave Source.** Suppose a transmitting medium (e.g., air) is stationary relative to an observer — there is no wind. Then any wave (e.g., the sound of a train whistle) that is transmitted in that medium will have the *same observed speed* at all points along the line of motion even if its source (e.g., the train) is in motion relative to the observer. This is seen from the "skipping stone" model of Figure (16.5.4). Although the source speed does not affect the wave speed, it does affect the measured frequency, the *Doppler effect* that produces a higher pitch for an approaching train and a lower pitch when the train recedes (*see Section* (16.6)).

(13.2.2) **Wave Speed *Is* Affected by Speed of the Medium.** As seen in (16.5.5), the observed speed of the transmitting medium (e.g., wind speed) *is* communicated to the speed of the transmitted wave (e.g., sound of a train whistle).

> **(13.2.2a)** **Since light needs no transmitting medium,** the speed of light in a vacuum is a constant c for all observers, whether they are in motion or not (1.5.2b).

> **(13.2.2b)** **However, when light *is* transmitted through a medium,** it is subject to the physics of (13.2.2) (*this is the democratization of light*), in which case, the speed of light is reduced to

(13.2.2c) $\quad v = \dfrac{c}{n}$

$$\left\{ \begin{array}{r|l} & n \ (unitless) \\ \hline Air & 1.0002926 \\ Diamond & 2.417 \\ Water & 1.33 \end{array} \right.$$

> where $n > 1$ is the unitless Refractive Index of the medium.

(13.2.3) **The Refractive Indices Are Approximate** since their values depend on the frequency of the transmitted light and the temperature of the transmitting medium. Consequently, the wave speed $v = c/n$ (13.2.2c) relative to the *stationary* medium is approximately equal to $c/1.33 \approx 140,000$ miles/second.

Due to (16.5.5), pg. 212, the observed speed of light in *moving* water is greater than (less than) $c/1.33$ if the water flow is in the same (opposite) direction of the light.

The Setup for Fizeau's Experiment

(13.2.4) **To describe how Fizeau designed his experiment** for measuring the pea-shooter effect, or the addition of speeds (1.13.2), we must define the reference frames for the observer, the water, and the impeded light beam traveling in water.

(13.2.5) • $\mathcal{F}^A$ is the frame of the *observer* (experimenter), which is analogous to Bernie's frame $\mathcal{F}^A$ in the figure of (13.1.1).

   • $\mathcal{F}^B$ is the frame of the transporting *water* which has speed w

relative to frame $\mathcal{F}^A$. This is analogous to frame $\mathcal{F}^B$ in (13.1.1), the frame of the transporting wagon. From (3.4.2), the water has zero speed relative to its own frame $\mathcal{F}^B$.

(13.2.6) *How Fizeau Measures Addition of Speeds*

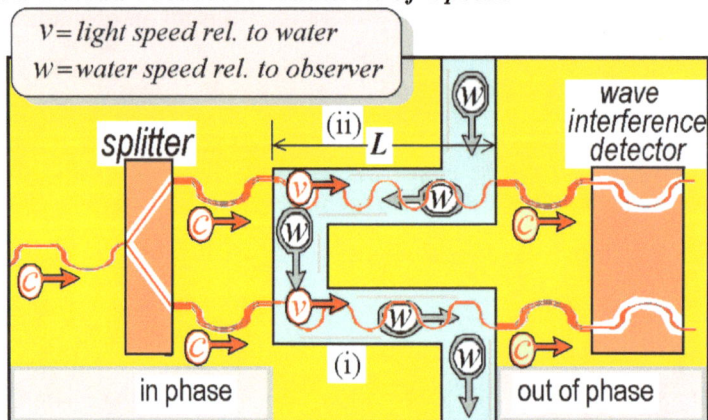

A *single light beam enters a splitter and exits as two identical beams that are directed to parallel tubes* (i) *and* (ii). *At the entry point of the tubes, the two beams are in phase. From the experimenter's viewpoint, water in tube* (ii) *has speed* $-w$ (*moving leftward*) *and* _diminishes_ *the speed of the opposing light beam. Similarly, water in tube* (i) *has speed* $+w$ (*moving rightward*) *and* _adds_ *to the speed of the light beam. This difference of speeds implies that at the exit point of the tubes, the light beams are out of phase as is seen by the interference patterns of their waves.*

(13.2.6a) Observed Speeds of Light in the Tubes. The speed of light relative to stationary water is $v = c/1.33$.

- As light travels in tube (i) in the same direction as the water, its observed speed is $v_{fast} > v = c/1.33$.

- As light travels in tube (ii) against the direction of the water, its observed speed is $v_{slow} < v = c/1.33$.

(13.2.6b) Therefore, there are *different numbers of wavelengths* in each tube. In Fizeau's experiment, it is observed that frequency of the light in each tube is unchanged. Such constant frequency tells us that a slower speed of the transmitted light (*upper tube* (ii)) implies a shorter wavelength and a faster speed for the transmitted light (*lower tube* (i)) implies a longer wavelength (*see Figure* (13.2.6) *and*

Definition (16.1.2)).

(13.2.6c) To see intuitively that increased wave speed implies increased wavelength, imagine a moving train (one wavelength) whose front and back ends must pass a fixed point on the platform exactly one second apart. The faster the speed of the train, the longer it must take to meet the one-second requirement.

To see this analytically, note that the formula $v = \lambda \cdot frequency$ (16.1.2) shows that wave speed v and wavelength λ are proportional. At any instant of time, the different wave speeds in the parallel tubes produce a different number of standing wavelengths (or cycles) as shown in Figure (13.2.6).

Because of the different speeds, the crests and troughs of the two light beams are not aligned when they exit the water-filled tubes. When they recombine, there are regions where they enhance each other and other regions where they cancel each other. The result is a pattern of light and dark fringes that changes as the speed w changes.

(13.2.7) Speeds of Light in Water. Since the speed of light relative to standing water is $c/1.33$, water traveling in the *same direction* at observed speed w adds to the speed of transmitted light (16.5.5). Non-relativistic Galilean physics holds that the augmented speed is

(13.2.7a) $\quad v_{fast} = c/1.33 + w.$
$$\qquad\qquad\qquad\qquad\qquad \textit{Galilean addition}$$
$$\textit{of speeds}$$
$$c/1.33 \textit{ and } w$$

Formula (13.2.7a) is not valid! What is observed instead is

(13.2.7b) $\quad v_{fast} = c/1.33 + \mathbf{f}w$
$$\textit{observed (true) addition}$$
$$\textit{of speeds}$$
$$c/1.33 \textit{ and } w$$

where $\mathbf{f}$ is the unitless FRESNEL COEFFICIENT OF DRAG, so-called for historical reasons.

Data obtained using different liquids gives the following relation between n, the Fresnel refractive index of (13.2.2c), and $\mathbf{f}$, the Fresnel drag of (13.2.7b):

(13.2.7c) $\quad \mathbf{f} = 1 - \left(\dfrac{1}{n}\right)^2.$ (*If $n \approx 1.33$ for water, then $\mathbf{f} \approx 0.43$.*)

This reasons for this empirical relation were not understood until Einstein provided the following theoretical basis:

(13.2.8) Einstein's Theoretical Basis. According to special relativity, if the speed of light with respect to the frame of reference where the medium (water) is at rest is c/n, and the speed of the medium with respect to the observer is w, then the speed of light with respect to the observer will be given by

(13.2.8a) $\quad c/n + \mathbf{f}w = c/n \oplus w$

Evaluating the Lorentzian sum above with $w << c$ gives $(c/n) + (1 - 1/n^2)w$ (see Exercise (13.3.2)). This prediction of the special theory then coincides precisely with the observations (13.2.7c)!

(13.3) Exercises

(13.3.1) **Wave Interference.** The symbols (m) and (m/s) indicate the units
Exercise of *meters* and *meters/second*. Given the Fizeau apparatus (13.2.6) where the known quantities are:

$L(m) =$ tube length,
$\lambda(m) = 570 \times 10^{-9}$ meters, the wave length of yellow light in a vacuum,
$w(m/s) =$ observed water speed,
$v(m/s) = c/1.33 = 1/1.33 =$ the speed of light relative to stationary water.

Suppose the tube length L and water speed w can be adjusted to respective values L_0 and w_0 such that wave interference is maximized as shown in Figure (13.2.6). This occurs when maximum cancellation of the interfering waves occurs. With these known values of L_0 and w_0, there is sufficient information to determine the observed speeds for light in the two tubes, v_{fast} in tube (i), and v_{slow} in tube (ii).

Hints:
- *From (16.1.2), we may set*

(13.3.1a) $\quad \phi_{yellow} = \dfrac{c(m/s)}{\lambda(m)} = \dfrac{1}{\lambda}, \qquad units\left(\dfrac{1}{seconds}\right)$

as the frequency of yellow light where the speed of light $c = 1$. The value of frequency ϕ_{yellow} is constant — the color (frequency) does not change as light is observed in running water.

- *Assuming the form (13.2.7b), write the speeds of light in the tube's running water as*

$$v_{fast} = c/1.33 + \mathbf{f}w$$

for the speed of light in tube (i) and

$$v_{slow} = c/1.33 - \mathbf{f}w$$

for the speed of light in tube (ii). The unknown constant $\mathbf{f}$ will be determined by our later calculations.

- *From (16.1.2), the wave length $\lambda_{fast} = v_{fast} \cdot \phi_{yellow}$ for the accelerated light beam in tube (i) and the wave length $\lambda_{slow} = v_{slow} \cdot \phi_{yellow}$ for the impeded light beam in tube (ii). Accordingly, the numbers of wavelengths of light in tubes (i) and (ii) respectively, are*

$$N_{fast} = \frac{L_0}{\lambda_{fast}} = \frac{L_0\phi_{yellow}}{v_{fast}} = \frac{L_0\phi_{yellow}}{(v + \mathbf{f}w_0)},$$
$$N_{slow} = \frac{L_0}{\lambda_{slow}} = \frac{L_0\phi_{yellow}}{v_{slow}} = \frac{L_0\phi_{yellow}}{(v - \mathbf{f}w_0)}.$$

- *The unitless number $(N_{slow} - N_{fast})$ is the <u>difference</u> between the number of waves in the two tubes. We find that*

$$(N_{slow} - N_{fast}) = 2L_0\phi_{yellow}\frac{\mathbf{f}w_0}{v^2}$$

$$= 2\frac{L_0}{\lambda}\frac{\mathbf{f}w_0}{v^2}. \qquad \begin{array}{l} set\ \phi_{yellow} = 1/\lambda\ from\ (13.3.1a) \\ and\ ignore\ (\mathbf{f}w_0)^2 \approx 0 \end{array}$$

- *In the equation above, set $(N_{slow} - N_{fast}) = 1/2$ so that wave interference is maximized. After L_0 and w_0 have been set to maximize wave interference, solve for the single unknown, $\mathbf{f}$, which then determines the desired speeds v_{fast} and v_{slow}.*

(13.3.2) **Deriving the Fresnel Drag Coefficient f.**   Show that (13.2.8)
Exercise   suffices to derive the empirical value of **f** (13.2.7c) for *all* refraction
indices n.

> **Hint:** *Set $c = 1$ so that $c/n \oplus w = 1/n \oplus w$. Using the form (7.3.1)
> for the Lorentz sum, equation (13.2.8) reduces to*
>
> $$\mathbf{f} = (1 - 1/n^2)/(1 + w/n).$$
>
> *Since water speed w is much less than light speed, 1 ($w \ll 1$), the
> $w/n \approx 0$ term in the denominator can be ignored.*

(14) The Bug-Rivet Paradox

NOTE: **To simplify calculations,** we set the speed of light $c = 1$.

(14.1) The Minkowski Diagram

(14.1.1)

animated online

In frame $\mathcal{F}^A$, two vertical plates, each with horizontally aligned holes, have left surfaces d_0 units apart. A rivet in $\mathcal{F}^B$ with at-rest shank length d_0 travels rightward towards the two aligned holes which are large enough so that the shank but not the head will pass through. A bug is in front of the hole in the right plate.

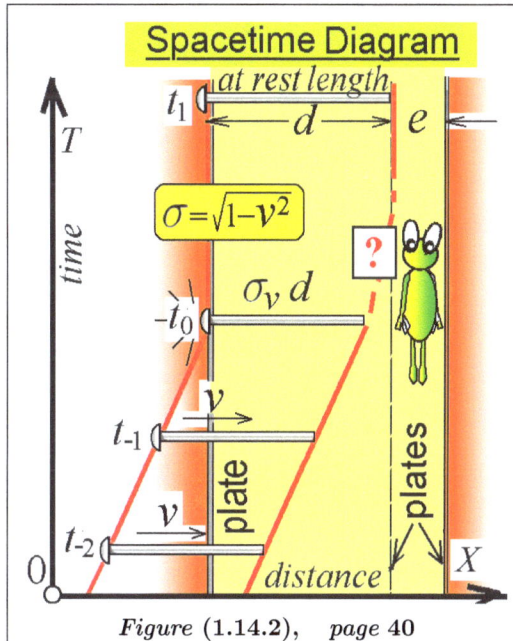

Figure (1.14.2),   *page* 40

Viewed from the (stationary) <u>plates</u>, the moving rivet has length $\sigma_v d_0 < d_0$ — hence, the bug is not injured when the rivet head collides with the left plate. **But viewed from the (stationary) <u>rivet</u>,** the moving plates are separated by only $\sigma_v d_0 < d_0$ units — the bug would be squashed after the rivet collides.

....

(14.1.2) **Question:** Is the bug squashed or not?

Answer: Analyzing the Minkowski diagram (14.1.3), we shall see why the bug cannot be spared.

(14.1.3) FIGURE: *Minkowski Diagram for Rivet Expanding Beyond Its At-rest Length*

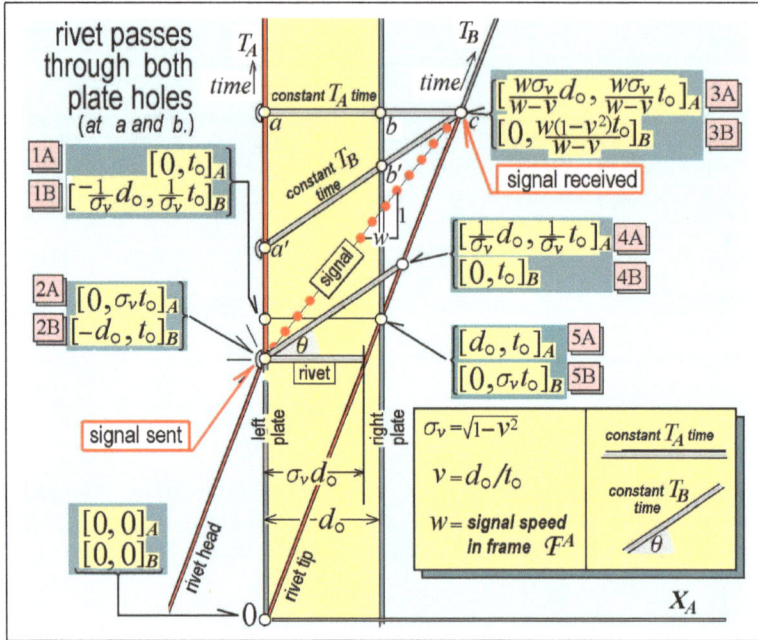

Just before the rightward-traveling rivet hits the left plate, all its atoms move at speed v. The moment the rivet head hits the left plate, it stops. But the rest of the rivet *cannot* stop simultaneously since the message "I've hit a plate" cannot travel instantaneously from the head of the rivet to the rightward tip. In fact, a signal is transmitted from the head to the rest of the rivet through the atomic bonds between the atoms of the rivet. Although the signal may travel quickly by human standards (*at speed w relative to the plates in $\mathcal{F}^A$*), it cannot travel faster than $c = 1$. The rivet tip will keep on moving at speed v until the signal finally reaches it.

Here is the analysis:

(14.1.4) Setup of the Minkowski Diagram (14.1.3):

    (14.1.4a)   Axes for $\mathcal{F}^A$ (*parallel plates frame*) are drawn in the Minkowski graph (14.1.3) so that its time axis, T_A, is vertical and its distance axis, X_A, is horizontal. The point $x = 0$ in $\mathcal{F}^A$ lies on the left surface of the left plate.

The units of the axes are normalized so that c, the speed of light in a vacuum in $\mathcal{F}^A$, equals 1.

....

(14.1.4b) Axes for $\mathcal{F}^B$ (*rivet frame*) are drawn in the Minkowski diagram (14.1.3) so that its time axis, T_B, forms an angle θ with the vertical. By symmetry (1.6.1), $-v$ is the speed of the plates in $\mathcal{F}^A$ as seen from the rivet's frame, $\mathcal{F}^B$.

The distance axis, X_B of $\mathcal{F}^B$, (not shown in Figure (14.1.3)), forms an angle θ with the horizontal. The angle θ is required since $v = \tan(\theta)$ is the speed of the rivet in $\mathcal{F}^B$ as seen from the plates' frame, $\mathcal{F}^A$.

The point $x = 0$ in $\mathcal{F}^B$ lies on the right hand tip of the rivet.

The units of the axes are normalized so that c, the speed of light in a vacuum in $\mathcal{F}^B$, equals 1.

....

(14.1.4c) Lines of simultaneity (*constant time*) in frame $\mathcal{F}^A$ lie on horizontal lines that are necessarily parallel to the X_A-axis.

All lines of simultaneity (*constant time*) of frame $\mathcal{F}^B$ lie on lines that form an angle θ with the horizontal that are necessarily parallel to the X_B axis.

....

(14.1.4d) Vertical plates (*in frame $\mathcal{F}^A$*) are parallel metal plates with horizontally aligned holes large enough to allow the shank of a rivet to pass through but not large enough for the rivet head. The at-rest distance between the left surfaces of the two plates is d_0.

....

(14.1.4e) The rivet (*in frame $\mathcal{F}^B$*) has a shank with at-rest length d_0. The speed of the inertial frame $\mathcal{F}^B$ of the rivet is $v = \tan(\theta)$ relative to the inertial frame $\mathcal{F}^A$ of the plates. Hence, from the symmetry principle, the speed of the frame $\mathcal{F}^A$ of the plates is $-v$ relative to the frame $\mathcal{F}^B$ of the rivet.

....

(14.1.4f) Synchronization: The left surface of the left hand plate in $\mathcal{F}^A$ and the right tip of the rivet in frame $\mathcal{F}^B$ are synchronized (3.4.5a) in Figure (14.1.3) at $[0,0]_A$ and $[0,0]_B$.

....

(14.1.4g)   **Time** $t_0 = d_0/v$ where v and d_0 are given in (14.1.4e). We can see that $v = d_0/t_0$ from the Minkowski diagram (14.1.3) since the slope of the worldline of the rivet is $1/v$. Specifically, the point $[d_0, t_0]_A$ $\boxed{\text{5A}}$ lies on the worldline as do the origins, $[0,0]_A$ and $[0,0]_B$, due to synchronization (14.1.4f).

....

(14.1.4h)   **The Lorentz Transformation,** L_v, connects the line-of-sight coordinates between frames $\mathcal{F}^A$ and $\mathcal{F}^B$ where frame $\mathcal{F}^B$ has speed v relative to frame $\mathcal{F}^A$ (7.1.3a), pg. 106, (7.2.2), pg. 109.

Specifically, for $\sigma_v = \sqrt{1 - v^2}$, an observer in $\mathcal{F}^A$ at time t with coordinates $[x, t]_A$ is linked (*via a line-of-sight corridor*) to an observer in $\mathcal{F}^B$ with coordinates $[x', t']_B$, if and only if

$$(*) \quad \underbrace{\frac{1}{\sigma_v}\begin{bmatrix} 1 & -v \\ -v & 1 \end{bmatrix}}_{L_v} * \begin{bmatrix} x \\ t \end{bmatrix}_A = \begin{bmatrix} x' \\ t' \end{bmatrix}_B \quad and \quad \underbrace{\frac{1}{\sigma_v}\begin{bmatrix} 1 & v \\ v & 1 \end{bmatrix}}_{L_{-v}} * \begin{bmatrix} x' \\ t' \end{bmatrix}_B = \begin{bmatrix} x \\ t \end{bmatrix}_A .$$

(14.2) Coordinates in the Minkowski Diagram

The coordinates $\boxed{\text{1A}}$, $\boxed{\text{1B}}$, $\boxed{\text{2A}}$, $\boxed{\text{2B}}$,...., $\boxed{\text{5A}}$, $\boxed{\text{5B}}$ of Figure (14.1.3) are justified through the following chain of ideas. We start with

$\boxed{\text{1A}}$ = $[0, t_0]_A$: On the T_A-axis of $\mathcal{F}^A$, choose t_0 in accordance with (14.1.4g). The Minkowski coordinate in frame $\mathcal{F}^A$ is $[0, t_0]_A$.

$\boxed{\text{1B}}$ = $[-\frac{1}{\sigma_v}d_0, \frac{1}{\sigma_v}t_0]_B = L_v([0, t_0]_A)$ in accordance with (14.1.4h).

....

$\boxed{\text{4B}}$ = $[0, t_0]_B$ are the coordinates in $\mathcal{F}^B$ of the intersection of the hyperbola $t_0^2 = t^2 - x^2$ and the T_B axis (the worldline of the rivet tip).

$\boxed{\text{4A}} = [\frac{1}{\sigma_v}d_0, \frac{1}{\sigma_v}t_0]_A = L_{-v}([0, t_0]_B)$ in accordance with (14.1.4h).

. . . .

$\boxed{\text{5A}} = [d_0, t_0]_A$ is the coordinate of the point on the left surface of the right plate which (at $\mathcal{F}^A$ time t_0) is d_0 units distance in $\mathcal{F}^A$ from $\boxed{\text{1A}}$, the point on the left surface of the left plate.

$\boxed{\text{5B}} = [0, \sigma_v t_0]_B = L_v[d_0, t_0]_A$ in accordance with (14.1.4h).

. . . .

$\boxed{\text{2B}} = [-d_0, t_0]_B$ is the coordinate of the point on the left surface of the left plate which (at $\mathcal{F}^B$ time t_0) is $-d_0$ units distance in $\mathcal{F}^B$ from $\boxed{\text{4B}}$, the point on the left surface of the right plate.

$\boxed{\text{2A}} = [0, \sigma_v t_0]_A = L_{-v}[-d_0, t_0]_B$ in accordance with (14.1.4h).

. . . .

$\boxed{\text{3A}} = \left[\frac{w\sigma_v}{w-v}d_0, \frac{w\sigma_v}{w-v}t_0 \right]_A$ are the spacetime coordinates of the rivet tip when it first receives the shock wave signal that has been traveling rightward along the shank at constant speed $w \leq c = 1$. The worldline of this signal (represented by red dots in (14.1.3)) starts at the point $\boxed{\text{2A}}$, $\boxed{\text{2B}}$ and has slope $1/w$.

$\boxed{\text{3B}} [0, \frac{w(1-v^2)}{w-v}t_0]_B = L_v \left[\frac{w\sigma_v}{w-v}d_0, \frac{w\sigma_v}{w-v}t_0 \right]_A$ in accordance with (14.1.4h).

(14.2.1)

> THEOREM: (***Rivet Overshoot Beyond the Second Plate***)
> *The rivet in frame $\mathcal{F}^B$ travels at speed v as measured from frame $\mathcal{F}^A$ of the vertical plates. Once the rivet head collides with the left plate at point* $\boxed{2A}$, $\boxed{2B}$ *of Figure (14.1.3), a shock wave is propagated through the rivet shank at speed w relative to the vertical plates, where $v < w \leq c$. Then*
>
> **(14.2.1a)** *The shock wave signal reaches the rivet tip at the point in Figure (14.1.3) denoted by $c =$* $\boxed{3A}$, $\boxed{3B}$.
>
> *Suppose the* extended shank length *is measured from the left surface of the right hand plate to the rivet tip. Then*
>
> **(14.2.1b)** $\dfrac{\text{extended shank length}}{\text{total shank length}} = 1 - \dfrac{1 - (v/w)}{\sigma_v},$
>
> *which is invariant for both frames, $\mathcal{F}^A$ and $\mathcal{F}^B$.*

PROOF: The coordinates for point $c =$ $\boxed{3A}$, $\boxed{3B}$ in Figure (14.1.3) have already been justified on page 175 as the point at which the shock wave arrives at the rivet tip. This establishes (14.2.1a).

Note that Figure (14.1.3) gives us both $\mathcal{F}^A$ and $\mathcal{F}^B$ coordinates of the event when the collision shock wave first arrives at the rivet tip. Using just the $\mathcal{F}^A$ coordinates for points a, b, and $c =$ $\boxed{3A}$, of Figure (14.1.3), we obtain

(14.2.1c) Time $t = (w\sigma_v)/(w-v)t_0$ is the common (simultaneous) time in frame $\mathcal{F}^A$ at points a, b, and $c =$ $\boxed{3A}$. From Figure (14.1.3), we see that their x coordinates a_x, b_x, and c_x are given by

point	x coordinate
c	$c_x = \dfrac{(w\sigma_v)}{(w-v)}d_0$
b	$b_x = d_0$
a	$a_x = 0$

which implies $\dfrac{c_x - b_x}{c_x - a_x} = 1 - \dfrac{1 - (v/w)}{\sigma_v}.$

This expression is exactly the ratio of the extended shank length to the total shank length. That is, (14.2.1b) is established and the theorem is proved. ∎

(14.2.2) N**ote:** ***The ratio (14.2.1b) is Frame-Invariant.*** *Table (14.2.1c) considers the ratio in* $\mathcal{F}^A$ *coordinates. Exercise (14.4.1) arrives at the same ratio using* $\mathcal{F}^B$ *coordinates.*

(14.2.3) N**ote:** ***Special relativity is incompatible with rigidity.*** *There are examples showing that rigidity forces certain particles to travel faster than c, the speed of light. One such example is a stiff rod pivoting between points A and B each of which rotates at speed $v = (2/3)c$. If the rod is rigid then the point C must rotate at twice the speed, viz., $2 \times (2/3)c = (4/3)c$. For the speed at C to be less than c, the rod has to bend so that point C "lags" behind.*

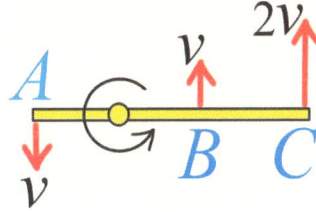

(14.2.4) N**ote:** ***The Future Affects the Present If Perfect Rigidity is Allowed.*** *If we assume that bodies remain rigid under acceleration, then the rivet can never expand beyond its at-rest length, d_0, and, consequently, the future must affect the present. Details are revealed in Exercise (14.4.4).*

If w is the speed of the shock wave as measured in frame $\mathcal{F}^A$ of the plates, then what is the shock wave speed as measured from $\mathcal{F}^B$, a passenger on the rivet?

We now present an alternate geometric proof that follows immediately from reading the coordinates of Figure (14.1.3).

(14.2.5)

> THEOREM: (***Relative Speeds of the Shockwave***): *The rivet in frame $\mathcal{F}^B$ travels at speed v as measured from frame $\mathcal{F}^A$ of the vertical plates. Once the rivet head collides with the left plate at point* 2A , 2B *of Figure (14.1.3), a shock wave is propagated through the rivet shank at speed w as measured in frame $\mathcal{F}^A$ of the vertical plates, where $v < w \le c$.*
>
> *Then*
>
> w_{rivet}, the speed of the shockwave, as measured in frame $\mathcal{F}^B$ of the rivet (or by a passenger on the rivet), is
>
> **(14.2.5a)** $$w_{\text{rivet}} = \frac{w + (-v)}{1 + w(-v)}.$$

PROOF: We compute the shock wave speed, $w_{\text{rivet}} = \Delta x_A / \Delta t_A$ in $\mathcal{F}^A$ coordinates and, redundantly, $w_{\text{rivet}} = \Delta x_B / \Delta t_B$ in $\mathcal{F}^B$ coordinates.

Using the orthogonal X_A and T_A axes in Figure (14.1.3),

$$\Delta x_A = (1/\sigma_v)d_0 \qquad \text{\textit{from}} \quad \boxed{4A} \,, \quad \boxed{2A} \,,$$
$$\Delta t_A = \left(\frac{w\sigma_v}{w - v} - \frac{1}{\sigma_v}\right)t_0 \quad \text{\textit{from}} \quad \boxed{3A} \,, \quad \boxed{4A} \,. \quad \Rightarrow \frac{\Delta x_A}{\Delta t_A} = \frac{w + (-v)}{1 + w(-v)}.$$

In Figure (14.1.3), the X_B-axis is parallel to the line formed between points $\boxed{2A,B}$ and $\boxed{4A,B}$. The T_B-axis is shown on the figure. Then

$$\Delta x_B = d_0 \qquad \text{\textit{from}} \quad \boxed{4B} \,, \quad \boxed{2B} \,,$$
$$\Delta t_B = \left(\frac{w(1 - v^2)}{w - v} - 1\right)t_0 \quad \text{\textit{from}} \quad \boxed{3B} \,, \quad \boxed{4B} \,. \quad \Rightarrow \frac{\Delta x_B}{\Delta t_B} = \frac{w + (-v)}{1 + w(-v)}.$$

In both frames $\mathcal{F}^A$ and $\mathcal{F}^B$ (14.2.5a) is confirmed, which ends the proof. ∎

(14.3) The Slinky Connection

The previous sections show that when a collision occurs between a plate and the head of a rivet, then information about the collision must travel from the rivet head along the shank and finally to the rivet tip before the tip "knows" that it must slow down.

In this section, we cite other situations in which a non-zero time is required for a message to propagate through a medium before a response to that message results.

- In the first instance (*Figure* (14.3.1)), a Slinky toy is suspended from its top. Once the top is released, the bottom of the Slinky will continue to feel an upward tension (*and remain fixed*) as long as the ring above it is stretched. Once the Slinky fully collapses, the bottom ring feels no more upward tension and the whole unit falls to the ground. (*Thanks to* DAVID ELLERMAN *for this example.*)

- The second example (14.3.2) illustrates that the loss of gravitational pull of the sun would not be felt on Earth for almost nine minutes.

(14.3.1) Fɪɢᴜʀᴇ: *Suspended Slinky is Released from the Top While Its Bottom Remains Fixed in Space*

(14.3.1a)   Question: What happens to the lower end of a suspended Slinky when the top end is released?[1]

(14.3.1b)   Answer: As indicated in Figure (14.3.1), [*Panels* ①, ②] the bottom of the Slinky remains suspended at the initial height h from the ground — *until* the collapsing wave from the falling Slinky top reaches the bottom of the Slinky®. At this moment [*Panel* ③], the following message reaches the Slinky bottom: "From now on, there is no longer an upward tension equal to the downward force of gravity. You must fall!" When this message is received [*Panel*④], the bottom of the Slinky joins the top in falling to the ground.

---

[1] For a video demonstration, search on YouTube.com for "Slinky Drop Answer"

(14.3.1c) **The Analogy of the Rivet and the Slinky Toy:**
In Figure (14.1.3), the rivet tip has constant speed $v > 0$ and is
in an inertial frame until the collision wave (message) reaches it at
point $\boxed{3A}$, $\boxed{3B}$. In Figure (14.3.1), the Slinky bottom also
has constant speed $v = 0$ in an inertial frame *until* the message that
there is no upward tension reaches from the top to the bottom at
point ③.

(14.3.2) **The Disappearing Sun:** *Time is required even for the "message" of gravity waves.* The earth is held in its orbit by centripetal force from the sun much as a pail of water, being swung by a rope, is held in its circular orbit by the centripetal pull of the rope. However, gravity waves from the sun require some eight minutes (at the speed of light) to reach the earth.

Now if the sun and all its mass were to instantly disappear, this would "cut the gravity rope," causing the earth to fly off its circular orbit on a straight-line tangent nine minutes *after* the sun's disappearance — this is the time required for the earth to "know" that there is no longer any gravitational pull from the sun. Afterwards, the earth would continue on its straight-line path, destined to travel forever into a void of cold darkness.

(14.4) **Exercises**

(14.4.1)
Exercise
(**Extending Theorem (14.2.1)**). Show that the ratio (14.2.1b)
which was shown for $\mathcal{F}^A$ coordinates is also valid for points a', b', c in
$\mathcal{F}^B$. Specifically, from Figure (14.1.3), show that in $\mathcal{F}^B$ coordinates,

(14.4.1a) $\qquad \dfrac{\textit{extended shank length}}{\textit{total shank length}} = \dfrac{c_x - b'_x}{c_x - a'_x} = 1 - \dfrac{1 - (v/w)}{\sigma_v}$

where a'_x, b'_x, c_x are the X_B coordinates of a, b, c. (*Equation* (14.4.1a)
can also be established using T_B *coordinates* a'_t, b'_t, c_t *in* $\mathcal{F}^B$.)

(14.4.2) Under the hypotheses of Theorem (14.2.1), show that for all shock
Exercise   wave speeds w, where $v \leq w \leq c = 1$, there is (at least) one moment
after the rivet-plate collision in which the rivet shank length exceeds
its at-rest length, d_0.

> **Hint:** *Formula (14.2.1b) is derived for the moment the shock wave
> reaches the rivet tip which finally gets the message to leave its inertial
> frame $\mathcal{F}^B$ and to decelerate. The question, then, is to produce a
> shock wave speed w for which (14.2.1b) is zero. But this implies that
> $w = (1/v) + \sqrt{(1/v)^2 - 1}$, which is greater than 1, the speed of light.*

(14.4.3) (***When Shock wave Speed = c=1***) Show that if either $w = 1$
Exercise   or $w_{\text{rivet}} = 1$ for the shock wave speed of Theorems (14.2.1) and
(14.2.5), then *both* values are equal to 1. (*This reflects the fact that
any wave traveling at the speed of light c in one frame has speed c in
<u>all</u> frames.*)

> **Hint:** *Apply Equation (14.2.5a).*

(14.4.4) (***When the Future Affects the Present***) Let the rivet shank
Exercise   have at-rest length d_0 as measured in the frame of the vertical plates,
$\mathcal{F}^A$. Using the Minkowski diagram coordinates (14.1.3), pg. 172,
show that if the rivet shank length is always less than or equal to its
at-rest length, d_0, then the future must affect the past.

> **Hint:**   *Figure (14.1.3) shows that the rivet collides with the left
> plate at point 2A ($\mathcal{F}^A$ coordinates) = 2B ($\mathcal{F}^A$ coordinates). From
> the Slinky principle, the collision signal information (the red dots in
> Figure (14.1.3)) require time to reach the rivet tip. In this case, the
> rivet must expand beyond its at-rest length until the collision signal
> arrives at the tip.*
>
> *Hence, the only way to maintain rigidity is to have the rivet and
> plates meet at point 2A =2B when their relative speed to each other
> is zero. This says the plates receive a rightward force or the rivet
> receives a leftward force in order to decelerate to relative speed zero
> before the actual collision. In other words, the <u>future</u> collision causes
> deceleration in the <u>present</u>.*

(14.4.5) (***Applying the Pea-Shooter Paradox***) Use the Lorentzian sum
Exercise $\oplus$ (7.3.1) directly to prove (14.2.5a).

> **Hint:** *Use the Pea-Shooter paradox as a template (see (1.13.2b)).*
> *Let w equal the speed of the shock wave (the pea) as observed from*
> $\mathcal{F}^A$, *the frame of the vertical plates (Bernie on solid ground). Then*
> w *is also equal to: The speed v of the rivet (Ashley on the train) as*
> *observed from* $\mathcal{F}^A$ *"plus" the speed w_{rivet} of the shock wave (the pea)*
> *as observed from* $\mathcal{F}^B$, *the frame of the rivet (Ashley on the train).*
> *Formally, then,*
>
> **(14.4.5a)** $w = v \oplus w_{\text{rivet}},$
>
> *so that using (1.13.3) to solve for w_{rivet} in (14.4.5a) yields (14.2.5a).*

VI. Energy and Mass

Chapter 15 $E=mc^2$ *This celebrated equation is almost an afterthought once the connection is made between the peashooter paradox and conservation of momentum.*

(15) $E = mc^2$

(15.1) How We Came to This Place

It might seem that our analyses of the paradoxes through Minkowski diagrams have little to do with $E = mc^2$, the relation of a mass m at rest and its energy equivalent E.

As it turns out, the structure underlying the pea-shooter paradox is also essential to showing that $E = mc^2$. The structure in question is the Lorentzian addition of speeds (7.3.1).

(15.1.1) FIGURE: (*Recapitulation: Three Steps to the Lorentz Sum*)

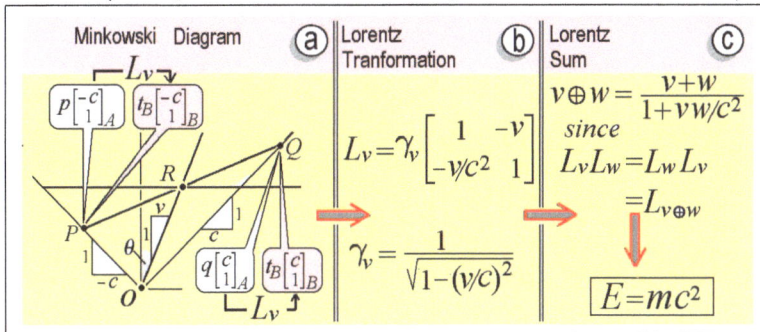

- PANEL (a): *Coordinates for the endpoints P and Q are found in two frames of reference: $P = p[-c, 1]$ and $Q = q[c, 1]$ relative to the frame $\mathcal{F}^A$ with perpendicular X and T axes, while $P = p[-c, 1]'$ and $Q = q[c, 1]'$ relative to the frame $\mathcal{F}^B$ with the T-axis lying on $\overline{OR}$.*

- PANEL (b): *The Lorentz transformation, a 2×2 matrix, is defined on the <u>basis</u> vectors (points) P, Q, and hence, is fully defined everywhere (the importance of the <u>basis</u> property for <u>linear</u> transformations follows from Theorem (A.4.2), pg. 343). The matrix has for its domain the frame of reference from which another frame is seen to be moving with relative speed v.*

- PANEL (c): *If frame $\mathcal{F}^A$ sees frame $\mathcal{F}^B$ move with speed v, then we have $L_v : \mathcal{F}^A \to \mathcal{F}^B$. If frame $\mathcal{F}^B$ sees frame $\mathcal{F}^C$ move with speed w, then $L_w : \mathcal{F}^B \to \mathcal{F}^C$. Since the matrix product $L_w * L_v : \mathcal{F}^A \to \mathcal{F}^C$*

$(L_w * L_v$ is defined in (A.4.3e)) is equivalent to $L_{v \oplus w} : \mathcal{F}^A \rightarrow \mathcal{F}^C$, where the Lorentz sum, $v \oplus w$, is given in Panel ©, then frame $\mathcal{F}^A$ sees frame $\mathcal{F}^C$ move with speed $v \oplus w$.

We will see in this chapter that if speeds must be added with the Lorentz sum, then necessarily $\boxed{E = mc^2}$.

....

(15.2) Speed-Dependent Mass: an Intuitive View

(15.2.1) **Speed-dependence in the notation.**   The mass of a particle viewed at speed zero (at rest) is denoted m_0. The mass of the *same* particle viewed at speed v is denoted m_v. Our Galilean/Newtonian intuition tells us that there is no need to change notation for a mass just because it happens to be in motion, since mass does not change with observed speed — or does it? (*In contrast, we accept the concept from introductory calculus that mass can change over* time *as is shown by the time-dependent "petri dish equation"* $m_t = m_0 e^{kt}$. *Einstein's brilliant insight revealed that mass also changes according to its observed* speed *v.*)[1] In any case, there is no harm in using the (ornamented) notation

(**15.2.1a**) $m_v \stackrel{def}{=\!=}$ *mass of particle at speed v.*

Hence, its rest mass, the mass at speed 0, is denoted m_0.

Intuition misleads us in another way. Let us empathize for a moment with a point mass that might ask itself, *If an observer measures my mass as equal to m_0 when I am observed at speed $v = 0$, then why is it that I* have a different *mass just because the observer now sees my speed to be $v > 0$?*

$$\boxed{\text{The Two-Particle Setup}}$$

(15.2.2) **Preliminaries:**   To acknowledge that a mass m is being observed

---

[1]At time $t = 0$, the initial population of bacteria in a petri dish has mass m_0. If at any time t, population growth is proportional to the present population, m_t, then in terms of differential equations, this statement translates to $dm_t/dt = km_t$ for all time $t > 0$ for some constant k. The solution is $m_t = m_0 e^k t$.

while it has speed v, we denote it by m_v. Therefore, m_0 denotes the **at-rest mass** at observed speed zero.

(15.2.3)

> **DEFINITION:** *The* MOMENTUM OF A PARTICLE *with rest mass* m_0^i *at velocity* $\vec{v}_i$ *is defined to be* $m_{\vec{v}_i}^i \cdot \vec{v}_i$ *(For clarity in equations such as (15.2.3a), we sometimes use the square bracket notation* $[m_{\vec{v}_i}^i] \cdot \vec{v}_i$). *The* MOMENTUM OF THE SYSTEM *of two particles with rest masses* m_0^1 *and* m_0^2 *when observed at respective velocities* $\vec{v}_1$ *and* $\vec{v}_2$ *is defined by*
>
> **(15.2.3a)** $\quad [m_{\vec{v}_1}^1] \cdot \vec{v}_1 + [m_{\vec{v}_2}^2] \cdot \vec{v}_2 \overset{\text{def}}{=} [m_{\vec{v}_1}^1 + m_{\vec{v}_2}^2] \cdot \vec{v}_{sys},$
>
> *where* $\vec{v}_{sys}$ *is the* SYSTEM VELOCITY *which, intuitively, treats the two particles as a single mass entity.*

If all motion occurs in the same direction or its opposite, then **conservation of momentum** gives us the

(15.2.3b) $\quad$ SYMMETRY PRINCIPLE, $\quad m_{-v} = m_v$

which holds that both speeds, $+v$ and $-v$, when applied to any mass m_0, produce the same observed speed-dependent mass m_v.

To see why this is so, imagine two ice-skaters, each of mass m_0 facing each other on frictionless ice. Since they each have speed $v = 0$, the system of two ice-skaters has total momentum zero. Next, they push each other away so that one ice-skater has speed $-v$ and the other has speed v. Now, the system momentum (*which is conserved to be zero*) has form $(m_{-v})(-v) + (m_v)(v) = (-m_v)(v) = (-m_{-v} + m_v)(v) = 0$ which implies (15.2.3b).

....

We will apply the momentum equation (15.2.3a) to the special case of two equal rest masses that initially have equal and opposite speeds $\pm v$. Accordingly, we shall use the the following figure that summarizes the states of these particles at various speeds. Note that we are obliged to add speeds relativistically according to the Lorentz formula (7.3.1) $v \oplus w = (v + w)/(1 + vw/c^2)$.

(15.2.4) FIGURE: *A Simplified Two-Particle System*

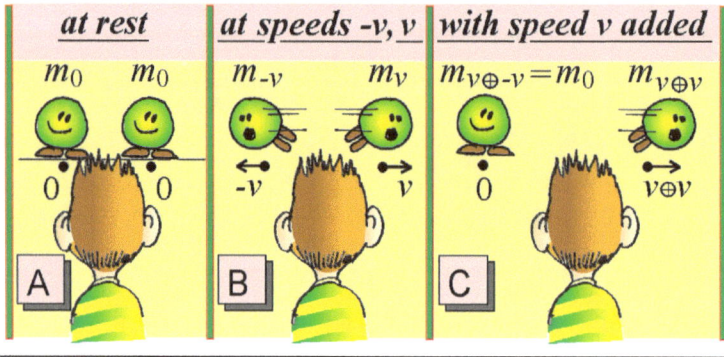

at rest	*at speeds -v, v*	*with speed v added*
$m_0 \qquad m_0$	$m_{-v} \qquad m_v$	$m_{v \oplus -v} = m_0 \qquad m_{v \oplus v}$
$0 \qquad 0$	$-v \qquad v$	$0 \qquad v \oplus v$
A	B	C

- PANEL ☐A☐ OF FIGURE (15.2.4). *Two particles are observed, each with the same at-rest mass m_0. Each particle has zero momentum mass $\cdot$ speed $= [m_0] \cdot 0 = 0$ so that from (15.2.3a), we compute the sum and obtain the system momentum and the system speed v_{sys} as follows:*

(15.2.4a) $\quad [m_0] \cdot 0 + [m_0] \cdot 0 = [m_0 + m_0] \cdot \underbrace{0}_{v_{sys}}$.

- PANEL ☐B☐ OF FIGURE (15.2.4). *The left and right masses, m_{-v} and m_v, say, have equal and opposite speeds, $-v$ and v. The respective values of their momentums are also equal and opposite. That is,*

(15.2.4b) $\qquad \underbrace{([m_{-v}] \cdot -v)}_{\substack{left \\ momentum}} = - \underbrace{([m_v] \cdot v)}_{\substack{right \\ momentum}}$.

We now obtain the following sum that produces the system momentum and the system speed v_{sys} (15.2.3a):

$$([m_{-v}] \cdot -v) + ([m_v] \cdot v) = [m_{-v} + m_v] v_{sys}.$$

From (15.2.4b), we may substitute $-([m_v] \cdot v)$ for $([m_{-v}] \cdot -v)$ above to obtain

(15.2.4c) $\quad \underbrace{-([m_v] \cdot v) + ([m_v] \cdot v)}_{0} = [m_{-v} + m_v] v_{sys}.$

Since $m_{-v} > 0$ and $m_v > 0$, (15.2.4c) implies $v_{sys} = 0$.

- PANEL ⬚C⬚ OF FIGURE (15.2.4). *Speed v is added to the entire 2-particle system of Panel* ⬚B⬚ *which means the system speed $v_{sys} = 0$ of (15.2.4c) is augmented by speed v to become*

(15.2.4d) $\quad new \ v_{sys} = (\underbrace{0}_{old \ v_{sys}} \oplus v) = v.$ $\qquad$ *from (7.4.3)*

Since the speed of each particle is augmented by v units, the left particle, formerly at speed $-v$, acquires the new speed $-v \oplus v = 0$ (see Proposition (7.4.3)) so its new momentum is $[m_0] \cdot (-v \oplus v) = [m_0] \cdot 0 = 0$. Similarly, the right particle, formerly with speed v, acquires added speed v to produce the new speed $v \oplus v$. This speed accounts for a new momentum of $[m_{v \oplus v}] \cdot (v \oplus v)$. In (15.2.3a), replace v_1 with 0 and v_2 with speed $v \oplus v$. This allows us to set $m_{v_1}^1 = m_0$ and $m_{v_2}^2 = m_{v \oplus v}$, which yields

(15.2.4e) $\quad \underbrace{[m_0] \cdot 0}_{0} + [m_{v \oplus v}] \cdot v \oplus v = [m_0 + m_{v \oplus v}] \cdot v_{sys}$

$$= [m_0 + m_{v \oplus v}] \cdot v. \quad \begin{array}{l} from \\ (15.2.4d) \end{array}$$

....

The stage is set for the following

(15.2.5)

THEOREM: ***Mass Increases with Increased Speed.*** *Given a particle with rest mass m_0. If m_w represents the measurement of this mass when observed at speed w, then*

(15.2.5a) $\qquad m_w > m_0.$

PROOF: Consider a system of two particles, each with identical rest mass m_0 so that all the equations of Figure (15.2.4) apply. In particular, among the three equal quantities in (15.2.4e), set only the first and last quantities equal to each other and solve for $m_{v \oplus v}$ to obtain

(15.2.5b) $\quad m_{v \oplus v} = m_0 \dfrac{v}{(v \oplus v) - v} = m_0 \dfrac{1 + (v/c)^2}{1 - (v/c)^2} \qquad$ *from (7.3.1)*

$$> m_0. \qquad since \ \frac{1 + (v/c)^2}{1 - (v/c)^2} > 1$$

Note that (15.2.5b) justifies (15.2.5a) once we set $w = v \oplus v$. This ends the proof. ∎

....

COROLLARY: *At speed* $w = v \oplus v$, *a particle with rest mass* m_0 *has increased mass* m_0 *growing to mass* m_w *where*

(15.2.5c) $m_w = m_0/\sqrt{1 - (w/c)^2}.$

PROOF: Exercise (15.5.3). ∎

_____

(15.3) Equivalence of Mass and Energy

(15.3.1)   **The power series expansion** for infinitely differentiable $f(x)$ is
Calculus
needed

(15.3.1a) $f(x) = f(0) + f'(0)x + \dfrac{f''(0)}{2}x^2 + \cdots + \dfrac{f^{(k)}(0)}{k!}x^k + \cdots,$

where $f'(0)$, $f''(0)$, $\ldots, f^{(k)}(0)$, $\ldots$, represent the first, second, $\ldots$, and kth derivative of the function f evaluated at 0. In particular, the kth derivative evaluated at 0 is

(15.3.1b) $f^{(k)}(0) = \dfrac{1 \cdot 3 \cdot 5 \cdots (2k-1)}{2^k}$ *when* $f(x) = \dfrac{1}{\sqrt{1-x}}.$

In (15.3.1a), set $x = (v/c)^2$ and replace the derivative expressions with (15.3.1b) to obtain the series

(15.3.1c) $\dfrac{1}{\sqrt{1 - (v/c)^2}} = 1 + \dfrac{1}{2}\left(\dfrac{v}{c}\right)^2 + \underbrace{\dfrac{3}{8}\left(\dfrac{v}{c}\right)^4 + \dfrac{5}{16}\left(\dfrac{v}{c}\right)^6 + \cdots}_{O(v^4)}$

where $O(v^4)$ indicates a sum of v^k terms where $k \geq 4$. (*At this point, units of speed v can be arbitrary (cm/sec, miles/hour) so it is not necessary that $c = 1$.*)

In (15.3.1a), replace $v < 1$ with $(v/c) < 1$ — this releases us from the special case where we write $c = 1$. Now recapture this new form of (15.3.1a) by multiplying (15.3.1c) through by $m_0 c^2$ to obtain the equalities

(15.3.1d)

$$m_v\, c^2 = \frac{m_0\, c^2}{\sqrt{1 - (v/c)^2}} = m_0\, c^2 + \underbrace{m_0\, \frac{v^2}{2}}_{\substack{\textit{Newtonian} \\ \textit{kinetic energy}}} + m_0\, O(v^4)$$

$$= m_0\, c^2 + \underbrace{E(v)}_{\substack{\textit{relativistic} \\ \textit{kinetic energy}}} \; .$$

(15.3.2) Interpreting (15.3.1d). Is the power series (15.3.1d) a mathematical invention designed to torture calculus students like some irrelevant chess game of the mind where symbols on paper only obey certain abstract rules? Or is there a valid connection with reality as we (think) we know it? Let us see how a certain marker term brings the series (15.3.1d) from the calculus textbooks into the world of physical reality.

Series (15.3.1d) represents energy. It is the classical kinetic energy term, $m_0\, v^2/2$, in (15.3.1d) that spreads its type (energy) to all other terms in the series. Informally, the energy attribute of $m_0\, v^2/2$ forces all other terms to have the same attribute.

Why $E = m_0\, c^2$. We write $E(v)$ to denote the sum of all the terms of (15.3.1d) on the right hand side except for the term, $m_0\, c^2$. This sum, $E(v)$, which depends on v, can only be interpreted as a general energy term that arises from the *energy required to bring the particle from a state of rest to a speed of v*. To see this, set $v = 0$ and $E(v)$ goes away. That is, $E(v)$ appears only when motion is induced, which means a force and distance, or energy, is invested into the particle in order to get it into motion.

Therefore, the left hand side of (15.3.1d) represents the sum of the at-rest energy, $m_0\, c^2$, which is clearly not zero, and $E(v)$, the energy necessary to bring the particle from speed $v = 0$ to speed $v \neq 0$. Since the term $m_0\, c^2$ does not change with v, we conclude

$$\boxed{E = m_0\, c^2 \;\; \text{is the rest energy} \\ \text{of a particle that has rest mass } m_0.}$$

(15.4) A Numerical Example

(15.4.1) *Masses Under the Effects of a Force Cannot Stay Constant.*
The following example shows how (15.2.3a), the definition of momentum using Lorentzian addition of speeds, will not tolerate constant mass with increased speed (due to a force).

(15.4.1a) Set *mass $m^1 = 1kg$ with speed $v_1 = -0.8$ and*
mass $m^2 = 2kg$ with speed $v_2 = 0.4$.
(*We need not write the units for speeds which are, after all, unitless ratios of form v/c*). Then from (15.2.3),

(15.4.1b)

$$\underbrace{1kg \cdot (-0.8) + 2kg \cdot (0.4)}_{0kg} = \underbrace{1kg + 2kg}_{3kg} \cdot w_{old}. \qquad \begin{array}{c} which\ implies \\ w_{old} = 0 \end{array}$$

Suppose the masses remain constant after the system speed has increased its value from $w_{old} = 0$ to $w = 0.8$. This means we must add $w = 0.8$ to the speeds $v_1 = -0.8$, $v_2 = 0.4$, and $w_{old} = 0$ in (15.4.1b) (*using relativistic addition $\oplus$ (7.3.1)*) to obtain

(15.4.1c)

$$\underbrace{1kg}_{m^1} \cdot \underbrace{(-0.8 \oplus 0.8)}_{0} + \underbrace{2kg}_{m^2} \cdot \underbrace{(0.4 \oplus 0.8)}_{0.909} = \underbrace{1kg + 2kg}_{3kg} \cdot \underbrace{(0 \oplus 0.8)}_{w_{new} = 0.8}$$

which gives us the absurd equation

(15.4.1d) $$\underbrace{1kg \cdot 0}_{0kg} + \underbrace{2kg \cdot 0.909}_{1.818kg} = \underbrace{3kg \cdot 0.8}_{2.4kg}.$$

Standard proof-by-contradiction (**??**) tells us we must reject our hypothesis that masses remain constant with augmented speed.

To say the same thing more formally, note that in reasoning from (15.4.1a) to (15.4.1d), we find that the constant-mass assumption leads to a false conclusion, namely,

(15.4.1e) $\{Constant\ mass\} \Rightarrow \{1.818 = 2.4\}\ (False)$.

We can never use valid reasoning where a true hypothesis leads to a false conclusion. Therefore, in (15.4.1e), our hypothesis $\{Constant\ mass\}$, which leads to a false conclusion, must be $(False)$.

. . . .

(15.4.2) **What is the non-absurd resolution to (15.4.1d)?**   The following table gives the values of

$$\widetilde{m}^1 = m^1_{v_1 \oplus w} \qquad and \qquad \widetilde{m}^2 = m^2_{v_2 \oplus w},$$

for system speed $w = 0.8$ and the resulting momentum.

(15.4.2a) Tᴀʙʟᴇ: *Momentum   P   at   system speeds $v_{sys}=0$ and $v_{sys}=0.8$*

	$v_{sys} = w = 0$	P	$v_{sys} = w = 4/5$	P
Speed	$v_1 = -4/5$	$-4/5$	$v_1 \oplus w = 0$	0
Mass	$m^1_{v_1} = 1kg$		$m^1_{v_1 \oplus w} = 3/5\ kg$	
Speed	$v_2 = 2/5$	$4/5$	$v_2 \oplus w = 10/11$	4.0
Mass	$m^2_{v_2} = 2\ \ kg$		$m^2_{v_2 \oplus w} = 22/5\ kg$	
	Total P	0	Total P	4.0

(15.4.3) **Two Computations for Momentum.**   When $v_{sys} = w = 4/5$, Table (15.4.2a) gives us the momentum as the sum of each particle's momentum, namely,

(15.4.3a) $M = \underbrace{m^1_{v_1 \oplus w}}_{(3/5)kg} \cdot \underbrace{v_1 \oplus w}_{0} + \underbrace{m^2_{v_2 \oplus w}}_{(22/5)kg} \cdot \underbrace{v_2 \oplus w}_{10/11} = 4.0.$

But from the same table, we also have the momentum of the "single point" particle, $m^1_{v_1 \oplus w} + m^2_{v_2 \oplus w}$, namely,

(15.4.3b) $M = \left[\underbrace{m^1_{v_1 \oplus w}}_{(3/5)kg} + \underbrace{m^2_{v_2 \oplus w}}_{(22/5)kg} \right] \cdot \underbrace{w}_{4/5} = 4.0.$

The rest masses are

$$m^1_0 = m^1_{v_1} \cdot \sqrt{1 - v^2_1} = 1kg \cdot \sqrt{1 - (-4/5)^2} = (3/5)kg$$

and

$$m_0^2 = m_{v_2}^2 \cdot \sqrt{1 - v_2^2} = 2kg \cdot \sqrt{1 - (2/5)^2} = (2/5)\sqrt{21}kg.$$

The increased masses at the speeds $v_i \oplus w$ are

$$m_{v_1 \oplus w}^1 = m_{-4/5 \oplus 4/5}^1 = m_0^1 = (3/5)kg$$

where Proposition (7.4.3) justifies $-4/5 \oplus 4/5 = 0$, and

$$m_{v_2 \oplus w}^2 = m_{2/5 \oplus 4/5}^2 = m_{10/11}^2 = (22/5)kg.$$

....

(15.4.4) The sum of the momentums of two particles equals the momentum of the sum (of particle masses), as is confirmed in (15.4.3a) and (15.4.3b). In Exercise (15.5.4), the reader constructs a program that generalizes Table (15.4.2a) by allowing two arbitrary mass inputs, an arbitrary speed $v_1 < 1$, and arbitrary system speed $w < 1$. Since the user has choice of masses and speeds, this program should give the reader a sense of how masses increase with speed.

(15.5) Exercises

(15.5.1) **An Alternate Proof for $E = mc^2$.**   The force F is defined as
Exercise   the rate of change of the momentum p:
Calculus

$$F = \frac{dp}{dt}.$$

The work $W_{i \to f}$ required from an initial position x_i to a final position x_f is defined as the line integral of the force [2]

$$W_{i \to f} = \int_{x_i}^{x_f} F\, dx$$

Then

---

[2]In more than one dimension this is the line integral along a path; in general the work will depend on the path. Forces for which the work is path-independent are called conservative forces.

$$
\begin{aligned}
W_{i \to f} &= \int_{x_i}^{x_f} F\, dx \\
&= \int_{x_i}^{x_f} \frac{dp}{dt}\, dx \\
&= \int_{p_i}^{p_f} v\, dp \qquad \left(set \ \ v = \frac{dx}{dt} \ \ in\ the\ previous\ integral \right) \\
&= pv\big|_i^f - \int_{v_i}^{v_f} p\, dv \qquad p = \frac{m_0 v}{\sqrt{1 - v^2/c^2}} \\
&= \left[\frac{m_0 v^2}{\sqrt{1 - v^2/c^2}} + m_0 c^2 \sqrt{1 - v^2/c^2} \right]_i^f \\
&= \left[\frac{m_0 c^2}{\sqrt{1 - v^2/c^2}} \right]_i^f \\
&= \left[m_v c^2 \right]_i^f .
\end{aligned}
$$

Finally, the work is also the change in the kinetic energy T:

$$
W_{i \to f} = T_f - T_i
$$

that implies $T = m_v c^2 + \text{constant}$. The kinetic energy vanishes at zero speed so

$$
T = m_v c^2 - m_0 c^2 \quad \Rightarrow \quad m_v c^2 = T + m_0 c^2
$$

that defines the rest energy $E_0 = m_0 c^2$ and the total energy $E = m_v c^2$:

$$
E = E_0 + T.
$$

Note that in terms of the momentum

$$
E = \sqrt{p^2 c^2 + m_0^2 c^4},
$$

which is valid even for particles with zero rest mass (the expression in terms of v is ambiguous because we let $m_0 \to 0$ while at the same time $v \to c$).

These definitions are useful because E is conserved. In any process where particles are free in the initial and final states the sum of the energies E (in terms of the momenta and rest masses) is conserved.

(15.5.2) **Derivatives.** Show that for $f(x) = (1 - x)^{-1/2}$, the kth derivative
Exercise $f^{(k)}(0)$ has the form

$$f^{(k)}(0) = \frac{(1 \cdot 3 \cdot 5 \cdots (2k - 1))}{2^k}.$$

(15.5.3) **Mass Increases with Speed.** Show that

animated
online

$$m_{v \oplus v} = m_0 \frac{1 + v^2}{1 - v^2} = m_0 \frac{1}{\sqrt{1 - (v \oplus v)^2}} \ .$$

Hint: *Definition (7.3.1) allows replacement of $v \oplus v$ with its equivalent, $v \oplus v = 2v/(1 + v^2)$.*

(15.5.4) **A Useful Spreadsheet.** Devise a spreadsheet, or a program in a
Exercise language of your choice (e.g., C, C^{++}, Matlab, Maple, Mathematica)
to generate tables of the form (15.4.2a).

Hint: *Construct your program with the following input/output properties:*

USER INPUT: ....

(I_1) The first mass $m_{v_1}^1$, and its unitless speed $v_1 < 1$.

(I_2) The second mass $m_{v_2}^2$. The program computes $v_2 < 1$ so that
momentum $m^1 v_1 + m^2 v_2 = 0$ is guaranteed (*see first column of
Table (15.4.2a)*).

(I_3) The arbitrary "add-on" speed $w < 1$.

PROGRAM OUTPUT: ....

(O_1) Speed v_2 which guarantees that for the two-particle system, the
initial momentum is zero (*mentioned in (I_2)*).

(O_2) The two at-rest masses, $m_0^i = m_{v_i}^i \sqrt{1 - v_i^2}$, where $i = 1, 2$.

(O_3) The Lorentz sums, $v_1 \oplus w < 1$ and $v_2 \oplus w < 1$.

(O_4) The two masses with initial speeds, v_1 and v_2, increased by w, namely, $m_{v_i \oplus w}^i = m_0^i / \sqrt{1 - (v_i \oplus w)^2}$, where $i = 1, 2$.

(O_5) The momentum sum of the two individual particles,
$[m_{v_1 \oplus w}^1] \cdot v_1 \oplus w + [m_{v_2 \oplus w}^2] \cdot v_2 \oplus w$.

(O_6) The system momentum (*due to system speed* $v_{sys} = w$),
$[m_{v_1 \oplus w}^1 + m_{v_2 \oplus w}^2] \cdot w$.

CHECK: ....
Outputs (O_5) and (O_6) should always be equal.

VII. The Mathematics of Waves and Light

Chapter 16 *develops the mathematical model of propagated waves since light exhibits the properties of a propagated wave under certain circumstances.*

Chapter ?? *offers details of some of early and ingenious methods for measuring c, the finite speed of light. As recently as the time of Galileo and Newton, it had not been proven that the speed of light was finite. But today, using a microwave oven at home, you can measure the speed of light in a 20-second experiment (Exercise (??), pg. ??).*

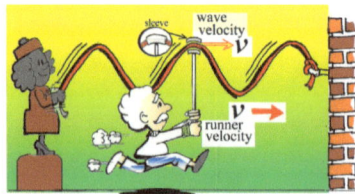

(16) The Nature of Waves

Preview

A fundamental starting point of Einstein's paper, [12], is that the speed of light is the same for all observers, whether or not the light source is moving. Equivalently (and non-intuitively), the speed of light is a constant 186,000 miles per second (in a vacuum) whether or not the observer is moving (*see* (1.5.4)).

Why should we believe such a non-intuitive idea? The following chapters provide the answer.

The constant speed of light, independent of frame of reference (point of view) was predicted decades before Einstein's paper, by Maxwell's equations which connected electricity, magnetism, and optics. These equations also predicted that the speed of all electromagnetic (*EM*) waves propagated in a vacuum was 3×10^8 meters/second — exactly the speed of light! These results are presented in Chapters (17) through (22).

(16.1) Propagated Waves(Calculus Required)

(16.1.1) Anatomy of a Wave: We consider a shape defined by a real-valued function, g, to be a **wave** traveling in one dimension if it has the following properties:

(1) **Wave length λ:** At any fixed time, t_0, its shape, $y = g(x,t)$ in the X-Y plane replicates itself over all successive intervals of length λ. For any x, such intervals take the form

(16.1.1a) $\underbrace{[\, x + k\lambda, \ x + (k+1)\lambda \,]}_{length\ \lambda}$, $k = 0, \pm 1, \pm 2, \ldots$.

Formally stated, for fixed time, t_0, for all x, and all integers k, we have $g(x, t_0) = g(x + k\lambda, t_0)$

(2) **Wave speed v:** The wave travels to the right, in the positive direction, with speed v. Consider a wave $g(\cdot, t_0)$ at fixed time t_0. As explained in Section (16.3), at wave speed v, after time Δt,

the full wave will shift to the right a distance $v\Delta t$. That is,

(16.1.1b)   *For all x, $g(x, t_0) = g(x + v\Delta t, t_0 + \Delta t)$.*

Equation (16.1.1b) says that for all x, the y-value of wave g at x_0 and time t_0 is the same as the y-value of g at $x + v\Delta t$ (*the rightward shifted distance*) at the later time $t_0 + \Delta t$.

(16.1.2)

> **DEFINITION:** *In a traveling wave with* WAVELENGTH λ *and* SPEED v, *we define the* PERIOD P *to be the time for one wavelength (one cycle) to pass an arbitrary point. The* FREQUENCY f *is the number of full cycles of function g that pass any fixed point in one second. Period P, frequency f, wavelength λ, and speed v, are related by the formulas*
>
> **(16.1.2a)** $P = \dfrac{\lambda}{v}, \qquad f = \dfrac{v}{\lambda}, \qquad$ *and* $\qquad P = \dfrac{1}{f}.$

‖ **NOTE** *how the units of (16.1.2a) are consistent with Appendix (??). Since the unit for λ is distance, d, and speed, v, is in units of d/t, the period, $P = \lambda/v$ is in units of t alone. This is the time for one full cycle of g to pass an arbitrary point. Also, frequency, f, is in units of $1/t$, or "per-second," which is the number of cycles of g that pass an arbitrary point per second.*

We analyze a traveling wave in a rope that is tied at one end.

(16.1.3) **FIGURE:** *Propagating a Wave in a Rope*

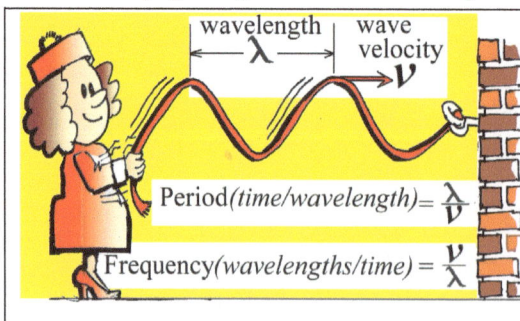

Ashley oscillates the loose end of a rope to generate a traveling wave that has velocity v. If Ashley keeps the tension constant, but increases the "hand speed," or energy to the rope by oscillating it faster, the wave will not travel any faster or slower. The wave length, λ, and the period, P, will decrease, but the wave speed, v, will be constant. Equivalently, the frequency, f, will increase, while the wave speed remains the same.

(16.1.4)

> THEOREM: (*Horizontal wave speed is independent of vertical "hand speed."*) *One end of a rope is tied to a wall while the other (loose) end oscillates vertically so as to generate a wave that travels toward the wall (as illustrated in Figure (16.1.3)). The rope*
>
> - *has linear density δ (ounces per inch, say) and*
> - *is pulled with constant horizontal tension T (pounds of force, say).*
>
> *Then, if gravitational pull on the rope is ignored, the speed v of the propagated wave is*
>
> **(16.1.4a)** $$v = \sqrt{T/\delta}.$$

PROOF: Assume the wave travels to the right as indicated in Figure (16.1.3). To measure the rightward wave speed v, it will be useful for Bernie, the runner, to keep pace with the wave, running to the right at identical speed, v. Hence, from Bernie's point of view, the rope coils are going leftward with speed $-v$ while the captured wave peak appears stationary with speed 0 as seen in Figures (16.1.5) and (16.1.6)).

(16.1.5) FIGURE: *The Wave Peak Captured in a Curved Sleeve.*

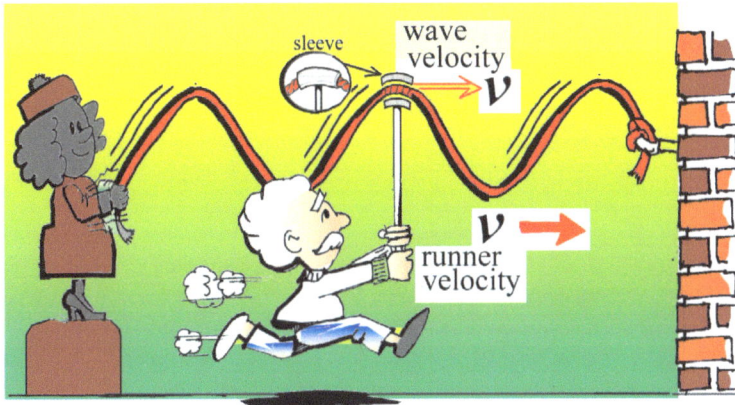

Bernie runs alongside the wave at the same speed of the propagated wave, capturing the wave peak in a curved tubular sleeve. The rope does not touch the sleeve's inside surface.

. . . .

There are two ways to calculate the centripetal (vertical downward) force illustrated in Figure (16.1.6): (i) with Newton's law $F = ma$ as seen in (16.1.6a)-(16.1.6c), or (ii) by decomposing the almost-horizontal tension force T as in (16.1.6d). The two analyses are merged to produce the conclusion (16.1.6e).

(16.1.6)
Trig
needed

FIGURE: *Two Ways to Measure Centripetal Force*

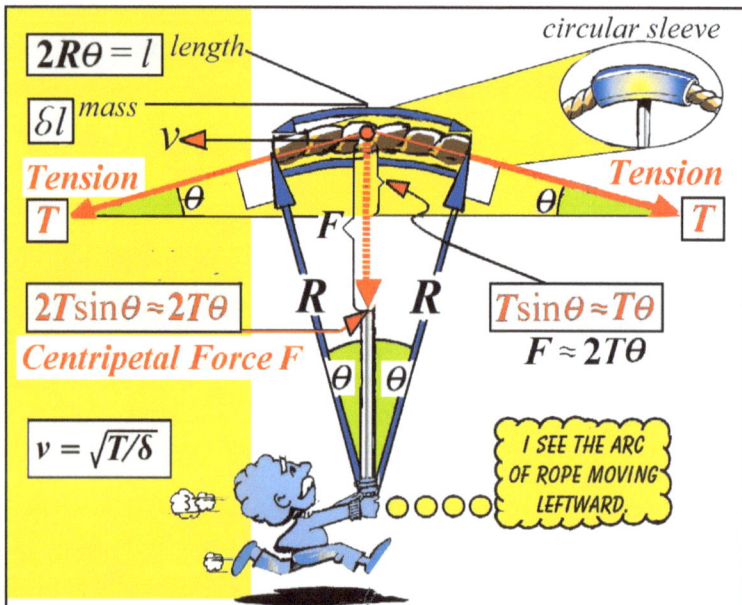

First Analysis of Centripetal Force, *F*. When the angle θ is small enough, both the rope segment and the sleeve approximate a small circular arc with radius R. At this small scale, the rope segment with length $l = 2R\theta$ and linear density δ gives us the following value of the rope segment mass:

(16.1.6a) $m = (l)\delta = (2R\theta)\delta$ {*mass of rope* }.

The rope in Figure (16.1.6) runs through the sleeve leftward on a circular path. From Newton's law, any mass on a circular arc of radius R with constant speed v is a result of a (centripetal) acceleration

(16.1.6b) $a = \dfrac{v^2}{R}$ $\begin{cases} see\ Exercise\ (16.7.1) \\ for\ a\ proof\ of\ this\ fact \end{cases}$.

If we take the mass m from equation (16.1.6a) and the (circular) acceleration a from equation (16.1.6b), then we can write the centripetal force, $F = ma$, as

(16.1.6c) $\qquad F = \underbrace{\dfrac{2\cancel{R}\,\theta\delta}{1}}_{m} \times \underbrace{\dfrac{v^2}{\cancel{R}}}_{a} = 2\,\theta\,\delta\,v^2 \,.$

Second Analysis of Centripetal Force, F. Now the centripetal force F of (16.1.6c) is also obtained from Figure (16.1.6) by decomposing the opposing tension vectors with length T into their horizontal and vertical components. The horizontal components are equal and opposite so they cancel.

The vertical components of the tension vectors, on the other hand, both point downward and reinforce each other. They each have length $T\sin(\theta)$, so their downward sum is $2T\sin(\theta)$. Since for small θ we have $\sin(\theta) \approx \theta$, we may write

(16.1.6d) $\qquad\qquad\qquad F = 2T\theta$

for the centripetal force on the rope segment where T is the initial tension on the ends of the rope.

Finally, equate F of (16.1.6c) with F of (16.1.6d) to obtain

(16.1.6e) $\qquad\qquad\qquad v = \sqrt{T/\delta}$

which is the equation (16.1.4a) we set out to prove. ■

NOTE: *Exercise (16.7.2) asks you to confirm that the dimensions of velocity — distance over time — result from an analysis of the right hand side of (16.1.4a) which is expressed only in terms of the tension force, T, and the linear mass-density, δ, of the rope.*

(16.2) Speed of Rope Wave is Constant

Your intuition might tell you that the speed of the wave propagated will increase if you oscillate the rope faster (increase the frequency). Not so. As long as the tension T is constant, the speed of
the traveling wave remains the same.

Light waves, or more generally, electromagnetic waves, share a similar
property. The speed of light in a vacuum is constant, no matter what
its frequency (color).

A rope is composed of matter and light is not. Nevertheless, they
share the common property that their speed of propagation does not
depend on the frequency carried by the wave.

A guitar string is an example of how the tension, T, affects the
frequency. According to (16.1.4a), tightening the string (increasing
tension) will increase the string's wave speed v which corresponds to
a higher frequency f, or a higher pitch.

(16.3) Shapes Traveling in One Dimension

Now that we see that a wave in a rope tends to maintain a constant
speed v, we characterize that differential equation, called the wave
equation (16.4), that describes shapes traveling at speed v in one
direction.

(16.3.1) **Notation for Curves Traveling to the Right:** What is the
proper notation to describe how a real-valued function, f, of one
variable, travels to the right along the X-axis over a distance Δx
units in time Δt? (The speed of rightward travel is then necessarily
$\Delta x/\Delta t = v$.)

We begin with the following graphs:

(16.3.2) FIGURE: *Rightward Traveling Curve:*
Angle View of X-Y-T Graph.

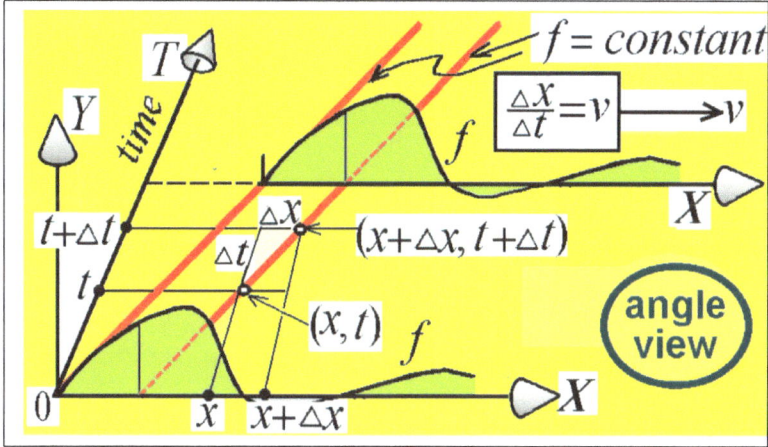

- *The Y values, $f(x,t)$, of the moving curve form a surface over the X-T SPACETIME plane. Since each point of the curve travels rightward on the X-axis with speed v, every vertical Y value of the surface is constant on a velocity line traced in the X-T plane. Two such velocity (red) lines of constant value are shown.*

(16.3.3) FIGURE: *Rightward Traveling Curve:*
Top View of X-Y-T Graph.

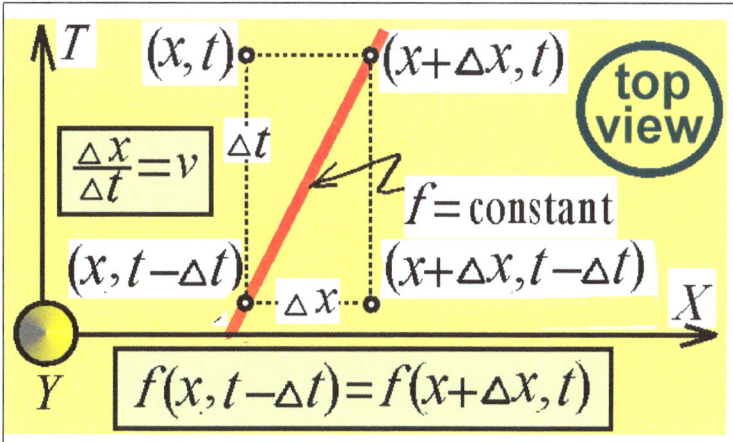

- *This is the "floor" shown in Figure (16.3.2). The Y-axis points up and out of the page. As we see from the graph, the lines in the X-T plane*

over which the surface values for f are constant, require that

(16.3.3a) $f(x, t - \Delta t) = f(x + \Delta x, t),$ *whenever* $\dfrac{\Delta x}{\Delta t} = v.$

NOTE: *To achieve the form of (16.3.3a) which appears in Figure (16.3.3), replace t with $t - \Delta t$ in Figure (16.3.2).*

(16.3.4)

> **DEFINITION:** *Given two-variable function $f(x, t)$. Then the limits*
>
> **(16.3.4a)** $\dfrac{\partial}{\partial x} f(x, t) \stackrel{def}{=} \lim\limits_{\Delta x \to 0} \dfrac{f(x + \Delta x, t) - f(x, t)}{\Delta x}$ *and*
>
> **(16.3.4b)** $\dfrac{\partial}{\partial t} f(x, t) \stackrel{def}{=} \lim\limits_{\Delta t \to 0} \dfrac{f(x, t + \Delta t) - f(x, t)}{\Delta t}$
>
> *are the* PARTIAL DERIVATIVES OF f *at point (x, t) with respect to the variables x and t, respectively.*

(16.3.5) **MOTIVATION:** *Definition (16.3.4) introduces symbols $\partial f / \partial t$ and $\partial f / \partial x$, the partial derivatives of $f(x, t)$, which are the usual derivatives, df/dx and df/dt, with all variables of f fixed except x and t, respectively. Use of the "curly" ∂ instead of the ordinary d is a reminder that function f has more than one variable.*

(16.3.6) **NOTE:** *Equivalent variations of symbol $\dfrac{\partial}{\partial x} f(x, t)$ (16.3.4a) are*

$$\frac{\partial f}{\partial x}, \quad \partial f / \partial x, \quad \partial f(x, t) / \partial x, \quad \partial / \partial x\ f(x, t).$$

(16.3.7)

Lᴇᴍᴍᴀ: *For the function (moving wave) $f(x,t)$ traveling rightward in time with speed v, the partial derivatives $\partial f/\partial x$ and $\partial f/\partial t$ are related as follows for all (x,t):*

(16.3.7a) $\dfrac{\partial}{\partial x}f(x,t) = -\dfrac{1}{v}\dfrac{\partial}{\partial t}f(x,t),$

or, in symbolic short-hand,

$$\frac{\partial}{\partial x} = -\frac{1}{v}\frac{\partial}{\partial t}$$

where existence of the rightward traveling function f, as t increasing, is assumed on both sides of the equal sign.

Pʀᴏᴏꜰ: For our two-variable function $f(x,t)$,

(16.3.7b) $\dfrac{\partial}{\partial x}f(x,t) \approx \dfrac{f(x+\Delta x,t) - f(x,t)}{\Delta x}$ *Def. (16.3.4a) of* $\dfrac{\partial}{\partial x}$

$= \dfrac{f(x,t-\Delta t) - f(x,t)}{\Delta x = v\Delta t}$ *Equation (16.3.3a)*

$= -\dfrac{1}{v} \cdot \dfrac{f(x,t-\Delta t) - f(x,t)}{-\Delta t}$ *Factor out* $-\dfrac{1}{v}$

$\approx -\dfrac{1}{v} \cdot \dfrac{\partial}{\partial t}f(x,t),$ *Def. (16.3.4b) of* $\dfrac{\partial}{\partial t}$

which ends the proof. ∎

Differences such as $f(x+\Delta x,t) - f(x,t)$, which we see in (16.3.7b), occur often enough to warrant a more compact notation, which we now present.

(16.3.8) NOTE: *For a real-valued function, $f(x,t)$ at a point, (x,t), in the spacetime plane, we use the symbols*

(16.3.8a) $\Delta f_x \overset{def}{=} f(x + \Delta x, t) - f(x,t)$ *when t is held fixed,*

$\Delta f_t \overset{def}{=} f(x, t + \Delta t) - f(x,t)$ *when x is held fixed.*

This notation is used in the following proposition that says: If f increases (or decreases) as x increases, then f necessarily decreases (or increases) as time t increases. More formally, we have

(16.3.9)

PROPOSITION: *Given a real-valued function traveling to the right over the X-axis with speed v. for any point (x,t) in spacetime,*

(16.3.9a) $\text{sign}(\Delta f_t) = -\text{sign}(\Delta f_x).$

That is,

$\Delta f_x > 0$ *implies* $\Delta f_t < 0$ *and* $\Delta f_x < 0$ *implies* $\Delta f_t > 0.$

PROOF: Write

$$\frac{\Delta f_x}{\Delta x} = \frac{f(x + \Delta x, t) - f(x,t)}{\Delta x} \qquad \text{Notation (16.3.8)}$$

$$\approx \frac{\partial}{\partial x} f(x,t) \qquad \text{Def. (16.3.4a) of } \frac{\partial}{\partial x}$$

$$= -\frac{1}{v}\frac{\partial}{\partial t} f(x,t) \qquad \text{Equation (16.3.7a)}$$

$$\approx -\frac{1}{v}\frac{f(x, t + \Delta t) - f(x,t)}{\Delta t} \qquad \text{Def. (16.3.4b) of } \frac{\partial}{\partial t}$$

$$= -\frac{1}{v}\frac{\Delta f_t}{\Delta t}. \qquad \text{Notation (16.3.8)}$$

Since $\Delta x > 0$ and $\Delta t > 0$, the first and last entries in this string of equalities tell us that $\text{sign}\Delta f_x = -\text{sign}\Delta f_t$. The proof is done. ∎

(16.3.10) NOTE: *The "arcane" mathematical fact, (16.3.9a), that the sign of the time change is opposite to that of the distance change, becomes an important physical fact when Maxwell uses it to determine c, the speed of light (Section (22.4)).*

(16.4)   The Wave Equation in One Dimension

(16.4.1)
Calculus
required

> **THEOREM:** (*One-dimensional wave equation.*) *Given f, a sufficiently differentiable real-valued function traveling rightward with constant speed v. Then*
>
> **(16.4.1a)**
> $$v^2 \frac{\partial^2}{\partial x^2} f(x,t) = \frac{\partial^2}{\partial t^2} f(x,t)$$
>
> *where* $\frac{\partial^2}{\partial w^2} f = \frac{\partial}{\partial w}\left(\frac{\partial}{\partial w} f\right)$ *for* $w = x$ *or* t. *This is the*
>
> ONE-DIMENSIONAL WAVE EQUATION.

PROOF: The equality

(16.4.1b) $\quad \dfrac{\partial}{\partial x}\left(\dfrac{\partial}{\partial t} f(x,t)\right) = -\dfrac{1}{v}\dfrac{\partial}{\partial t}\left(\dfrac{\partial}{\partial t} f(x,t)\right)$

is a rewriting of (16.3.7a) in which the function $f(x,t)$ has been replaced with another function, $\dfrac{\partial}{\partial t} f(x,t)$. If, in (16.4.1b), we replace $\dfrac{\partial}{\partial t}$ on the left hand side with $-v\dfrac{\partial}{\partial x}$ (again, from (16.3.7a)), the result is

(16.4.1c) $\quad -v\dfrac{\partial}{\partial x}\left(\dfrac{\partial}{\partial x} f(x,t)\right) = -\dfrac{1}{v}\dfrac{\partial}{\partial t}\left(\dfrac{\partial}{\partial t} f(x,t)\right),$

which, when multiplied through by $-v$, yields the desired wave equation (16.4.1a). ∎

(16.4.2) **Solutions to the Wave Equation.** The preceding development assumes only that the real-valued function, f, travels to the right over the 1-dimensional X-axis. If we impose periodicity (the shape of f repeats over every interval of length L), or certain initial conditions (specifying the shape of f over the $[0, l]$ interval at time $t = 0$), or specific boundary conditions (for all time t, $f(x) = 0$ for $x = 0$, L, $2L, \dots$), then specialized techniques exist for finding the appropriate periodic solutions.

(16.5) Wave Propagation: The Skipping Stone Model

Suppose an individual, the "source" of a ball, is stationary (source speed $= v_s = 0$). The source individual throws the ball at a speed of 10 mph relative to herself (ball speed $= v_w = 10$ mph). A stationary observer reckons the speed of the ball to be $w_w = 10$ mph.

Now suppose the source moves at speed 3 mph (source speed $= v_s = 3$ mph). In this case, the stationary observer sees the ball moving with speed (source speed + ball speed $= v_s + v_w = 3$ mph + 10 mph) $= 13$ mph. In other words, the speed of a particle is affected by the speed of its emitting source.

In contrast, the speed of a propagated wave is not affected by the speed of its emitting source.

. . . .

Before developing this invariant property of propagated waves, we introduce the following notation.

(16.5.1)

> **NOTATION:** *We will use the following notation:*
> - v_w *is the speed of a* wave *relative to its carrying medium.*
> - v_s *is the speed of a* wave's source *relative to the medium.*
> - v_m *is the speed of the carrying* medium *of the wave relative to an observer.*

Now instead of a ball being thrown from a moving platform, consider a sound wave (a train whistle) being "thrown" from a moving train whistle. Assume there is no wind so the speed of air, the carrying medium, is $v_m = 0$.

The source of the sound wave, the train whistle, has speed $v_s = 100$ mph relative to an observer. The speed of sound waves in air, relative to the observer, is $v_w = 750$ mph. Will the sound of a train's whistle reach us at $v_s + v_w = 850$ mph? The answer is no.

Unlike particles, wave speeds are not affected by the speed of the wave source. As long as the observer and the air (*the wave's medium*) are not moving relative to each other, the speed, v_s, of the sound's *source* does not affect the speed, v_w, of the wave relative to the medium. In other words,

(16.5.2)

> **P**ROPOSITION: **The speed of a wave** *through a stationary medium is constant no matter what the speed of its source is.*

To see why this is so, consider the "skipping stone model" for propagated waves.

(16.5.3) **The speed of a wave source v_s relative to the medium is not transmitted to the speed of the propagated wave v_m relative to the medium.** For motivational purposes, Figure (16.5.4) presents a simplified model of circular ripples or wave fronts (*instead of spherical ripples*) propagated on a 2-dimensional space or "table-top" by a dancer whose every step produces a new expanding ripple from the point of contact.

(16.5.4) **F**IGURE: *The Skipping Propagator of Waves.*

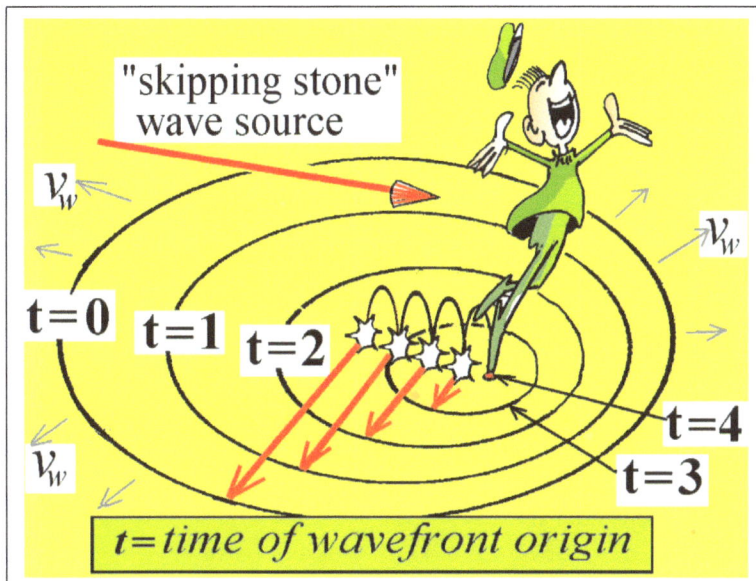

I*magine the dancer's floor with infinitely many uniformly distributed buttons, each of which, when pressed, generates one expanding ripple. All ripples in the medium (table-top) expand at the same speed, v_w from the center. Whenever the dancer taps any one of the fixed buttons, that button*

doesn't care how fast the dancer is prancing overhead — the expansion of the radius of each disk relative to its generating button is always v_w.

(16.5.5) **The speed of the medium — not the source — *is* transmitted to the speed of the propagated wave.** We have noted in (16.5.3) that to an outside observer, the speed of a wave, v_w, is always the same regardless of the speed of the wave source, v_s — provided that v_m, the speed of the transmitting medium (air) is zero relative to the observer.

Speeds, V_w and V_s, are relative to the medium.

Now if the medium is moving at speed $v_m \neq 0$ relative to the observer, then the entire medium-wave-source "unit" is, likewise, carried along at speed v_m relative to the observer. Thus, $v_m + v_s$ is the source speed, and $v_m + v_w$ is the wave speed, relative to the observer. The relative speed of the medium *is* transmitted to the speed of the wave (and to the speed of the source).

(16.5.5a)

NOTE: *Since the medium speed v_m and the source speed v_s are tiny compared to the speed of light, c, there is a miniscule difference between the Galilean/Newtonian arithmetic sum $v_m + v_s$ and the true Lorentzian or relativistic sum, $v_m \oplus v_w$ (1.13.3), (7.3.1), whose value, as we shall see, was experimentally confirmed by Fizeau (13.2).*

(16.6) The Doppler Effect in Spacetime

(16.6.1) A Qualitative View of Frequency Shift. If the speed of the waves (ripples) does not change with the speed of the source, then does the source speed have any observable effect at all?

To fix ideas, and for convenience, we may think of the waves as sound waves traveling through air. As Figures (16.5.4) and (16.6.2) illustrate, an observer who sees (more exactly, hears) the source approaching will perceive an increase of the frequency (the ripples bunch up) while an observer who sees the wave source departing, will hear a decreased frequency (the ripples spread out).

(16.6.2) Fɪɢᴜʀᴇ: *Shifting Ripple-Centers Affect Frequency*

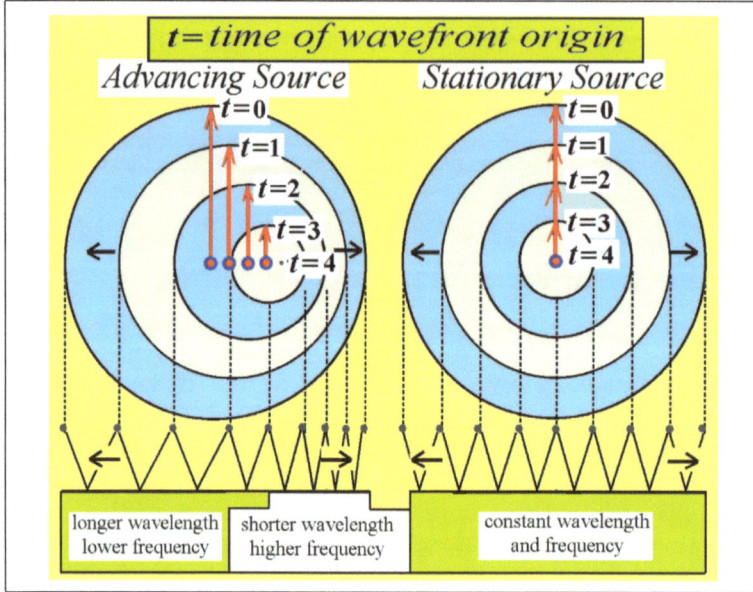

From experience, we know that as a train's whistle approaches, we hear one pitch, but as soon as the train passes by, we hear another, lower pitch.

(16.6.3)

> Dᴇꜰɪɴɪᴛɪᴏɴ: *If f is the frequency of a wave with a stationary source, then the increase or decrease of wave frequency f induced by relative speed of the source is known as the* Dᴏᴘᴘʟᴇʀ Eꜰꜰᴇᴄᴛ.

(16.6.4) **As we shall see, the listener will hear only two distinct pitches,** (16.6.6a), from the whistle if the train has constant speed on a 1-dimensional track and the observer is standing in the train's path. (*To ensure that the observer actually gets to hear the second tone, we suppose he or she stands very close to, but not on, the track. However, this deviation from the 1-dimensional ideal will produce two almost-constant tones separated by a brief, continuous slide from the higher to the lower pitch.*)

(16.6.5) **A Quantitative View of Frequency Shift.** The following theorem quantitifies the value of the two tones that are heard by a stationary observer from a wave-emitting source traveling at constant speed, v_s.

(16.6.6)

THEOREM: **(Doppler Shift: *Fixed Observer, Moving Source*)** *A wave has speed v_m relative to a transmitting medium that is stationary with respect to an observer. If the observer measures the wave source to have speed $v_s = 0$, then the observed frequency is f_0. If the observer measures the wave source to have speed $v_s > 0$, then the observed frequency is*

(16.6.6a)
$$f' = \begin{cases} f_0 \cdot \dfrac{v_m}{v_m - v_s} & \text{while the source} \\ & \text{approaches the observer} \\[1em] f_0 \cdot \dfrac{v_m}{v_m + v_s} & \text{while the source} \\ & \text{recedes from the observer.} \end{cases}$$

PROOF: In Figure (16.6.6c), $P = 1/f_0$ is the period of the wave (*time for one wavelength to pass a given point*).

At time $t = 0$, the wave crest (*wave maximum*) emanates from source Point (0).

At time $t = P$, as shown in the top view in Figure (16.6.6c), the wave crest, traveling with speed v_m, has spread from Point (0) to a circle with radius *distance = speed × time = $v_m \times P$*. Points (a) and (a') lie on this circle.

Also at time $t = P$, the source Point(0) has traveled to Point (b) with speed v_s. The source will now transmit another wave crest (*maximum*). The distance traveled by the source is *distance = speed × time = $v_s \times P$*.

Figure (16.6.6c) shows a full wave length between Points (b) and (a), and another full wave length between Points (b) and (a') since these points represent the positions of the wave crests at time $t = P$. As shown in Figure (16.6.6c), the distance between Points (b) and (a), i.e., a wavelength, is $\lambda_{right} = (v_m - v_s) \times P$. From Definition (16.1.2), the frequency as seen by the right observer, is

(16.6.6b)

$$f'_{right} = \frac{wave\ speed}{wavelength} = \frac{v_m}{\lambda_{right}} = \frac{v_m}{(v_m - v_s)}\frac{1}{P} = \frac{v_m}{(v_m - v_s)}f_0.$$

This proves the first equation of (16.6.6a).

(16.6.6c) FIGURE: *Top View of Wave Crest Propagated After P units of Time*

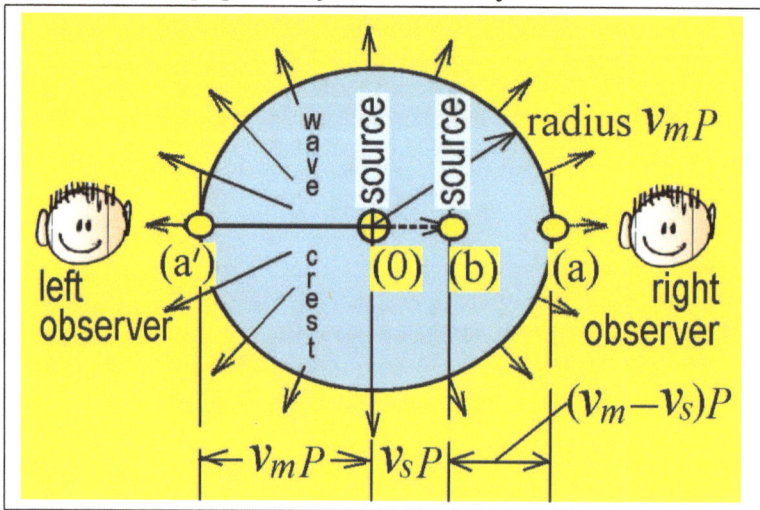

To prove the second equation of (16.6.6a), we establish that the wavelength $\lambda_{left} = (v_m + v_s) \times P$ and then argue as we did in (16.6.6b). This is left as Exercise (16.7.4). The proof is done. ■

(16.7) Exercises

(16.7.1)
Exercise
v^2/R = **Centripetal Acceleration.** A particle travels clockwise with speed $v = v_a = v_b$ on an arc with radius R as it subtends an angle of 2θ. It takes t seconds to travel from point a to point b. If the acceleration is defined as $[\text{vector}(v_b) - \text{vector}(v_a)]/t$, then show that the acceleration is directed toward the center and has magnitude v^2/R.

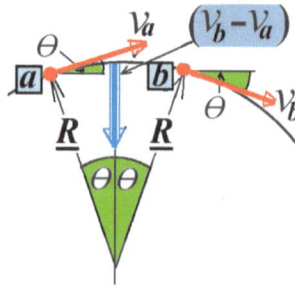

(16.7.2) **Dimensional Consistency.** Show that (16.1.4a) is dimensionally
Exercise consistent. That is, both quantities,

$$v \quad \text{and} \quad \sqrt{T/\delta},$$

have common dimension, namely, *distance/time*.
Hint: *The dimension of the force of tension, T, is* (mass × acceleration) *and the dimension of linear density, δ, is* (mass/length).

(16.7.3) **Speed and rpm.** Given that a wheel has radius R and is turning
Exercise at ρ rpm (revolutions per minute). What is the speed of a point on
the circumference?

(16.7.3a)Why, in Exercise (16.7.3), do we use symbols R and
ρ rather than specific numbers like "7.3 feet" or "2,300 rpm?" The
answer is that whether symbols or specific numbers are used, the
time and energy required to solve this problem are identical since
the manipulations in each case are identical. The advantage of using
symbols is that one solution provides a valid template for infinitely
many cases, each case corresponding to one of the infinitely many
values for R and ρ. But a solution for particular numbers such as
"7.3 feet" and "2,300 rpm", gives you that solution and no other.

(16.7.4) Following the pattern of (16.6.6b), prove the second equation of
Exercise (16.6.6a).

(17) Maxwell's Mathematical Toolkit

(17.1) Preface

(17.1.1)
Calculus
will be
required

Things to Come: We present the mathematical machinery used by Maxwell to develop his four equations (20.7.3a)-(20.7.3d). This requires a significant amount of calculus, much of which will be reviewed in this chapter.

Use of calculus may seem daunting but the rewards are great. Maxwell's equations have stood the test of time, as they describe not only the motion of charged particles, properties of magnets and electromagnetic waves at *all* frequencies, but they apply to phenomena in the quantum regime as well.

The language of mathematics is essential — and there does not seem to be a simpler path.

(17.1.2)

Note: *Maxwell's equations (20.7.3), pg. 280 will be and should be applauded as a great achievement in mathematical modeling. However, we must be aware that sometimes, apparently reliable models can be based on false premises. See (??), pg. ?? for a spectacular example.*

(17.2) Language and Proportionality

(17.2.1)

> **DEFINITION:** *Given a set of numbers X, each of which is associated with a number Y. We say number X is* <u>PROPORTIONAL</u> *to its related number Y if there is a number k that is constant with respect to X and Y, where for all X and its related Y,*
>
> **(17.2.1a)** $\qquad X = kY.$
>
> (k *may depend on variables other than X and Y.*)

Motivation for (17.2.1). To say "X is proportional to Y," is to say if X doubles, then so does its related number Y; if X triples, then so does Y, and so on. In other words, the quotient X/Y is a constant k for all X and its related Y. But

(17.2.1b) $\quad X/Y = k$ is algebraically equivalent to $X = kY$.

From equations (17.2.1a) and (17.2.1b), we can say more:

(17.2.2)

> **THEOREM:** *If a quantity X is proportional to both Y and Z, then X is proportional to the product YZ. (For example, the area A of a rectangle is proportional to both length L and width W. Hence, A is proportional to the product LW.)*

PROOF: Exercise (17.9.1). ∎

NOTE: Theorem (17.2.2) will prove to lie at the foundation of Coulomb's Law (18.2.4b).

(17.2.3)

> **DEFINITION:** *We say X is* <u>INVERSELY PROPORTIONAL</u> *to Y whenever X is proportional (17.2.1) to $1/Y$, the inverse of Y. That is,*
>
> **(17.2.3a)** $\qquad X = k/Y$ *for some constant k*
>
> *that is called the* <u>CONSTANT OF PROPORTIONALITY</u>.

(17.2.4) Examples:

(17.2.4a)
We observe the set of related numbers, (F, m), where force F is used to accelerate mass m by a fixed amount, a ft/sec^2. To say that the force F is always proportional to mass m is to state Newton's Second Law, $F = ma$, where a is a constant of acceleration.

(17.2.4b)
To say the force F is *inversely* proportional to the square of distance r, is to say F is proportional to $(1/r^2)$. That is, $F = k/r^2$. Such a proportionality is called an INVERSE SQUARE LAW.

(17.2.4c)
Consider the list of related numbers, (A, L) and (A, W), where A is the area of a rectangle with length L and width W. Since $A = LW$, then, consistent with Definition (17.2.1), we declare that the area A of a rectangle is proportional to both the length L and the width W. However, area A is also proportional to a third number, the product, LW. In other words, $A = kLW$. (See Exercise (17.9.1).)

(17.3) 1D Lengths & 2D Areas as 3D Vectors

(17.3.1) A new interpretation of vectors. We first encountered ordered triples, $[x_1, x_2, x_3]$ in (A.2.3) as a way to numerically locate points in 3-space within a 3D (*three-dimensional*) frame of reference. Arrows can be used to represent such ordered triples, or vectors.

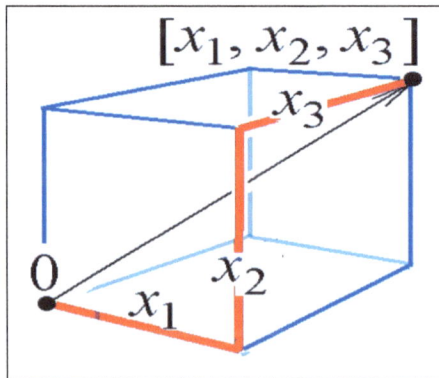

In the following, we use arrows to graphically represent physical quantities of *lengths* and *areas* in three-dimensional space.

(17.3.2) Tiny, (i.e., differential) pieces of 1D arc lengths and 2D surfaces can be described by 3D vectors only if there is sufficient smoothness — 1D arcs are not severely crinkled and 2D surfaces are not very crumpled. Here are the details:

- LEFT PANEL OF FIGURE (17.3.3): LENGTHS AS VECTORS. Arc segment of length Δl is assigned a 3-dimensional arrow, $\Delta \vec{l}$, whose tail is placed at a point in the segment, whose length is Δl, and whose direction is tangent to the arc. Either this orientation or its negative (a rotation of 180°) is declared to define the *positive* direction of $\Delta \vec{l}$.

- RIGHT PANEL OF FIGURE (17.3.3): AREAS AS VECTORS. A region of area ΔA is taken small enough so that it is essentially flat — much the way we view the curved surface of a lake, a small area of Earth's surface, as flat. The area ΔA is assigned a 3-dimensional arrow, $\Delta \vec{A}$, whose tail is placed at a point in the region, whose length is ΔA, and whose direction is NORMAL, or perpendicular to the area. Either this orientation or its negative (a rotation of 180°) is declared to define the *positive* direction of $\Delta \vec{A}$. (*Vectors describing surface orientation will be defined in* (17.4.7).)

(17.3.3) FIGURE: *1D-Lengths and 2D-Areas as 3D-Vectors, $\Delta \vec{l}$ and $\Delta \vec{A}$.*

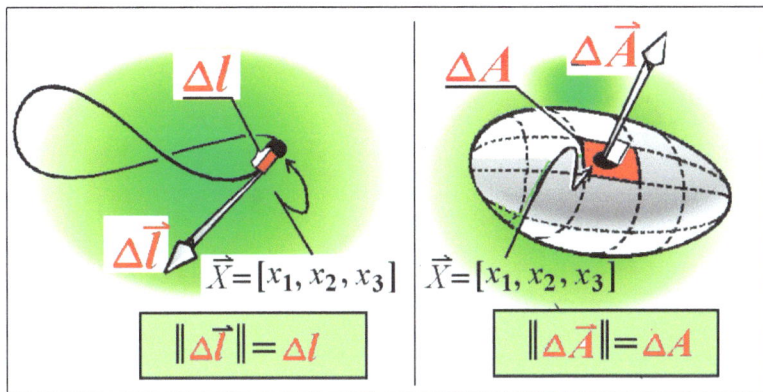

The notations for vectors and their lengths are given by Table (17.3.4).

(17.3.4) *Vector Representations of Length and Area*

	Vector	Vector Magnitude	Vector Direction
Length	$\Delta\vec{l}$	$\Delta l = \|\Delta\vec{l}\|$	tangent to curve
Area	$\Delta\vec{A}$	$\Delta A = \|\Delta\vec{A}\|$	normal to surface

Physical experiments show that for vectors $\vec{V}$ in $\boldsymbol{R}^3$ (em See Def. (A.2.1), pg. 335) emanating from points on a surface or an arc length, it is useful to consider their components or projections onto the vectors $\Delta\vec{A}$ or $\Delta\vec{l}$ of (17.3.4). This motivates the following discussion:

. . . .

The setup: The vector $\vec{V}_A$ in $\boldsymbol{R}^3$ emanates from a point on the surface element ΔA and forms an angle θ_A with the normal vector, $\Delta\vec{A}$. Similarly, the vector $\vec{V}_l$ is situated at a point on the differential line segment Δl, forming an angle θ_l with the tangent vector $\Delta\vec{l}$.

(17.3.5) Trig

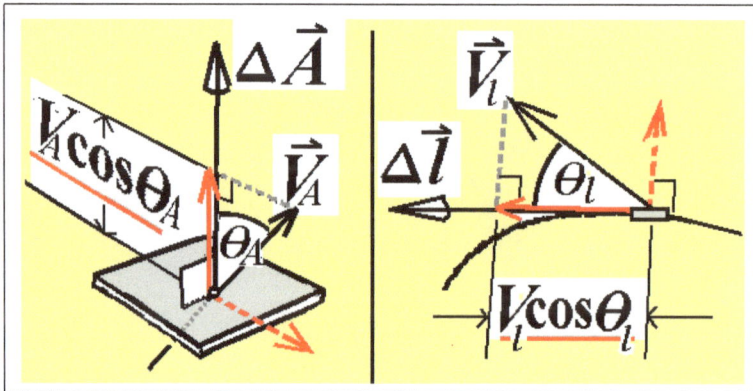

NOTE: *The components of arbitrary vectors, $\vec{V}_A$ and $\vec{V}_l$, on the area vector, $\Delta\vec{A}$, and length vector, $\Delta\vec{l}$, (17.3.3) are called the* PROJECTION *of vectors $\vec{V}_A$ and $\vec{V}_l$, respectively.*

The *lengths* of the projection vectors, as illustrated in Definition (17.3.5), are given by the table:

(17.3.5a)

$\vec{V}_A$	$\vec{V}_l$	{Vector}
$V_A \cos \theta_A$	$V_l \cos \theta_l$	{Length of Projection}

(17.4) Orientations of Lines and Surfaces

(17.4.1) Motivation: There will follow a considerable number of results about lines and surfaces which culminate in the four famous equations of Maxwell (20.7.3a)-(20.7.3d). The importance of these equations is indicated by the fact that Einstein acknowledged and applauded Maxwell's work in the very first paragraph of his paper of 1905 [12] in which the ideas of special relativity were first introduced.

At first glance, these mathematical results might seem to have little to do with special relativity. However, the time spent in understanding the mathematics in this chapter will bear fruit when used in subsequent chapters. In addition, we hope the reader will recognize that there is an intrinsic elegance that, in itself, justifies the effort to understand the tools of this chapter.

(17.4.2) Inside and Outside Regions. Where Section (17.3) described the lengths of a tiny line segment at a *point*, and the area of an infinitesimal surface element at a *point*, we now describe the orientation in space of the entire line and surface.

First, we are interested in lines and surfaces that are *closed*, a term we now define in an intuitive, non-rigorous way:

(17.4.3)
Calculus
required

> DEFINITION: *A non-intersecting 1-dimensional line in space that closes back on itself, like a string whose two ends meet, is called a* CLOSED LINE. *Similarly, a non-intersecting 2-dimensional surface that closes in on itself, like a potato skin that covers a finite-volume potato, is called a* CLOSED SURFACE.

∥ *See Figure (17.3.3) for an illustration of a closed line and a closed*
∥ *surface.*

(17.4.4) **Closed**   non-intersecting 1-dimensional lines (loops) are called JORDAN CURVES, after the French mathematician MARIE ENNEMOND CAMILLE JORDAN (1838-1922). Like a closed 1-dimensional circle, 1-dimensional Jordan curves define an infinite *exterior* (outside region) and a finite *interior* (inside region). This may seem obvious, but a rigorous proof is a rather tough mathematical nut. In fact, the proof by Jordan himself was in error.

Like the two-dimensional surface of a sphere, any closed non-intersecting "potato-skin" area in space defines an infinite outside region and a finite inside region.

(17.4.5) FIGURE: *Wheel-Axle Orientation of Closed Wire Loops.*

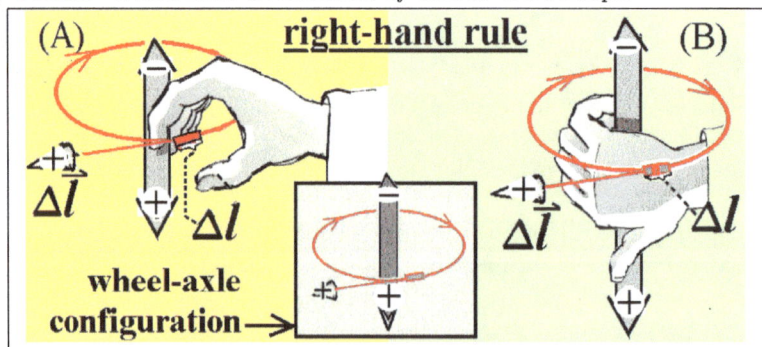

A *closed loop and a line that enters its enclosed area is a* **wheel-axle** *configuration. It is an experimental fact that a magnet moving through a loop generates a current through the loop. The direction of the current depends on whether the magnet's north pole or the south pole enters first. The right hand rule applied to the wheel-axle configuration provides a mathematical means of describing the leading pole of the magnet (north =t humb) and the*

direction of the current (clockwise = direction of the right hand fingers).

(17.4.6) The Right Hand Rule: If the positive direction of one component (wheel or axle) is known, then the positive direction of the other is uniquely determined according to the following:

- PANEL (A) OF (17.4.5) **Four Fingers are Inside the Loop:** *The four fingers point in the positive direction of the straight wire (axle) while the thumb points in the positive direction of the loop (wheel).*

- PANEL (B) OF (17.4.5) **Thumb is Aligned with the Straight Wire:** *The thumb points in the positive direction of the wire (axle) while the four fingers point in the positive direction of the loop (wheel).*

Motivation: The Right Hand Rule is devised to uniquely relate the direction in which we travel along a *closed* curve (*defined in* (17.4.3)) to the "direction" of the enclosed area. If we imagine that we travel in a certain direction on the closed curve — the boundary of a small area ΔA — then we can uniquely define a positive (right hand) direction of the vector $\Delta\vec{A}$ whose length $\|\Delta\vec{A}\|$ numerically coincides with the area of ΔA as is shown in the next section.

. . . .

(17.4.7)

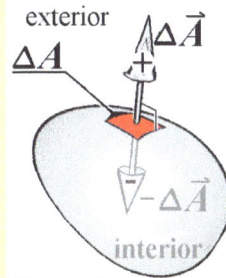

DEFINITION: *At each point on a closed surface, the* POSITIVE DIRECTION *of the three-dimensional area vector,* $\Delta\vec{A}$

- *is perpendicular to the tangent plane at that point of the surface and*

- *points away from the enclosed interior, that is, toward the exterior of the closed surface from which it emanates.*

NOTE: Table (17.3.4) marks the distinction between Δl and $\Delta\vec{l}$ in Figure (17.4.5), and between ΔA and $\Delta\vec{A}$ in Figure (17.4.7).

(17.5) **Vectors Modeling Reality**

(17.5.1) **Perception and Mathematics.**   Vectors and their components
are observed in real life.

For example, observers at different positions will each see different
vector components of an automobile's velocity vector, $\vec{X}$, as Figure
(17.5.1a) shows.

- A viewer above the auto (i.e., the reader) sees a 50 mph speed for
  the skidding car,

- Bernie perceives the car to be moving 40 mph to his left, and

- Ashley observes a 30 mph drift to her right.

In Figure (17.5.1a), Bernie and Ashley each see perpendicular *components* (A.2.1), $\vec{X}_1$ and $\vec{X}_2$, respectively, of the vehicle's velocity
vector, $\vec{X}$, which, by the Pythagorean theorem, has value $\vec{X} = 50$
mph. It is the overhead viewer, the reader, who sees the full velocity
vector $\vec{X}$ and its components.

(17.5.1a)   **F**IGURE: *Perceived Vector Components of Velocity*

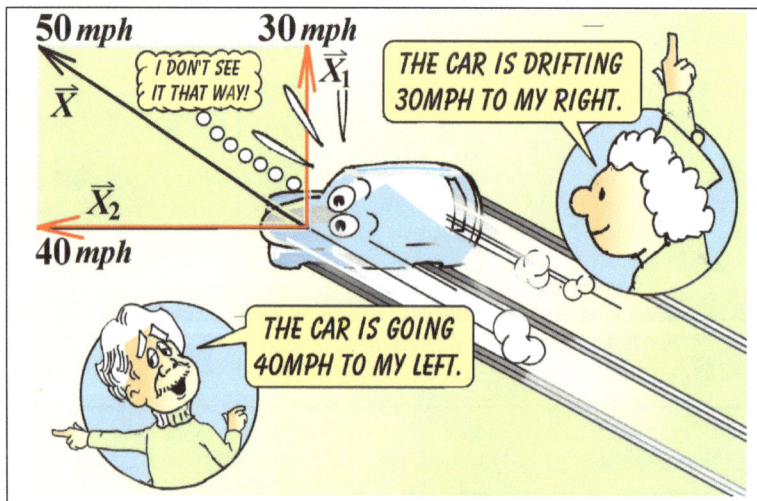

Figure (17.5.1a), representing a 2-dimensional situation, illustrates a
mathematical structure — **vector decomposition** — that accounts
for three differently perceived speeds of 30 mph, 40 mph, and 50
mph, for the same automobile. We state (without proof) that any

2-dimensional vector $\vec{X}$ has two component vectors, $\vec{X}_1$ and $\vec{X}_2$ (*see* (A.2.4b)) such that $\vec{X} = \vec{X}_1 + \vec{X}_2$, where $\vec{X}_1$ and $\vec{X}_2$ form a 90° angle with each other.

In Figure (17.5.1a), the lengths of the vector $\vec{X}$ and its mutually perpendicular (A.2.6b) pair of components, $\vec{X}_1$, and $\vec{X}_2$, are

$$X = 50 \text{ mph, } X_1 = 30 \text{ mph, and } X_2 = 40 \text{ mph.}$$

(17.5.2) **Vector Fields:** Given a region X in space $\boldsymbol{R}^2$ or $\boldsymbol{R}^3$. At each point (or vector) $\vec{X}_i$ in the region X, the state of the point $\vec{X}_i$ may be described by an associated vector $\vec{v}(\vec{X}_i)$ that also lies in $\boldsymbol{R}^2$ or $\boldsymbol{R}^3$.

For example, the state of a particle at point $\vec{X}_i$ might be described by an associated vector $\vec{v}(\vec{X}_i)$ which could be the *velocity* vector of that particle, or a *force* vector acting on that particle due to gravity or magnetism.

A region X with this structure $\vec{X}_i \rightarrow \vec{v}(\vec{X}_i)$ is called a *field*. More formally,

(17.5.3)

DEFINITION: *Given a region X in either $\boldsymbol{R}^2$ or $\boldsymbol{R}^3$ where for each point (vector) $\vec{X}_i$ of space X, there is an associated vector $\vec{v}(\vec{X}_i)$ that also lies in $\boldsymbol{R}^2$ or $\boldsymbol{R}^3$. Then the region X is called a* VECTOR FIELD. *Vectors $\vec{v}(\vec{X}_i)$ are called the* FIELD VECTORS *of X that are tangent to (red dotted) lines that are called* FIELD LINES.

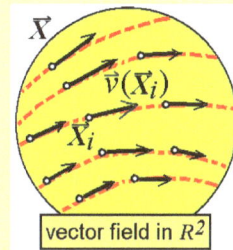

vector field in R^2

(17.5.4) NOTE: The norms $\|\vec{v}(\vec{X}_i)\|$ of the field vectors $\vec{v}(\vec{X}_i)$ are not necessarily constant along red-dotted field lines.

(17.6) Inner and Cross Products

We first encountered the inner product $\langle \vec{X}, \vec{Y} \rangle$ as the coordinate-dependent summation given by Definition (A.2.6). We now present an equivalent definition that is more geometrically based in that $\langle \vec{X}, \vec{Y} \rangle$ depends on lengths of vectors $\vec{X}, \vec{Y} \in \mathbf{R}^3$ and the angle between them. Consistent with the notation of Table (17.3.4), $X = \|\vec{X}\|$ and $Y = \|\vec{Y}\|$, we have

$(17.6.1)$

DEFINITION: *Given   the 3-dimensional   vectors $\vec{X}$ and $\vec{Y}$ with   respective lengths X and Y. Then the* INNER PRODUCT, $\langle \vec{X}, \vec{Y} \rangle$, *is the number*

(17.6.1a)

$$\langle \vec{X}, \vec{Y} \rangle = X \cdot \cos\theta \cdot Y$$

where θ is the angle between vector $\vec{X}$ and vector $\vec{Y}$.

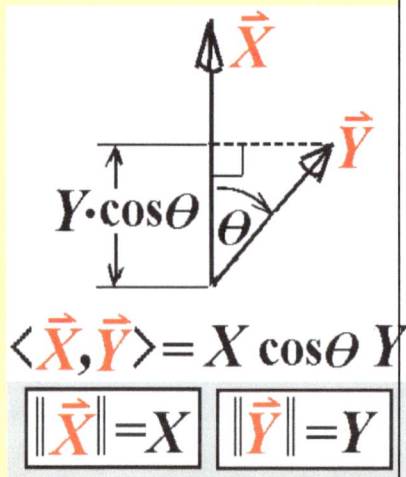

$$\langle \vec{X}, \vec{Y} \rangle = X \cos\theta \ Y$$

$$\|\vec{X}\| = X \qquad \|\vec{Y}\| = Y$$

Beyond 3-Dimensional Reality. Classical trigonometry defines the *cosine* of an angle (*formed by a side and the hypotenuse of a right triangle*) to be the ratio of the adjacent side length to the hypotenuse length.

Now, thanks to the general definition of inner product (*the left hand side of* (17.6.1a)), which is equivalent to the coordinate-based definition (A.2.6), we can go beyond familiar 3-dimensional space to define the cosine of angles formed by vectors in higher dimensions.

For example, angles that are formed "between" vectors in spaces $\mathbf{R}^k$ of *any* dimension, $k = 1, 2, 3, \ldots$ also have a cosine — this extended definition of cosine does not *contradict*, but *agrees with* the familiar 2-dimensional version (A.2.6).

Using (17.6.1a) we unambiguously define

(17.6.1b) $\quad \cos\theta \overset{def}{=\!=} \dfrac{\langle \vec{X}, \vec{Y} \rangle}{\|X\| \|Y\|}$

for any pair of vectors $\vec{X}, \vec{Y} \in \boldsymbol{R}^k$. Using this number $-1 \leq \cos\theta \leq 1$, we find the angle θ by using the usual 2-dimensional cosine. For example, if (17.6.1b) tells us that for vectors $\vec{X}$ and $\vec{Y}$ in 400-dimensional space, the cosine of the included angle is $\cos\theta = 0.5$, then we would declare that the angle between these vectors is $\theta = 30^o$.

A special case. As shown in Definition (A.2.6b) and Figure (A.2.6d), vectors $\vec{X}$ and $\vec{Y}$ are *perpendicular (form an angle of* 90°*)* if and only if $\langle \vec{X}, \vec{Y} \rangle = 0$.

(17.6.2) NOTE: **Two Right Hand Rules** The Right Hand Rule of (17.4.6) in three-dimensional space $\boldsymbol{R}^3$ defines the "direction" of an enclosed area to be the *direction of travel* along its closed boundary.

In (17.6.3) following, we will define a different right hand rule, one that determines the positive direction for a new product $\vec{A} \times \vec{B}$ in $\boldsymbol{R}^3$ for pairs of vectors $\vec{A}$ and $\vec{B}$ in $\boldsymbol{R}^3$.

(17.6.3)

DEFINITION: *Right Hand Rule and Cross Product* $\vec{A} \times \vec{B}$.

Vectors $\vec{A}$ and $\vec{B}$ form an angle θ.

1. *Point the index finger of your right hand in the direction of $\vec{A}$, a vector with length A.*

2. *Face your palm toward vector $\vec{B}$ whose length is B.*

3. *The cross-product, $\vec{C} = \vec{A} \times \vec{B}$, is the vector that is parallel to your thumb and perpendicular to the plane formed by $\vec{A}$ and $\vec{B}$.*

(17.6.3a)   The length $\|\vec{C}\|$ is $C = A\,\sin(\theta)\,B$.

. . . .

(17.6.4) **Notes on the Angle θ:**

Trig $\underline{\theta = 90^o}$ When the angle $\theta = 90^o$, we have $\sin(\theta) = 1$ which, from (17.6.3a), says that the length$(\vec{A} \times \vec{B}) = A \cdot B$, its maximum value.

$\underline{\theta = 0}$ Geometrically, when $\theta = 0$ then $\vec{A}$ and $\vec{B}$ are colinear. Also, since $\sin(\theta) = \sin(0) = 0$, the cross-product $\vec{A} \times \vec{B} = \vec{0}$.

(17.6.5) **The previous statement tells us:** The cross product of non-zero factors (vectors) *can* produce a zero result (when $\theta = 0^o$). Contrast this with the fact that the product of ordinary non-zero numbers *can never* be zero.

(17.6.6) **The cross product and the inner product** are both products of a pair of vectors. However, only for vectors $\vec{X}$ and $\vec{Y}$ in three dimensional space $\boldsymbol{R}^3$, is the *cross product* $\vec{X} \times \vec{Y}$ defined. Moreover, this cross product is another vector in $\boldsymbol{R}^3$. But for a vector pair $\{\vec{X}, \vec{Y}\}$ in any finite dimensional space (and even for infinite dimensions), the *inner product*, $\langle \vec{X}, \vec{Y} \rangle$, is always a number, i.e., a vector in $\boldsymbol{R}^1$.

(17.7) Riemann Sums and Integrals

One of the most interesting results derived from Maxwell's equations is the deep connection between magnetic and electric phenomena. For example, as mentioned in (17.4.5), moving a magnet through a wire loop will generate an electric current in the wire. Quantitatively, the current in the wire is determined by the combination of the magnetic field changes on each section of the wire loop.

The mathematical tool for carrying out this "combination" along the wire is called an *integral*. In (17.7.4), we will provide a brief introduction to the definition and properties of integrals.

(17.7.1)

> **DEFINITION:** *Let $\mathcal{U}$ be a region in space — $\mathcal{U}$ is either a surface area, A, or an arc, L. Now let $\mathcal{U}$ be decomposed into n separate, mutually exclusive pieces. Then*
>
> **(17.7.1a)** $\qquad \Pi_{\mathcal{U}, \Delta u_i} = \{\Delta u_1, \Delta u_2, \ldots, \Delta u_i, \ldots, \Delta u_n\}$
>
> *is the notation for this* <u>PARTITION</u> *of the region $\mathcal{U}$.*

NOTE: The construction of (17.7.1a) is an approximation of a curve with a series of short line segments which, if very short, will closely approximate the original curve. Similarly, with surfaces, we use smaller and smaller polygons to approximate a surface (*see Figure (17.7.1c)*).

At each point in the region $\mathcal{U}$, there is defined a real-valued function, $f : \mathcal{U} \to \boldsymbol{R}$ (A.3.2) where each Δu_i stands for either an area segment, ΔA_i, or a line segment, Δl_i, accordingly.

Assumption: We will suppose that the Δu_i's of (17.7.1a), are so small that the function f is virtually constant over each piece, Δu_i — the values of f do not significantly vary over any single Δu_i, or

(17.7.1b) For any two points, u_i and u_i', in Δu_i, $f(u_i) \approx f(u_i')$.

Accordingly, at each of the n differential (small) segments, Δu_i, of partition $\Pi_{\mathcal{U}, \Delta u_i}$ (17.7.1a), we may construct the single, unambiguous product,

(17.7.1c)

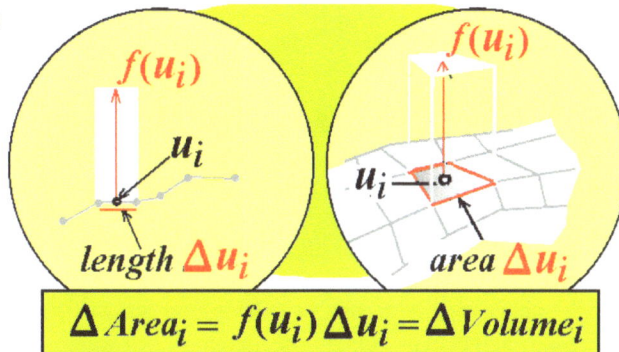

$$\Delta Area_i = f(u_i)\, \Delta u_i = \Delta Volume_i$$

which we may interpret geometrically as (*see Figure (17.7.1c)*)

- The *ith differential area* of a rectangle over the line segment, Δu_i, whenever region $\mathcal{U}$ is an arc in space, or as

- The *ith differential volume* of a 3-dimensional "box" over surface segment, Δu_i, whenever region $\mathcal{U}$ is a surface in space.

We now have the structure to set the following definition in place:

(17.7.2)

> **DEFINITION:** *Given a region $\mathcal{U}$, which is either a 1-dimensional arc, L, or a 2-dimensional surface A. Let a real-valued function, $f : \mathcal{U} \to \mathbf{R}$ be defined on $\mathcal{U}$, which is partitioned into n parts as described in (17.7.1a). Then the* RIEMANN SUM *relative to this partition, $\Pi_{\mathcal{U}, \Delta u_i}$, is denoted by*
>
> **(17.7.2a)** $\displaystyle\sum_{\Delta u_i} f(u_i)\Delta u_i$ *or, equivalently,* $\displaystyle\sum_{i=1}^{n} f(u_i)\Delta u_i.$

(17.7.3) We need proper notation for the result of taking the limit of Riemann sums (17.7.2a) as the size of the biggest Δu_i goes to zero. (*The limit will not depend on the particular partition — all partitions (17.7.1a) will yield the same result (17.7.2a).*)

Now if the sum of terms (17.7.2) does not depend on the partition, but only on the region, $\mathcal{U}$, and the function, f, then what notation is suitable for this two-variable limit? For example, we might choose to represent the Riemann sum

$$\lim_{max\ \Delta u_i \to 0} \sum_{\Delta u_i} f(u_i)\Delta u_i$$

with a symbol such as $\Lambda(\mathcal{U}, f)$. However, we reject this form since it conceals the summation structure of the Riemann sum (17.7.2a). The standard notation for the Riemann sum is

(17.7.3a) $\displaystyle\int_{\mathcal{U}} f(u)du,$

which echoes the structure of a Riemann sum in the sense that the symbol $\int$ suggests a "continuous" form of the sum $\sum$ (17.7.2a) where du represents the presence of Δu in the sum.

(17.7.4)
Calculus
needed

> **DEFINITION:** *The limit of the sequence of Riemann sums* (17.7.2a), *as the size of the largest partition piece goes to zero, that is, as* $max\{\Delta u_i\} \to 0$, *is the* DEFINITE RIEMANN INTEGRAL,
>
> **(17.7.4a)**
> $$\lim_{\substack{max\,\Delta u_i \\ \to\, 0}} \sum_{\Delta u_i} f(u_i)\Delta u_i \quad \text{which is denoted by} \quad \int_{\mathcal{U}} f(u)du.$$
>
> *If the region* $\mathcal{U}$ *is a closed arc (loop) or a closed surface (potato skin), then the integral* (17.7.4a) *is written with an embellished circle notation,*
>
> **(17.7.4b)** $\qquad \oint_{\mathcal{U}} f(u)du$
>
> *and is called the* CLOSED INTEGRAL *over* $\mathcal{U}$.

(For the definition of a closed line or surface, see (17.4.3), pg. 223.)

(17.7.5) **For certain special cases, shortcuts have been invented** to avoid calculation of the actual limit (17.7.4a).

For example, consider the power functions, $f(x) = x^k$, for a fixed scalar $k \neq -1$. We can compute $\int f(x)dx$ without wrestling with the Riemann sum of (17.7.4a). Thanks to the Fundamental Theorem of Calculus, we remove ourselves from this definition altogether by following precise rules of mechanical manipulation of the fixed exponent, k. In particular, if the region $\mathcal{U}$ is the interval $[a, b]$, this technique produces the tidy answer,

$$\int_{\mathcal{U}} f(x)dx = \int_a^b x^k dx = \frac{b^{k+1} - a^{k+1}}{k+1}.$$

For more general functions, we are obliged to use the unwieldy limit-based definition (17.7.4a) and stoically deal directly with the Riemann sums (17.7.2a). The desire to devise calculations that defy round-off and accumulation errors (that are due to limitations of computer architecture) is the engine that drives the field of NUMERICAL ANALYSIS.

(17.8) Integrals of the Inner Product

Integrals in the Laboratory. In the study of electromagnetism, there arise integrals that take the form

(17.8.1) $$\int_L \langle \vec{V}_l, d\vec{l}\,\rangle \quad \text{or} \quad \int_A \langle \vec{V}_l, d\vec{A}\,\rangle,$$

where the 3-dimensional vector $\vec{V}_l$, a function of $\vec{l}$, represents either a magnetic field vector $\vec{B}$ or an electric field vector $\vec{E}$. (*Formal definitions follow in Chapter* 18.)

But the integrals of (17.8.1) do not seem to resemble the Riemann integral form, $\int f(u)du$, given by (17.7.4). That the integrals are, in fact, disguised forms of the Riemann integral (17.7.4a), will now be shown.

The Physical Background for (17.8.1):
Figure (17.8.2), which illustrates this behavior of $\vec{V}$ relative to arcs and surfaces, is the framework for the following discussion that resolves the apparently different forms, (17.7.4a) and (17.8.1), for the Riemann integral.

(17.8.2) FIGURE: *Projection Component of $\vec{V}$ and its Inner Products*

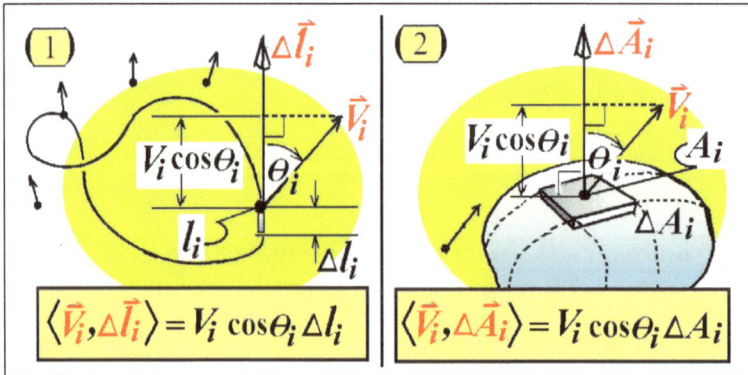

(17.8.3) The Setup for $\int \langle \vec{V}_l, d\vec{l}\,\rangle$ of (17.8.1). *See* (17.3.3)-(17.3.5).
Choose any point u_i inside the ith differential arc segment Δu_i. The length of the arc is also denoted by the symbol Δu_i, which, in this

case, is also written Δl_i. Through this point u_i, there passes the ith (electric/magnetic) field vector, $\vec{V}(u_i)$, (*abbreviated as V_i*) which forms an angle $\theta(u_i)$ (*abbreviated as θ_i*) with respect to the tangent vector, $\Delta \vec{u}_i = \Delta \vec{l}_i$. For physical reasons, we are interested in the product of the following two quantities, (17.8.3a) and (17.8.3b):

(17.8.3a) $f(u_i) = V(u_i)\cos(\theta(u_i)) = V_i\cos\theta_i,$

the projection component (17.3.5) of the field vector, $\vec{V}_i$, and

(17.8.3b) $\Delta u_i = \Delta l_i,$

the differential arc length. But the product of (17.8.3a) with (17.8.3b) produces a familiar scalar — the coordinate-free inner product (17.6.1)

(17.8.3c) $f(u_i)\Delta u_i = V_i \cdot \cos\cdot\theta_i \cdot \Delta l_i = \langle \vec{V}_i, \Delta \vec{l}_i \rangle.$

Summing on the first and third quantities of (17.8.3c) over all $\Delta u_i = \Delta l_i$ of the arc L gives us two forms of the Riemann sum (17.7.2)

(17.8.3d) $\displaystyle\sum_{\Delta u_i \ in \ L} f(u_i)\Delta u_i = \sum_{\Delta l_i \ in \ L} \langle \vec{V}_i, \Delta \vec{l}_i \rangle$

whose respective limits, as $\max\Delta u_i \to 0$ (as $\max\Delta l_i \to 0$) are written

(17.8.3e) $\displaystyle\int_{arc\,L} f(u)du = \int_{arc\,L} \langle \vec{V}_l, d\vec{l} \rangle.$

We conclude, therefore, that the right hand side of (17.8.3e), the integral of inner products, is a *bona-fide* Riemann integral consistent with Definition (17.7.4).

(17.8.4) The Setup for $\int \langle \vec{V}_l, d\vec{A} \rangle$ of (17.8.1).
We follow (17.8.3a)-(17.8.3e) almost word for word, except that

- The arc length is replaced by a surface area. Instead of Δu_i representing the ith differential arc length (*Panel (1) of Figure (17.8.2)*), the quantity $\Delta u_i = \Delta A_i$ now represents the ith differential surface element (*Panel (2) of Figure (17.8.2)*).

- Instead of being *tangent* to the ith differential arc length $\Delta \vec{l}_i$ (*Panel (1) of Figure (17.8.2)*), the projection component of vector $\vec{V}_i$ is now *perpendicular* to the ith surface element ΔA_i (*Panel (2) of Figure (17.8.2)*). Note that relative to both cases, $\Delta u_i = \Delta l_i$ and $\Delta u_i = \Delta A_i$, the magnitude of the projection component of $\vec{V}_i$ is $V_i \cdot \cos\theta_i$.

(17.8.5)

> **DEFINITION:** *The* INTEGRAL OF THE INNER PRODUCT $\langle \vec{V_l}, d\vec{l}\rangle$ *over an arc L is defined by,*
>
> **(17.8.5a)** $\quad \displaystyle\int_{arc\,L} \langle \vec{V_l}, d\vec{l}\rangle \stackrel{def}{=} \lim_{\substack{max\,\Delta l_i \\ \to\,0}} \left[\sum_{\Delta l_i} \langle \vec{V_i}, \Delta \vec{l_i}\rangle\right].$
>
> *The scalar $\Delta l_i > 0$ is the length of the ith segment in the partition $\Pi_{L,\Delta l_i}$ (17.7.1) of the arc L. The vector $\vec{V_i} = \vec{V}(l_i)$ is the vector $\vec{V}$ evaluated at some point, l_i, in the segment Δl_i. Similarly, the integral of the inner product $\langle \vec{V_l}, d\vec{A}\rangle$ over the area A is defined by*
>
> **(17.8.5b)** $\quad \displaystyle\int_{A} \langle \vec{V_l}, d\vec{A}\rangle \stackrel{def}{=} \lim_{\substack{max\,\Delta A_i \\ \to\,0}} \left[\sum_{\Delta A_i} \langle \vec{V_i}, \Delta \vec{A_i}\rangle\right].$
>
> *The scalar $\Delta A_i > 0$ is the area of the ith segment in the partition $\Pi_{A,\Delta A_i}$ (17.7.1) of the surface A and the vector $\vec{V_i} = \vec{V}(l_i)$ is the vector $\vec{V}$ evaluated at some point l_i in the area segment, ΔA_i.*

(17.8.6) **EXAMPLE:** ***Integrating the inner product over an interval.*** We show how Definition (17.8.5a) applies to determine that

(17.8.6a) $\qquad \displaystyle\int_{x=a}^{x=b} \langle \vec{V_0}, d\vec{l}\rangle = V_0 \cos\theta_0 (b-a)$

where the *arc L* in (17.8.5a) is the interval $[\,a\,,\,b\,]$ on the *X*-axis in $\boldsymbol{R}^3$, and the vector $\vec{V_0} = V_l$ in $\boldsymbol{R}^3$ is constant at every point l in the interval $[\,a\,,\,b\,]$.

> **NOTE:** *Integrating over the interval $[\,a\,,\,b\,]$ (17.8.6a), we may use any one of the three (equivalent) forms,*
>
> **(17.8.6b)** $\quad \displaystyle\int_{x=a}^{x=b} \langle \vec{V}, d\vec{l}\rangle, \ \int_{[a,b]} \langle \vec{V}, d\vec{l}\rangle, \ or \ \int_{a}^{b} \langle \vec{V}, d\vec{l}\rangle.$

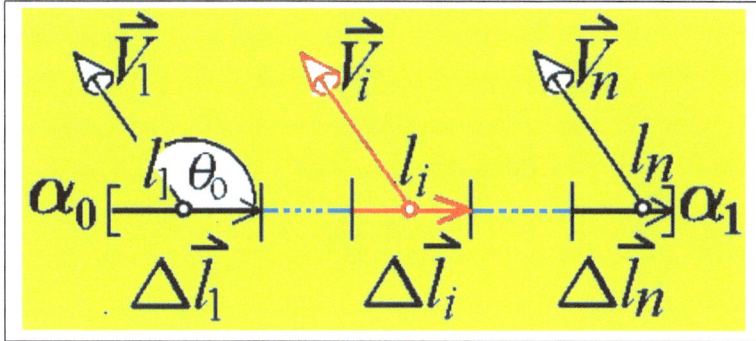

The partition: Following Definition (17.8.5), the interval $[\alpha_0, \alpha_1]$ is partitioned or cut (17.7.1) into n parts that are represented by the rightward-oriented vectors,

(17.8.6c) $\qquad \Delta\vec{l}_1, \ldots, \Delta\vec{l}_i, \ldots, \Delta\vec{l}_n,$

so that

(17.8.6d) $\qquad \displaystyle\sum_{\Delta l_i} \Delta l_i = |\alpha_1 - \alpha_0| > 0. \qquad \begin{cases} \textit{the length of the} \\ \textit{interval } [\alpha_0, \alpha_1] \end{cases}$

‖ **NOTE:** *We use the absolute value notation, $|\alpha_1 - \alpha_0| > 0$, since it may be the case that $\alpha_1 - \alpha_0 < 0$, that is, $\alpha_0 > \alpha_1$.*

‖ **NOTE:** *We use notation Δl_i to represent both the line segment and the positive length of the line segment.*

Now

$$\int_{x=\alpha_0}^{x=\alpha_1} \langle \vec{V}_0, d\vec{l} \rangle = \lim_{\substack{max\, \Delta l_i \\ \to\, 0}} \sum_{\Delta l_i} \langle \vec{V}_0, \Delta\vec{l}_i \rangle \qquad \begin{cases} \textit{from} \\ \textit{Definition (17.8.5a)} \end{cases}$$

$$= \lim_{\substack{max\, \Delta l_i \\ \to\, 0}} \sum_{\Delta l_i} V_0 \cos\theta_0 \Delta l_i \qquad \begin{cases} \textit{from} \\ \textit{Definition (17.6.1a)} \end{cases}$$

$$= V_0 \cos\theta_0 \left[\lim_{\substack{max\, \Delta l_i \\ \to\, 0}} \sum_{\Delta l_i} \Delta l_i \right] \qquad \begin{cases} \textit{factor out the} \\ \textit{constant } V_0 \cos\theta \end{cases}$$

$$= V_0 \cos \theta_0 \quad \lim_{\substack{max\, \Delta l_i \\ \to\, 0}} |\alpha_1 - \alpha_0| \qquad \left\{ from\ (17.8.6d) \right.$$

$$= V_0 \cos \theta_0 |\alpha_1 - \alpha_0| \quad \left\{ \begin{array}{l} for\ constant\ |\alpha_1 - \alpha_0|, \\ \lim |\alpha_1 - \alpha_0| = |\alpha_1 - \alpha_0|. \end{array} \right.$$

Equating the first and last terms in this chain of equalities establishes (17.8.6a). It is important to recall that well-defined integrals require that the result is independent of the particular partition, that is, all partitions must yield the same result. ∎

(17.9) Exercises

(17.9.1) **If X is proportional to numbers Y and Z,** then show that X
<u>Exercise</u>   is proportional to their product, YZ.

Hints:

• **The Defining Equations:** *To say that the variable X is proportional to the variable Y is to say that*

$$X = kY \text{ for some constant } k.$$

For example, if X doubles then Y doubles. If X triples, then so does Y, and so on.

But if there is a third variable, Z, say, then the "constant" k_1 relating X and Y may depend on Z and the constant k_2 relating X and Z may depend on Y. Hence, we denote these "constants" by k_1^Z and k_2^Y, respectively. That is, to say X is proportional to both Y and to Z is to say

(17.9.1a) $X = k_1^Z Y \quad and \quad X = k_2^Y Z.$

where k_1^Z is constant with respect to X and Y, and k_2^Y is constant *with respect to X and Z.*

- **A Proof:** Divide (17.9.1a) through by YZ to obtain

(17.9.1b) $$\frac{X}{YZ} = \underbrace{\frac{k_1^Z}{Z}}_{f(Z)} = \underbrace{\frac{k_2^Y}{Y}}_{g(Y)} .$$

Now the functions $f(Z)$ and $g(Y)$ cannot be equal for all choices of independent variables Z and Y unless they are both equal to the same <u>constant</u> d_0, say. To see this, note that

$$\frac{d}{dZ}f(Z) = \frac{d}{dZ}g(Y) = 0 \Longrightarrow f(Z) = g(Y) = d_0 = a\ constant,$$

which, from (17.9.1b), implies $X = d_0 YZ$. ∎

(17.9.2)
Exercise
Inner Product Equivalence. Show that definitions (A.2.6) and (17.6.1) are equivalent for vectors $\vec{X}$ and $\vec{Y}$ in 2-dimensional space $\mathbf{R}^2$.

(17.9.3)
Exercise
Wheel of Balanced Vectors: (*validating vector decomposition*). The apparatus illustrated will verify that force vectors decompose into their horizontal and vertical components. Hang the weights with a low-friction line that attaches to the center ring, draping the line over the peg to achieve the desired angles. If the forces actually decompose as the mathematical structure indicates, then a stable, balanced center ring occurs when the sum of all forces on it equals zero — that is, when the sum of all *horizontal* components and the sum of all *vertical* components each equals zero.

(17.9.3a) Given Fixed Angles. Find the ratio of the weights w_1, w_2, and w_3 that produce equilibrium, i.e., the sum of the forces on the center ring is zero.

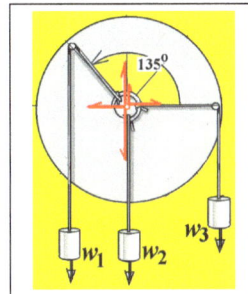

(17.9.3b) The 3-4-5 Triangle.   If the three forces, 30 *lbs.*, 40 *lbs.*, and 50 *lbs.*, decompose according to the geometry of the 3-4-5 triangle, then equilibrium on the central ring is achieved.

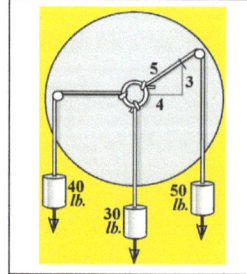

(17.9.3c) Forbidden Angles.   At what angles, if any, in Figure (17.9.3a), is it impossible to achieve equilibrium on the central ring, regardless of the value of the weight w_1? (**Note:**   *All weights w_1, w_2, w_3 are positive.*)

(17.9.4)   **Integration and Electromagnetism.**   Show that

Exercise

(17.9.4a) $\int_A \langle \vec{V}, d\vec{A} \rangle$

is a Riemann integral (17.7.4). **Hint:** *Model your response on the similar argument that follows Figure (17.8.2).*

VIII.   Maxwell's Equations

Chapters (17) - (22) *present the mathematical modeling techniques that Maxwell used in developing his four equations that unified the theories of electricity, optics, and magnetism.*

Section (22.4) *of chapter (22) shows how Maxwell, using just two laboratory measurements, predicted the value of c, the speed of electromagnetic radiation in a vacuum, and also identified light as a particular type of electromagnetic radiation. These predictions have now been verified experimentally to enormous accuracy, though it took decades to develop the technology to provide such accurately measured and convincing evidence.*

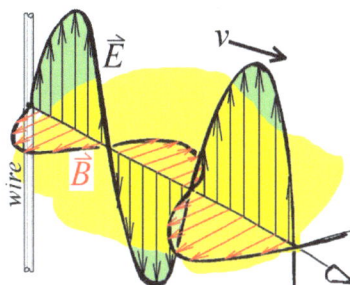

(18) Electric and Magnetic Fields

(18.1) Background

In 1873, the Scottish physicist JAMES CLERK MAXWELL (1831-1879) published *A Treatise on Electricity and Magnetism* in which he extended and unified discoveries of CHARLES-AUGUSTIN DE COULOMB CARL FRIEDRICH GAUSS (1777-1855),　ANDRÉ MARIE AMPÈRE (1775-1836), and MICHAEL FARADAY (1791-1867).

With only four fundamental equations (listed in Section (20.7)), Maxwell was able to integrate the theories of electricity, magnetism, and optics. Moreover, these equations enabled him to make the armchair calculation that the speed of electromagnetic waves in a vacuum was $c = 3 \times 10^8$ meters/second, a number that exactly coincided with the experimentally measured speed of light.

Merely a coincidence? Maxwell thought not.

Maxwell suspected, but could not prove, that light was as much an electromagnetic phenomenon as the magnetic and electric fields he studied in his laboratory — later experiments were to fully confirm his suspicions.

....

The catalog of electromagnetic (EM) waves was extended in 1888 when HEINRICH RUDOLPH HERTZ (1857-1894) (*whose name "hertz," abbreviated as (Hz), became used for radio and electrical frequencies*) showed that oscillating electrical charges produced electromagnetic waves (radio waves). Although Hertz failed to measure the speed of electromagnetic waves, later investigators were to confirm that the waves move at the speed of light c.

But what about light whose speed is also c? Is light a special electromagnetic wave? If oscillating electrical charges produced electromagnetic waves, then which oscillating charges were responsible for the production of light? In 1895, Dutch physicist HENDRIK ANTOON LORENTZ (1853-1928) suggested that the emitted light was generated by oscillating charges within the atoms. In 1896, Lorentz's student, PIETER ZEEMAN (1865-1943), confirmed this hypothesis. At this point, light became a valid entry in the catalog of electromagnetic waves. (See also [8], pp. 267-269.)

This chapter will describe ideas of electricity and magnetism that eventually led to Maxwell's marriage of observation and mathematics that united the theory of electricity, magnetism, and optics.

(18.2) Electric Forces: Coulomb's Law

Electrically charged particles have either a plus or a minus charge. An electron, a negatively charged particle, that is placed between a plus and a minus charged particle will experience a force toward the positive particle. (*The positive/negative terminology for electrical charges is due to* BENJAMIN FRANKLIN *(1706-1790).*)

(18.2.1) NOTE: *By convention, we use the single symbol, q, to represent both the physical particle and the charge on that particle.*

(18.2.2)

> DEFINITION: *The collection of all forces in a region experienced by an electron in that region is the* FORCE FIELD *created by a set of charged particles of the region. The paths within the region that the electrons take as they yield to the attractive force of the positive particle(s) are called the* LINES OF FORCE.

(18.2.3) FIGURE: *Rules for Drawing Electric Field Lines*

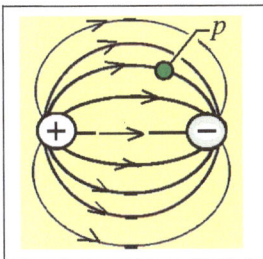

(1) *Lines run from + charges to − charges whenever there are ± pairs of electrically charged particles.*
(2) *Lines never cross.*
(3) *The number of lines emanating from any particle is proportional to the charge of that particle.*
(4) *A charged particle positioned on a field line experiences a force that is tangent to that field line. Two charges with identical signs repel each other, and two particles of opposite sign attract each other.*

A graphical representation. Figure (18.2.3) shows two charged particles with opposite charges. By convention, lines of force (within the force field) always emanate from positively charged particles. An analytical description of the lines of force will follow.

(18.2.4) FIGURE: *Forces Between Charges: Coulomb's law*

T*wo electrically charged particles — q_0 at point p_0 in space and q_1 — are r units apart. Experimental measurements confirm that a test particle q_0 (being held) exerts a resisting force, $\vec{F}_{0,1}$, whose magnitude $F_{0,1}$ is proportional to the charge, q_0, the charge q_1, and to $1/r^2$, the inverse square of the distance r. These proportionality relationships, as per Theorem (17.2.2), are expressed mathematically as*

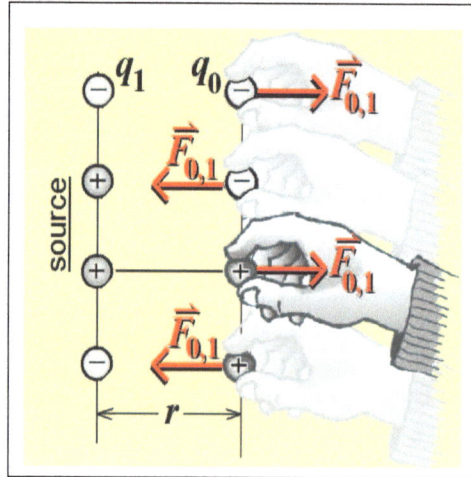

(18.2.4a) $F_{0,1} = k_1 q_0, \quad F_{0,1} = k_2 q_1, \quad$ and $\quad F_{0,1} = k_3(1/r^2)$
for certain proportionality constants k_1, k_2, k_3.

Since (18.2.4a) tells us that $F_{0,1}$ is proportional to q_0, q_1 and $(1/r^2)$, Theorem (17.2.2) will guarantee that $F_{0,1}$ is proportional to their product, i.e., $F_{0,1} = k_0(q_0)(q_1)(1/r^2)$ for a certain proportionality constant k_0. This is COULOMB'S LAW   which has form

(18.2.4b) $$\vec{F}_{0,1} = \underbrace{\left(\frac{1}{4\pi\varepsilon_0}\right)}_{k_0} \frac{q_0\, q_1}{r^2}\, \vec{e}_{0,1}$$

where $k_0 = \left(\dfrac{1}{4\pi\varepsilon_0}\right)$ is the constant of proportionality and

(18.2.4c) $\vec{e}_{0,1} = \dfrac{\vec{r}_1 - \vec{r}_0}{\|\vec{r}_1 - \vec{r}_0\|}$ *is a unit vector ($\|\vec{e}_{0,1}\| = 1$) pointing* *from q_0 to q_1 (see (18.2.4)).*

From (A.2.13a), pg. 339, the length of the vector $\vec{F}_{0,1}$ of (18.2.4b) is

$$(\textbf{18.2.4d}) \quad F_{0,1} = \| \vec{F}_{0,1} \| = \left| \frac{1}{4\pi\varepsilon_0} \frac{q_0\, q_1}{r^2} \right| \underbrace{\| \vec{e}_{0,1} \|}_{1} = \frac{1}{4\pi\varepsilon_0} \frac{q_0\, q_1}{r^2}.$$

(18.2.5) Permittivity. If we place known and measurable quantities, q_0, q_1, π, r, and $F_{0,1}$, on one side of the equation (18.2.4d), we are left with the single "unknown" quantity ε_0, the electrical PERMITTIVITY, which computes to be

$$(\textbf{18.2.5a}) \qquad \varepsilon_0 \overset{def}{=} \frac{1}{4\pi} \frac{q_0\, q_1}{r^2\, F_{0,1}}.$$

(18.2.6) The sign of the product, $q_0\, q_1$, in (18.2.4b) will algebraically point the unit vector $\vec{e}_{0,1}$ in the correct direction: That is, $q_0 q_1 < 0$ implies that the vector $-\vec{e}_{0,1}$ points *toward* the source particle (*particles with different-signed charges will attract*) while $q_0 q_1 > 0$ implies $+\vec{e}_{0,1}$ points *away* from the source particle (*particles with same-signed charges will repel*).

(18.2.7) Permittivity, ε_0 (with the 0 subscript), depends on the *vacuum* in which the force $\vec{F}_{0,1}$ of (18.2.4b) is measured. Measurements of $\vec{F}_{0,1}$ taken in other environments — nitrogen, air, water, etc. produce their own permittivity constants, ε.

The units of permittivity, ε_0, are

$$(\textbf{18.2.7a}) \qquad \frac{q^2}{l^2 \times F} \approx \frac{\text{charge}^2}{\text{length}^2 \times \text{Force}}.$$

What Are the Specific Units for (18.2.7a)? At this point it is not necessary to state a particular system of units since the relation is valid for all unit systems. However, when we later calculate the speed of light in (22.4.3), we will use MKS units, namely,

    (1) charge $q = C$ (Coulombs),
    (2) length $l = \text{mtr}$ (meters),
    (3) force $F = N$ (Newtons).

(18.3) Electric Fields

(18.3.1) **Multiple Source Charges.** Coulomb's law (18.2.4b) gives us the force between two particles q_1 and q_0. But what force is produced by n source charges $q_1, q_2, \cdots, q_n$ on a single "target" or test charge q_0 that is fixed at position p_0 (*see Figure* (18.3.4)) ?

The *experimental* result is that for n charges, the force on charge q_0 at position p_0 is the *vector sum* (A.2.4b) of the electric forces generated by the source charges q_i, each acting independently of the other charges — as if the other charges were not present. This experimentally confirmed vector-sum behavior does not follow automatically from Coulomb's law (18.2.4b).

(18.3.2) **Question:** How can we measure the total force exerted by several source charges $q_1, q_2, \ldots, q_n$ on a single test charge q_0?

Answer: Since each force $\vec{F}_{0,i}$ (18.2.4b) between q_i, $i = 1, 2, \ldots, n$, and the test particle q_0 is proportional to both charges q_i and q_0, we normalize the situation by setting the test particle charge $q_0 = 1$. This normalization ensures that the force computed at point p_0 does not depend on q_0, which is always equal to 1. Equivalently, we can always divide equation (18.2.4b) through by the value of the target test charge q_0 which then yields the force per unit charge due to a single source charge q_1, namely,

(18.3.3)

> DEFINITION: *A* TEST CHARGE q_0 *is located at a point* p_0. *The force-per-unit-charge, or the* ELECTRIC FIELD VECTOR *at a point* p_0 *due to a single charge* q_1 *that lies* r *units from* p_0 *is the vector*
>
> (18.3.3a) $$ \vec{E}_{0,1} \overset{def}{=} \frac{\vec{F}_{0,1}}{q_0} = \frac{q_1}{(4\varepsilon_0 \pi r^2)} \vec{e}_{0,1} $$
>
> *where the direction of the unit vector* $\vec{e}_{0,1}$ *is determined by* q_0 *and* q_1 *in accordance with* (18.2.6).

We now extend the quotient of (18.3.3a), which is defined for a single source charge q_1, to several test charges $q_1, q_2, \ldots, q_n$.

(18.3.4) FIGURE: **Electric Field Vector due to Many Charges**

The Setup: Given n charged (source) particles,

$$q_1, q_2, \ldots, q_i, \ldots, q_n,$$

with respective distance vectors,

$$\vec{r}_1, \vec{r}_2, \ldots, \vec{r}_i, \ldots, \vec{r}_n$$

emanating from a charged test particle q_0 situated in space at point p_0. Note that $r_i = \|\vec{r}_i\|$ as declared in (A.2.13a).

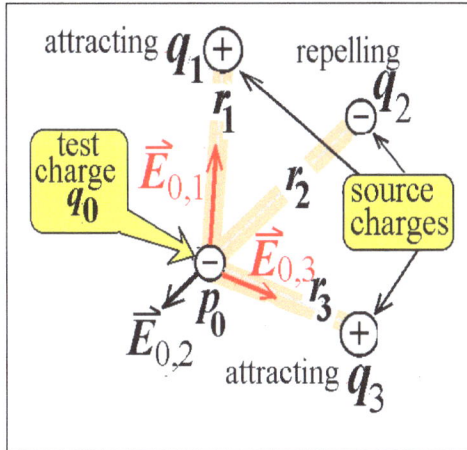

Definition (18.3.3) describes the electric field vector $\vec{E}_{0,1}$ at point p_0 due to a single source charge q_1 that lies r_1 units distant. Definition (18.3.5) will describe the "composite" electric field vector $\vec{E}_0$ at point p_0 due to several source charges, $q_i, i = 1, 2, \ldots, n$, where the source charge q_i lies r_i units from point p_0.

(18.3.5)

> **DEFINITION:** *The* ELECTRIC FIELD VECTOR, $\vec{E}_0$ *at a point* p_0 *due to* n *source particles* $\{q_i : i = 1, 2, \ldots, n\}$ *is the vector sum (A.2.4b) of all (force-per-unit-charge) field vectors* $\vec{E}_{0,i}$ *(18.3.3a) generated by* all *the source charges in the region, namely,*
>
> **(18.3.5a)** $\quad \vec{E}_0 \stackrel{def}{=} \sum_{i=1}^{n} \vec{E}_{0,i} = \sum_{i=1}^{n} \frac{\vec{F}_{0,i}}{q_0} = \frac{1}{4\pi\varepsilon_0} \sum_{i=1}^{n} \frac{q_i}{r_i^2} \vec{e}_{0,i}.$
>
> *An* ELECTRIC FIELD $\vec{E}$ *(no subscript) is the set of all electric field vectors of a region.*

> NOTE: The construction of electric field vectors in (18.3.5a) is but one example in nature where mathematical vector addition (A.2.4b) applies.

(18.4) Magnetic Fields

Just as we have an electric field $\vec{E}$ (18.3.5) of vectors in 3-space that exert forces on electrically charged particles, there are magnetic fields $\vec{B}$, represented by vectors in 3-space, that exert forces on *moving* charged particles.

Magnetic fields are confirmed by their effects on moving charges and on other magnets. By using a compass needle as the "other magnet," we can see the effects of a magnetic field.

(18.4.1) In sum, an <u>electric</u> field $\vec{E}$ is a vector field that results from nearby charged particles — a MAGNETIC FIELD $\vec{B}$ is a vector field created by moving charged particles or a permanent magnet.

(18.4.2) FIGURE: *Magnetic Field Lines*

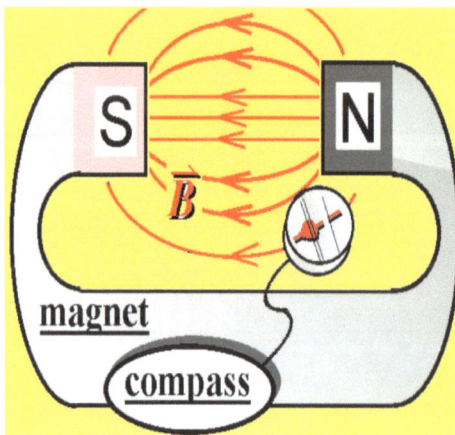

A *magnet, with ends marked "S" (South) and "N" (North) is shown with invisible lines, $\vec{B}$, of magnetic force running from "N" to "S". The magnetic field lines are revealed when the needle of a (south-seeking) compass is observed to follow the lines as indicated in the diagram. The "N" and "S" notations are arbitrary but universally accepted.*

(18.4.3) The following rules for drawing magnetic field lines are similar to the rules (18.2.3) for drawing electric field lines:

(1) Lines run from North to South poles.

(2) Lines never cross.

(3) The number of lines emanating from any North pole, or ending at any South pole, is proportional to the strength of that pole.

(4) A magnetic needle experiences a force tangent to the field line. Since like poles repel and opposite poles attract, the North pole of the needle is attracted to an external South pole.

(18.4.4) FIGURE: (*No Magnetic Monopoles Exist.*)

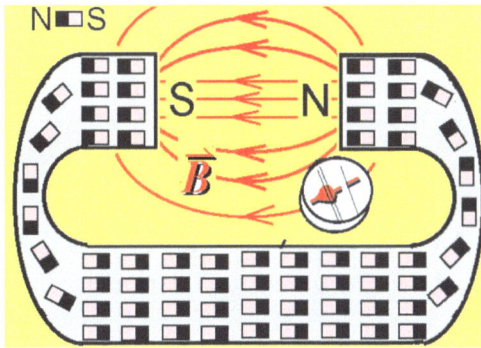

An object is a magnetic MONOPOLE if it exhibits the properties of either a South pole or a North pole, but not both. However, no magnetic monopoles have yet been observed. No matter how small a piece is taken from a magnet, there is always both a North end and a South end. Unlike positive and negative electric charges that can exist in isolation from each other, magnetic materials can *not* be decomposed so as to isolate a North pole from a South pole. As Figure (18.4.4) indicates, the magnet is composed of atomic "micro" magnets which, due to their alignment, produce a "macro" magnet.

NOTE: *To reveal the direction of the magnetic field, we use a south-seeking compass, as shown in Figure (18.4.2). A standard north-seeking compass becomes south-seeking after the arrowhead of the needle is transferred to the opposite end.*

(18.5) Magnetic Forces: Lorentz Forces

(18.5.1) How Nature requires the cross product (17.6.3). We know from Coulomb's law (18.2.4) that forces are exerted on charged particles in an *electric* field, whether they are in motion or not. In contrast to this, forces are exerted on charged particles in a *magnetic* field only when they are in motion.

Moreover, the particle motion and the resulting force must both be against the "grain" or lines of the magnetic field. That is, experiments show that the force on a moving charge is perpendicular to both the velocity $\vec{v}$ and the $\vec{B}$ field which provides an example of the cross product (17.6.3) in Nature.

Here are the details.

IN THE WORLD OF MECHANICS: If you jump into a stream whose current runs right to left, then you would feel a force from the right pushing you leftward (downstream).

IN AN ELECTRIC FIELD: If the "stream" is an electric field running right (+) to left (−), and you are a positively charged particle, then, similar to the mechanical condition above, you would feel a "downstream" force pushing you along the field lines in the (+)-to-(−) direction.

HOWEVER, IF YOU ARE A CHARGED PARTICLE IN A MAGNETIC FIELD or stream, and you run (motion is necessary) across the magnetic field lines which run right (N) to left (S), then you would not be pushed leftward, or downstream. Instead, consistent with the geometry of the cross-product, you would be pushed upward, toward the heavens. (*You must be positively charged to experience an upward thrust. A negatively charged particle is destined to be pushed downward, toward — well — the other place.*)

$$\|\vec{v} \times \vec{B}\| = v \sin\theta\, B$$

These ideas are codified in (18.5.2a) of the following discussion:

(18.5.2) F<small>IGURE</small>: *Magnetic Force on a Moving Charge*
Trig

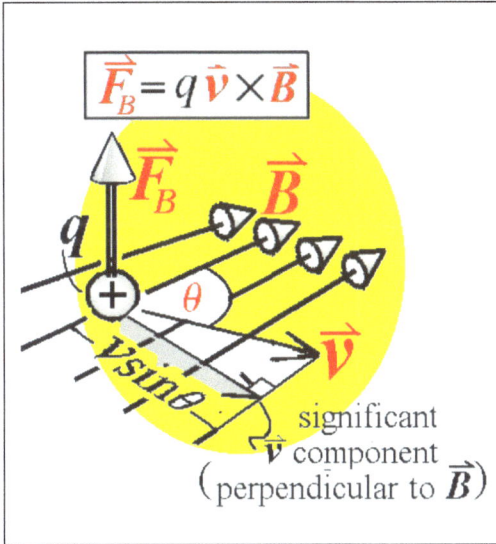

$$\boxed{\vec{F}_B = q\,\hat{v} \times \vec{B}}$$

significant
$\vec{v}$ component
$\left(\text{perpendicular to } \vec{B}\right)$

A particle with charge q travels with velocity $\vec{v}$ at an angle θ with respect to the magnetic field lines. Only the component of $\vec{v}$ that is perpendicular to the magnetic field lines contributes to the induced force, $\vec{F}_B$, whose direction is determined by the cross product, $\vec{v} \times \vec{B}$. The length of the perpendicular part of $\vec{v}$ is $v \cdot \sin\theta$ (17.6.3) (*Figure at left*).

In sum, the magnetic force $\vec{F}_B$ induced by a particle with charge q and velocity $\vec{v}$, passing through a magnetic field line $\vec{B}$ is

(18.5.2a) $\qquad \vec{F}_B = q\,\vec{v} \times \vec{B} \qquad$ so that, as per (17.6.3),

(18.5.2b) $\qquad \|\vec{F}_B\| = \|q\,\vec{v} \times \vec{B}\| = q \cdot v \cdot \sin\theta \cdot B$

where $v = \|\vec{v}\|$, $B = \|\vec{B}\|$ (17.3.4) and particle with charge q travels at an angle θ with respect to the magnetic field line $\vec{B}$.

. . . .

As a consequence of (18.5.2a), an electrically charged particle moving through a magnetic field $\vec{B}$ experiences no force if its path is parallel to the field lines of Figure (18.4.2). In this case $\theta = 0$ in Figure (18.5.2) so that using $\sin\theta = 0$ in (18.5.2b), we see that force $\vec{F}_B = 0$. Similarly, the force $\vec{F}_B$ is *maximal* when the path is perpendicular to the field lines, i.e., $\theta = 90^o$ *implies* $\sin\theta = 1$ in (18.5.2b).

. . . .

> NOTE: *In contrast to the magnetic force on a charged particle*
> *(18.5.2a), which depends on the speed of the particle, the electrical*
> *force on charged particles, as given by Coulomb's law, (18.2.4b),*
> *is not affected by particle speeds at all. For two particles, the elec-*
> *trical force between them at any instant of time depends only on*
> *their charges, q_0 and q_1, and the instantaneous distance r between*
> *them.*
>
> *The force between charged particles at a distance r changes in time*
> *when the distance r changes. This raises the question of how fast*
> *(or instantaneous) the signal must be between particles to effect*
> *the force change.*

(18.5.3) The Lorentz Force. Forces on a charged particle are due to (1) *electric fields* generated by neighboring charged particles (Coulomb's law (18.2.4b)) and (2) *magnetic fields* due to the speed of the particle. Since both electric and magnetic fields can exist simultaneously, we have the LORENTZ LAW that combines (1) and (2) as the vector sum (A.2.4b):

(18.5.3a) $$\vec{F} = q\vec{v} \times \vec{B} + q\vec{E},$$

where q is the charge moving through the electric field $\vec{E}$ and the magnetic field $\vec{B}$ with (vector) velocity $\vec{v}$.

> NOTE: *The coexistence of both electric and magnetic fields allowed*
> *for Thomson's discovery of the electron. See Section (18.6).*

In the following section, we shall see how J. J. Thomson, inferring from effects alone, manages to "see", and discover, the *electron*.

(18.6) How Thomson Discovers the Electron

The interplay of the forces induced by the electric field (18.2.4b) and the magnetic field (18.5.2a) led to the discovery of the electron. Here is how it happened.

In 1897, J. J. THOMSON (1856-1940) conducted an experiment at Cambridge University that involved an evacuated cathode ray tube (CRT), much like the picture tube in a television receiver. At one end of the tube, a hot filament emitted charged particles, later determined by Thomson to be electrons. The stream of particles was channeled through a slit to form a narrow beam that collided with and illuminated a spot on the phosphorescent screen at the other end.

Here follows a description, with schematics, of the crossed-fields apparatus used in Thomson's experiment:

(18.6.1) F<small>IGURE</small>: ***Electric and Magnetic Forces on an Electron***

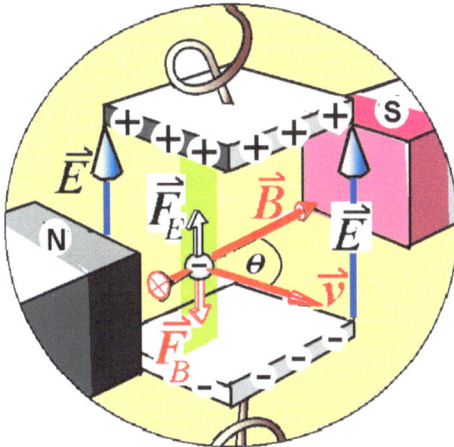

- A magnetic field, $\vec{B}$, runs horizontally ($\rightarrow$) from the N pole to the S pole.

- An electric field, $\vec{E}$, runs vertically ($\uparrow$) from the negative ($-$) plate to the positive ($+$) plate.

- A negatively charged electron, whose path forms an angle θ with the magnetic field, travels horizontally, at speed v. The angle θ is adjusted by rotating the magnet. In this way, the

magnetic force on particles with charge q varies with θ according to the formula, $\vec{F}_B = q\vec{v} \times \vec{B} = q \cdot v \sin\theta \cdot B$ (18.5.2a). Due to the Right Hand Rule (18.5.2) for the magnetic field, the force $\vec{F}_B$ will direct *negatively charged* electrons downward. The electric field $\vec{E}$ on the other hand, drives each electron upwards with force $\vec{F}_E$.

. . . .

Figure (18.6.2) shows how the trajectory of the electron beam, whose particles have speed v, is affected by the magnetic and electric fields separately.

(18.6.2) Fɪɢᴜʀᴇ: *Beam Deflection in Electric and Magnetic Fields*

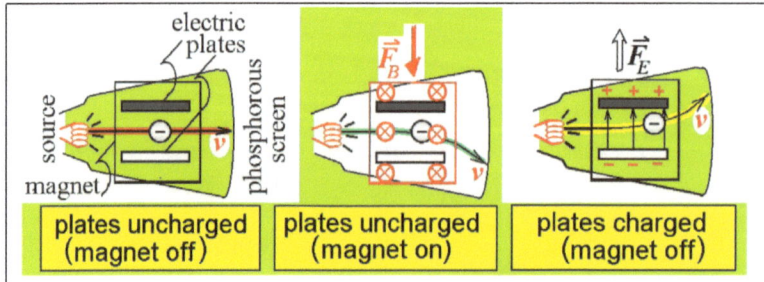

- Lᴇғᴛ Pᴀɴᴇʟ: Both the electric and magnetic fields are off. The beam travels horizontally to its reference point on the screen.

- Mɪᴅᴅʟᴇ Pᴀɴᴇʟ: The electric field is off while the magnetic field, at an angle θ to the path of the beam, is turned on. As in Figure (18.6.1), the "x"s indicate the tail of the magnetic $\vec{B}$ arrow — the magnetic field is directed into the page. From (18.5.2a), the magnetic force $\vec{F}_B$ on the *negatively* charged particle is downward with magnitude $q \cdot v \cdot \sin\theta \cdot B$.

- Rɪɢʜᴛ Pᴀɴᴇʟ: The magnetic field is off and the electric field is turned on. The beam is forced upward with force $\vec{F}_E$ due to the attractive force of the positive plate at the top and the repelling force of the negative plate at the bottom.

(18.6.3) **Thomson used the apparatus as follows:** With the two fields turned off (left panel), the position of the undeflected beam on the screen is marked. Once the electric field $\vec{E}$ is turned on (right panel), the position of the deflected beam is measured. Then, the magnetic field is turned on and adjusted until the magnetic force equals that of the electric field and returns the beam to its original, undeflected, position. From the geometry of the cathode ray tube, the deflection distance and the combined forces of the fields $\vec{E}$ and $\vec{B}$, as governed by the <u>Lorentz equation</u> (18.5.3), Thomson calculated

(18.6.3a) $v = E/B$,   the *speed v* of the negative charges, and

m/q,   the *ratio* of mass to electric charge.

The low value of m/q meant that each negatively charged particle, now called an *electron*, had mass much less than $1/10$ of one percent of the mass of the smallest known atom, hydrogen. The existence of such a tiny mass meant that the atom (which means "indivisible") was not indivisible at all. In fact, Thomson asserted, these ultra-light particles, must be common to, and part of, every atom of every element.

These inferences of Thomson, made in 1897, mark the discovery of the (apparently) invisible electron. Thomson, following the principles of the scientific method (**??**), embraced its reality through its effects.

Thomson received the Nobel prize for physics in 1937.

(19) Electricity and Magnetism: Gauss' Laws

(19.1) Flux of Vector Fields

We consider a vector field $\vec{E}$ (17.5.3), pg. 227, which is composed of field vectors including $\vec{E}_i$, $i = 1, 2, \ldots, k$. For example, the individual field vectors $\vec{E}_i$ might be water speeds, electric field vectors, magnetic field vectors, or gravitational field vectors.

As per (17.3.3), we can represent a small element with area ΔA as a vector $\Delta \vec{A}$ whose length is ΔA. That is, $\|\Delta \vec{A}\| = \Delta A$.

(19.1.1) FIGURE: **Passing Through a Surface Element.** Our objective is to measure the flow of "*E*-stuff" or so-called *flux* passing through a surface. To do this, we only measure the perpendicular component of $\vec{E}_i$ which has length $\|\vec{E}_i \cos(\theta)\|$ where θ is the angle between vectors $\vec{E}_i$ and ΔA_i.

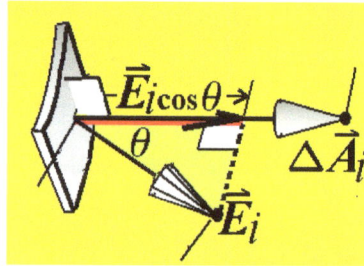

Flow along the surface cannot be said to *pass through* the surface.

(19.1.2) EXAMPLE: *Varying Area and Constant Flux of a Water Stream*

A 3-dimensional velocity vector field $\vec{E}$ describing the flux (*amount of water volume per unit time that flows in the direction of flow*) of incompressible water flows through a hose

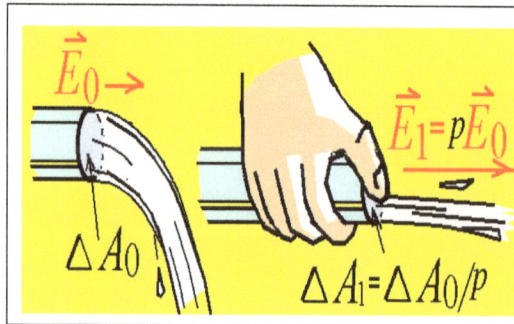

with cross-sectional area $\Delta A_0 = 12$ cm^2. The average velocity vector $\vec{E}_0$ runs perpendicularly to the cross-sectional area with magnitude $\|\vec{E}_0\| = 8$ cm/sec. The flux defined in (19.1.3) gives us the rate of volume flow through ΔA_0, namely,

(19.1.2a) $\quad \Phi_E(\Delta A_0) = <\vec{E}_0, \Delta \vec{A}_0>$

$$= \|\vec{E}_0\| \cos(0) \|\Delta \vec{A}_0\| = 8\frac{cm}{sec} \times 1 \times 12cm^2 = 96\frac{cm^3}{sec}.$$

Therefore, as shown in Figure (19.1.2), when the exit area ΔA_0 is *decreased* by a factor of p, as when a thumb is held over the hose end, then the speed of flow, $\vec{E}_0$, *increases* to $\vec{E}_1 = p\vec{E}_0$. Conversely, if the exit area *increases* by a factor p, then the water speed *decreases* (is divided by) the factor p.

(19.1.3)

DEFINITION OF FLUX: *Given vector field $\vec{E}$ (17.5.3) containing a surface area A that is partitioned into k small disjoint surface elements $\Delta A_i, i = 1, 2, \ldots, k$. Sub-area ΔA_i is tiny enough so that an individual field vector $\vec{E}_i$ passing through it can be considered to be constant. Then the <u>APPROXIMATE FLUX</u> <u>through the i-th area element ΔA_i</u> is the inner product (A.2.6a)*

(19.1.3a) $\quad \Phi_E(\Delta A_i) \approx <\vec{E}_i, \Delta \vec{A}_i> = \|\vec{E}_i\| \cos\theta \|\Delta \vec{A}_i\|$

where $\|\vec{E}_i\|$ and $\|\Delta \vec{A}_i\|$ are the lengths of vectors $\vec{E}_i$ and $\Delta \vec{A}_i$. The <u>APPROXIMATE FLUX</u> <u>through</u> <u>the</u> <u>entire</u> <u>area</u> <u>A</u> is the sum of the inner products

(19.1.3b) $\quad \Phi_E(A) \approx \sum_{i=1}^{k} <\vec{E}_i, \Delta \vec{A}_i> = \sum_{i=1}^{k} \|\vec{E}_i\| \cos\theta \|\Delta \vec{A}_i\|.$

Through the limit process, we define the <u>TOTAL FLUX</u> through the area A to be

(19.1.3c) $\quad \Phi_E(A) \overset{\text{def}}{=} \int_A <\vec{E}, d\vec{A}>$

$$= \lim_{max\|\Delta A_i\| \to 0} \left[\sum_{\Delta A_i} \|\vec{E}_i\| \cos\theta \|\Delta \vec{A}_i\| \right].$$

(19.1.4) The following example (19.1.5) illustrates a field $\vec{E}$ constrained in a "tubular" region in which flux is constant through each cross-section. This constancy of flux prevails with fields of incompressible fluids, magnetic fields, and electric fields. We may think of our flowing substance acting like incompressible water flowing through a volume.

(19.1.5) Fɪɢᴜʀᴇ: ***Constant Flux Through Parallel Cross-Sections***

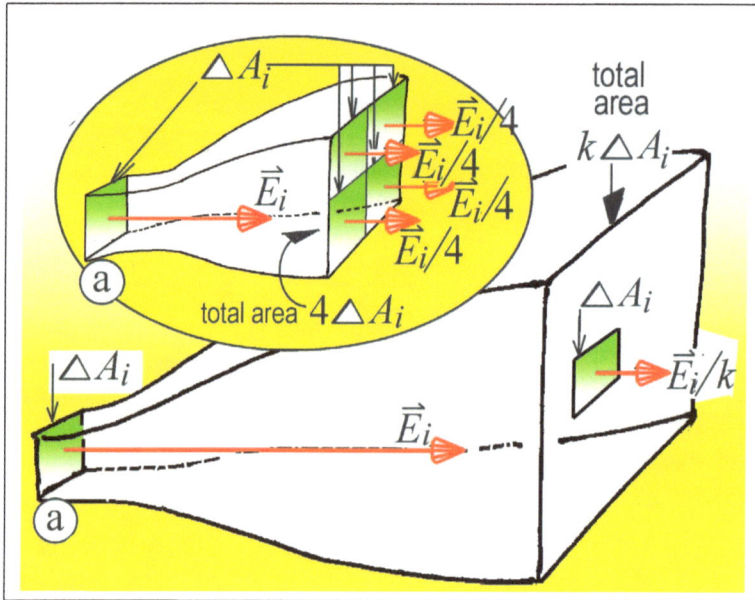

Over a sample area ΔA at ⓐ, field E has field (flow) vector with average value $\vec{E}_i$. Hence, from (19.1.3a), the flux at the entry area cross-section ⓐ is

(19.1.5a) $\Phi_E(\Delta A_i) = \|\vec{E}_i\| \cos 0° \|\Delta \vec{A}\| = \|\vec{E}_i\| \cdot 1 \cdot \Delta A.$

Small Inset Figure (19.1.5).   Since the exit area cross-section (*to the right of the entry area* ⓐ) is the sum of four areas

$$\Delta A_i + \Delta A_i + \Delta A_i + \Delta A_i = 4\Delta A_i,$$

the flow vector $\vec{E}_i$ must distribute itself over the four pieces of this larger cross-sectional area. (*Recall from (19.1.4) we use water flow as our model and hence assume there is no source or sink inside the volume and that the flowing substance is incompressible.*)

This means that each of the four areas of size ΔA_i accepts only $(1/4)$th of the flow, or $\frac{1}{4}\vec{E}_i$. The flux at each quarter of the exit cross-sectional area is $\frac{1}{4}\|\vec{E}_i\|\Delta A$ which implies that the total flux over all four regions is the 4-term sum

(19.1.5b)

$$\underbrace{\frac{1}{4}\|\vec{E}_i\|\Delta A_i) + \frac{1}{4}\|\vec{E}_i\|\Delta A_i) + \frac{1}{4}\|\vec{E}_i\|\Delta A_i + \frac{1}{4}\|\vec{E}_i\|\Delta A_i}_{Total\ Flux = \sum_{1=1}^{4}\frac{1}{4}\|\vec{E}_i\|\Delta A_i} = \|\vec{E}_i\|\Delta A_i.$$

Larger Diagram of Figure (19.1.5). Since the exit area cross-section (*to the right of the entry area* ⓐ) is $k\Delta A_i$, the flow vector $\vec{E}_i$ must distribute itself over the k pieces of this larger area. This means that each of the k areas of size ΔA_i accepts only $(1/k)$th of the flow, or $(1/k)\vec{E}_i$. The flux at each smaller piece of the exit area is $(1/k)(\vec{E}_i) \times \Delta A$ which implies that the total flux over all k regions is the k-term sum

(19.1.5c)

$$\underbrace{\frac{1}{k}\|\vec{E}_i\|\Delta A_i + \frac{1}{k}\|\vec{E}_i\|\Delta A_i + \ldots + \frac{1}{k}\|\vec{E}_i\|\Delta A_i}_{Total\ Flux = \sum_{1=1}^{k}\frac{1}{k}\|\vec{E}_i\|\Delta A_i} = \|\vec{E}_i\|\Delta A_i$$

(19.2) Electric and Magnetic Flux

As per Definition (19.1.3), electrical "throughput," or *flux* is defined for an arbitrary surface area, A as follows:

(19.2.1)
Calc

> **DEFINITION:** *Given a surface area A in space that lies within an electric field $\vec{E}$. Then the integral*
>
> **(19.2.1a)** $\qquad \Phi_E(A) = \displaystyle\int_A \langle \vec{E}, d\vec{A} \rangle$
>
> *is called the* ELECTRIC FLUX *through the area A.*

> NOTE: *If the integral (19.2.1a) is taken over a closed surface area A (like the "potato surface" shown in the Figure of (17.4.7), pg. 225), then the ornamented integral symbol, $\oint_A$, is used in place of $\int_A$.*

> NOTE: *the integral (19.2.1a) has been defined in (17.8.5b) of Definition (17.8.5).*

Similar to Definition (19.2.1), magnetic flux will be defined by (19.2.2) following.

(19.2.2)
Calculus
needed

> DEFINITION: *Given a surface area A in space that lies within a magnetic field, $\vec{B}$. Then the integral*
>
> **(19.2.2a)** $\qquad \Phi_B(A) = \int_A \langle \vec{B}, d\vec{A} \rangle$
>
> *is called the* MAGNETIC FLUX *through the area A.*

(19.3) Gauss' Law for Electricity

At all points of the surface of a sphere with radius r, $\| \vec{E}_r \| = E_r$ (18.3.5) is constant. This gives us an easy proof of Gauss' law of electricity of the special case where a single source charge q_0 is enclosed by an r-sphere.

(19.3.1)
Calculus
needed

> THEOREM: (*Gauss' Law for electricity.*) *Given a closed surface, A, which encloses charged particles whose total charge is q. Then the total enclosed electric charge q_{enc} can be determined from the following form of the electrical flux $\Phi_E(A)$:*
>
> **(19.3.1a)** $\qquad \Phi_E(A) = \oint_A \langle \vec{E}, d\vec{A} \rangle = \dfrac{q_{enc}}{\varepsilon_0}.$

Proof: We only prove the result for the simplest case where A is the closed surface of a sphere of radius r_0 and there is only one charged particle, $q_{enc} = q_0$, at its center. As per (18.3.3a), electric field vectors $\vec{E}_{r,0}$ point radially outward, parallel to the area vector $\Delta\vec{A}$. The electric flux $\Phi_E(A)$ (19.2.1) through this r_0-sphere has the form

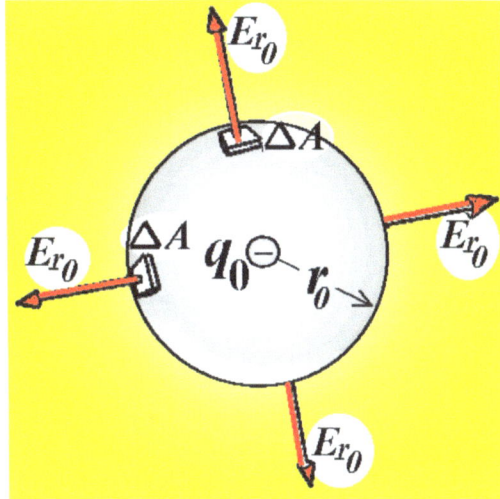

$$\oint_A \langle \vec{E}_{r_0}, d\vec{A} \rangle = \lim_{\max \Delta A_i \to 0} \left[\sum_{\Delta A_i} \langle \vec{E}_{r_0}, \Delta\vec{A}_i \rangle \right] \qquad \text{from (17.8.5b),}$$

$$= \lim_{\max \Delta A_i \to 0} \left[\sum_{\Delta A_i} E_{r_0}(\cos 0^{\mathrm{o}}) \Delta A_i \right] \qquad \text{from (17.6.1)}$$

$$= E_{r_0} \cdot \left[\lim_{\max \Delta A_i \to 0} \sum_{\Delta A_i} \Delta A_i \right] \qquad \begin{cases} constant\ E_{r_0}\ factors \\ out\ and\ \cos(0^{\mathrm{o}}) = 1 \end{cases}$$

$$= \frac{q_0}{4\pi\varepsilon_0 r_0^2} \cdot \left[4\pi r_0^2 \right] = \frac{q_0}{\varepsilon_0}. \qquad \begin{cases} from\ (18.3.5a)\ with\ n = 1 \\ and\ \lim \sum \Delta A_i = 4\pi r_0^2 \end{cases}$$

This chain of equalities establishes (19.3.1a) and the theorem is proved for the special case where the closed area A is the surface of the r_0-sphere. ∎

. . . .

(19.3.2) Note: **Coulomb ⇒ Gauss** *Coulomb's Law* (18.2.4b) *implies Gauss' Law of electricity, Theorem* (19.3.1), *since the last equality of the proof above depends on* (18.3.5a) *which depends on Coulomb's Law. Exercise* (19.5.2) *shows the converse,* **Gauss ⇒ Coulomb.**

(19.3.3) $\|$ NOTE: *Theorem* (19.3.1) *states that Equation* (19.3.1a) — *Gauss' law for electricity — holds true even when the closed surface, A, does not define a sphere, and there may be several charged particles at arbitrary points within the closed surface. Exercise* (19.5.5) *takes you through the computation when the closed surface defines a right circular cylinder and there is a single charged particle at* an arbitrary *point on the central axis.*

(19.4) Gauss' Law for Magnetism

(19.4.1) POSTULATE: (*Gauss' Law for magnetism*). *Given a closed surface*
Calc   *A that encloses a magnetic source. Then the magnetic flux* $\Phi_B(A)$ *is zero. That is,*

(19.4.1a) $$\Phi_B(A) = \oint_A \langle \vec{B}, d\vec{A} \rangle = 0 \, .$$

Empirical justification: This postulate, which has never been experimentally contradicted, is equivalent to saying there are no monopoles (*see Figure* (18.4.4)). That is, as the figure on the right shows, any magnetic source comes equipped with both a North pole and a South pole. Hence, every magnetic field line $\vec{B}$ gives a *positive* reading as it exits the surface

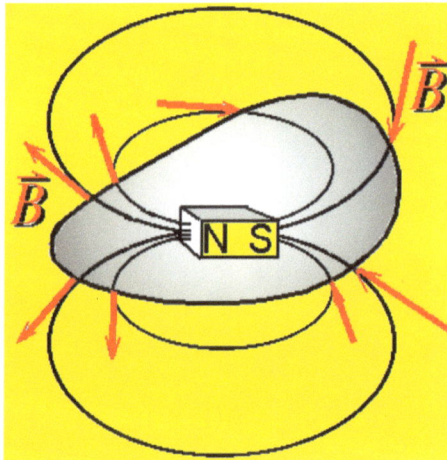

and produces a *negative* reading on its return when it re-enters the surface. The positive and negative orientations are a result of Definition (17.4.7) for surface orientation. ∎

(19.5) **Exercises**

(19.5.1) **Surface Areas of Maxwell Cones**

Exercise

Given the cone of length r with rectangular top with dimensions Δx and Δy.

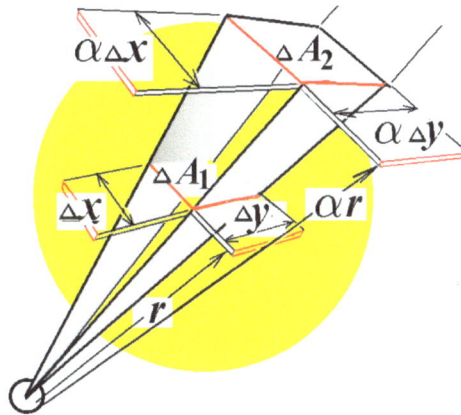

....

- Show that the extension of the cone to a length αr produces rectangular top dimensions $\alpha \Delta x$ and $\alpha \Delta y$.

....

- Conclude from this that the area,

$$\Delta A = \Delta x \Delta y,$$

of the rectangular top is proportional to the square of length r. (See (17.2.1) for definition of proportionality.)

(19.5.2) **Gauss $\Rightarrow$ Coulomb.** Show that Gauss' law (19.3.1) implies

Exercise Coulomb's law (18.2.4b). We have already shown the converse, **Coulomb $\Rightarrow$ Gauss** (*see* (19.3.2)).

Hint: *Gauss' Law* $\displaystyle\oint_A < \vec{E}, d\vec{A} >= \frac{q_{enc}}{\varepsilon_0}$.

Since $4\pi r^2$ is the surface area A of a sphere of radius r, and the charge q_{enc} is at its center, physical symmetry requires that $\|\vec{E}\|$, the length of $\vec{E}$, is constant at each point of the sphere's surface. Since $\vec{E}$ and differential area vector $d\vec{A}$ are both perpendicular to the sphere's surface (see Figure (17.3.3)),

$$| < \vec{E}, d\vec{A} > | = \|\vec{E}\| \cdot |d\vec{A}|.$$

The constant $\|\vec{E}\|$ then factors out of the integral implying that

$$\|\vec{E}\| \times \text{(sphere area)} = q_{enc}/\varepsilon_o,$$

which is a form of Coulomb's Law.

(19.5.3) **Motivation for Decomposing E_{r_0} into exactly k^2 Shorter Pieces:**

Exercise   At each distance, kr_0, from the source charge, q_0, the field line of strength E_{r_0} could have been decomposed into one, three, seven, or *any* number of pieces. The choice to cut E_{r_0} into k^2 equal pieces is motivated by the fact that at radius r_0, the area is multiplied by a factor of k^2.

In fact, at distance r_0, the field line pierces an initial area of size ΔA_{r_0} (see Figure (19.1.5)). Then, at distance $2r_0$, the area grows to size $4\Delta A_{r_0} = 2^2\Delta A_{r_0}$; at distance $3r_0$, the area grows to size $3^2\Delta A_{r_0}$. In general, at distance kr_0, the area grows to size $k^2\Delta A_{r_0}$.

To provide each area segment with a field line, then, consistent with rules (18.2.3), we cut the field line of magnitude E_{r_0} into k^2 (shorter) pieces and give one of these pieces to each area segment of size ΔA.

(19.5.3a) **FIGURE:** *A Cone Intersecting a Sphere.*

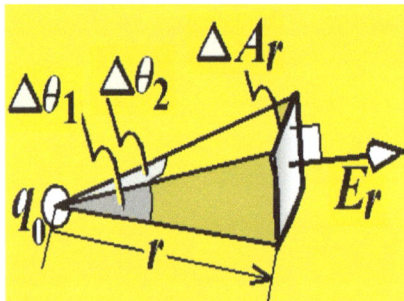

A charge q is at the center of a sphere with radius r. Four rays from the center define a four-faced cone whose intersection with the sphere's surface forms a four-sided rectangle with area ΔA_r The angles θ_1 and θ_2 are sufficiently small so that the surface element approximates a flat rectangle. Moreover, the field vector $\vec{E}_r$ is perpendicular to the surface element.

With the construction of Figure (19.5.3a), the product $(E_r \cdot \Delta A_r)$ is independent of r — the product is the same for any cone that is a multiple (or stretching) of the original cone centered at q. This is the content of the next theorem.

(19.5.4) T<small>HEOREM</small>: (***Product $E_r \cdot \Delta A_r$ independent of r.***)   *Given the*
Trig   *intersection of the sphere centered at charge q_0 and the cone of Figure*
(19.5.3a). Then the product

(19.5.4a) $E_r \cdot \Delta A_r = \dfrac{q_0 \cdot \Delta\theta_1 \cdot \Delta\theta_2}{4\pi\varepsilon_0}$,

which depends on the vertex angles, $\Delta\theta_1$, $\Delta\theta_2$, the charge q_0, and the constant ε_0, is, therefore, independent of r.

P<small>ROOF</small>: For very small angles, $\Delta\theta_1$, $\Delta\theta_2$, the intersection of the cone and the sphere in Figure (19.5.3a) closely approximates a rectangle with sides $r \cdot \Delta\theta_1$ and $r \cdot \Delta\theta_2$. Hence, the area of this surface element is $\Delta A_r = r^2 \cdot \Delta\theta_1 \cdot \Delta\theta_2$. Now use the form of Coulomb's law given by (18.3.5) to see that $E_{\mathbf{p}} = E_r = q_0/4\pi\varepsilon_0 r^2$. The product of E_r and ΔA_r gives us (19.5.4a), which proves the theorem. ■

Particular radius r_0. Choose as a reference the cylinder intersecting the sphere with radius r_0. At this point, the electric field has magnitude E_{r_0} and the surface has value ΔA_{r_0}. The product of these two quantities, therefore, computes to be

(19.5.4b) $E_{r_0} \cdot \Delta A_{r_0}$.

General radius r. The electric field is proportional to $1/r^2$ (17.2.1). Thus, at every point (area) that is r units from the charge has value

(19.5.4c) $(k_1)/r^2 = \underbrace{(r_0^2 E_{r_0})}_{k_1} /r^2$.

The surface area r units from the charge is proportional to r^2. Hence, the area has value

(19.5.4d) $(k_2)r^2 = \underbrace{(\Delta A_{r_0}/r_0^2)}_{k_2} r^2$.

Therefore, the product of the field magnitude and the surface area at an arbitrary radius r computes to be $E_{r_0} \cdot \Delta A_{r_0}$, exactly the same as (19.5.4b), the product at our starting radius, r_0.

From (18.3.5), we see that the electric field E_r *decreases* algebraically in proportion (17.2.1) to $1/r^2$. At the same time, the surface area A_r

of a sphere of radius r *increases* with r^2. More generally, the area, ΔA_r, say, of any subsection of the sphere's surface also increases in proportion to r^2.

The (area × field) product is constant at all distances r from the center. This follows since the area of the cone is proportional to r^2 and the field is proportional to $1/r^2$. Hence, their product $(\delta A_r \cdot E_r)$ is constant for all $r > 0$.

(19.5.5) **Gauss' Law for Electricity in a Cylinder.**   By direct compu-
<u>Exercise</u>   tation, verify Gauss' law for electricity over the surface of a cylinder of radius R that encloses a point charge q_0. Specifically, show that

$$\oint \langle \vec{E}, d\vec{A} \rangle = \frac{q_0}{\varepsilon_0}$$

when the enclosed charge lies H units from the top and H' units from the bottom on the radial axis. (*The cylinder has total height* $H + H'$.)

(19.5.5a) FIGURE: *"Upper" Half-cylinder*

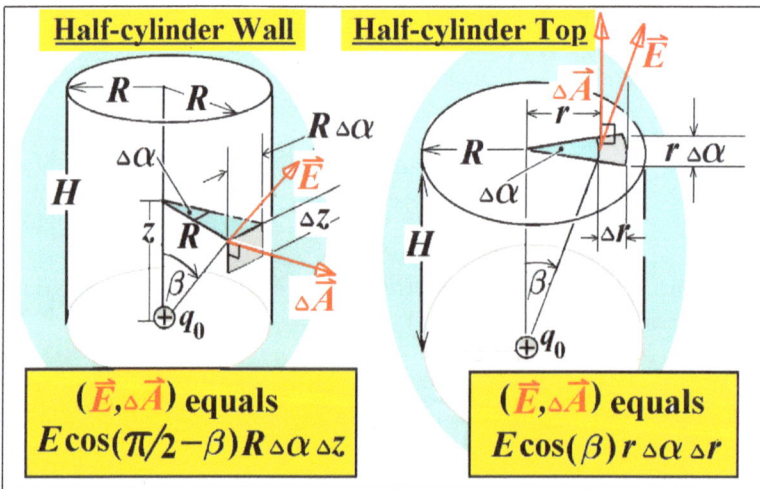

Half-cylinder Wall   **Half-cylinder Top**

$(\vec{E}, {}_\Delta\vec{A})$ **equals** $(\vec{E}, {}_\Delta\vec{A})$ **equals**
$E\cos(\pi/2-\beta)\,R\,{}_\Delta\alpha\,{}_\Delta z$ $E\cos(\beta)\,r\,{}_\Delta\alpha\,{}_\Delta r$

Hints: *First, compute the surface integral $\oint \langle \vec{E}, d\vec{A} \rangle$ over the "upper half surface" of Figure (19.5.5a) that consists of the top H units of the cylinder wall and the top circle. Then integrate over the "lower half surface" which is the bottom H' units of the cylinder's wall along with its bottom circle.*

Figure (19.5.5a) shows the values of the electric field vector $\vec{E}$, the included angle β, and the differential surface element $\Delta\vec{A}$ that we need to know before computing the closed Riemann integral $\oint \langle \vec{E}, d\vec{A} \rangle = \lim_{\Delta A \to 0} \sum_{\Delta A \to 0} (\vec{E}, \Delta\vec{A})$. (To recall the definition of the Riemann integral, see (17.7.4a).)

(19.5.5b) **F**IGURE: *Distances from $\vec{E}$ to q_0.*

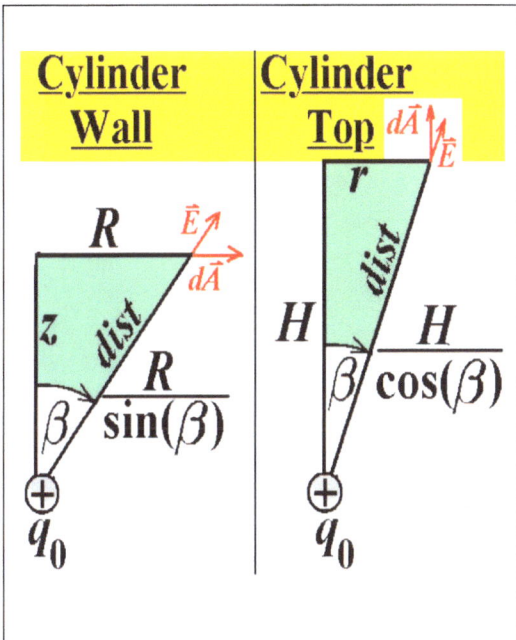

- **Value of E.** Formula (18.3.5), pg. 247, tells us that the magnitude of the electric field is $E = (q_0/4\pi\varepsilon_0)(1/dist^2)$ where *dist* is the distance from the surface element to the single enclosed charge q_0. On the cylinder wall (left triangle of (19.5.5b)), $dist = R/\sin(\beta)$ and on the circular top (right side of (19.5.5b)), $dist = H/\cos(\beta)$.

- **Value of the Included Angle.** As can be read from Figures (19.5.5a) and (19.5.5b), the angle between the electric field vector $\vec{E}$, and the *vertical* area vector $\Delta\vec{A}$ on the top circular surface is β. But on the cylinder wall, between $\vec{E}$ and the *horizontal* area vector $\Delta\vec{A}$ lies the included angle $(\pi/2 - \beta)$.

- **Value of ΔA.** On the cylinder wall (left cylinder in (19.5.5a)), the differential area element is $\Delta A = \Delta z \, R\Delta\alpha$. On the top circular surface (right cylinder of (19.5.5a)), the differential area element is $\Delta A = \Delta r \, r \, \Delta\alpha$.

- **The Final Integrals.** With the values of E, β, and ΔA, the calculated values for the inner products $\langle \vec{E}, \Delta\vec{A} \rangle$ in the boxes shown in Figure (19.5.5a) are justified. Now take the sum $\sum_{\Delta A} \langle \vec{E}, \vec{\Delta A} \rangle$ as $\Delta\vec{A}$ runs over the entire cylinder wall (19.5.5c), and then over the entire circular top (19.5.5d). In each case, take the limit as the maximum value of $\Delta A \to 0$ to produce the Riemann integrals $\oint \langle \vec{E}, d\vec{A} \rangle$ which assume respective forms

(19.5.5c) $\int E \cos(\pi/2 - \beta) R \, d\alpha \, dz$ (cylinder wall), and

(19.5.5d) $\int E \cos(\beta) r \, d\alpha \, dr$ (top surface).

- **Eliminating the r, dr, and z, dz Variables.** The integrals above can be simplified by eliminating the dependent r and z variables and integrating only with respect to β and α. Specifically, as Figure (19.5.5b) illustrates, $r = H \tan(\beta)$ so that $dr = [H/\cos(\beta)] d\beta$. Similarly, $z = R/\tan(\beta)$ which, when differentiated with respect to β, yields

$$dz = -[R/\sin^2(\beta)] d\beta.$$

- **An Efficient Computation.** Once the variables r, dr, z, and dz are eliminated (above), the disagreeable surface integrals, (19.5.5c) and (19.5.5d), assume the simple $\int \sin(\beta) d\beta$ form, whose sum produces the total surface integral (wall + top) of the upper cylinder:

(19.5.5e)

$$\underbrace{\frac{q_0}{2\varepsilon_0} \int_0^{\beta_0} \sin(\beta) d\beta}_{(19.5.5d)} + \underbrace{\frac{q_0}{2\varepsilon_0} \int_{\beta_0}^{\pi/2} \sin(\beta) d\beta}_{(19.5.5c)} = \frac{q_0}{2\varepsilon_0} \int_0^{\pi/2} \sin(\beta) d\beta.$$

The angle β_0 is formed at the meeting point of the cylinder wall and the circular surface, that is, when both triangles in Figure (19.5.5b) are the same right triangle with sides $\{H, R, \sqrt{H^2 + R^2}\}$. Equivalently, β_0 is the unique angle for which $\tan(\beta_0) = R/H$.

(19.5.5f) How to Measure Angle β.   The angle β, formed by a line originating from the point charge q_0, is measured from the vertical axis of the cylinder. Angle β increases as the line rotates away from the vertical as shown in Figures (19.5.5a) and (19.5.5b). Thus, a vertical line pointing up from q_0 is at an angle $\beta = 0$ and at an angle $\beta = \pi/2$ when the line points horizontally.

- **The Other Half of the Cylinder.** The surface integral above covers only the upper H units of the cylinder. Since neither the radius R nor the wall height H appears in the integral, the identical integral, $(q_0/2\varepsilon_0) \int_0^{\pi/2} \sin(\beta) d\beta$, will result when integrating over the lower H' units of the cylinder.

(20) Towards Maxwell's Equations

Preview

The Source of Magnetism.

- Magnetism is generated by moving electrical charges and (*in a quantum setting that has nothing to do with moving electrons*) by spin on particles such as the neutron and the neutrino.

- Some atoms become tiny magnets because of their moving electrons and also because of the spin of the atom's electrons, protons, and neutrons. Not all atoms have this property.

In 1819, HANS CHRISTIAN OERSTED noticed that a compass needle was deflected by a wire carrying a current. Then, in 1820, JEAN-BAPTISTE BIOT (1774-1862) and FÉLIX SAVART (1791-1841) conducted experiments that measured the force exerted on a compass by the electric current.

(20.1) Biot-Savart Law: Magnetism from Electricity

The numerical value of current, I, is measured as the number of positive coulombs, Δq, passing through a cross-section of the wire over a time period Δt. This definition is motivated by the fact that in the presence of an electric field, the stronger the field, the more electrons move per unit time through a wire. Specifically,

$$(\textbf{20.1.1})\; I = \frac{\Delta q}{\Delta t} \;\left(\lim_{\Delta t \to 0} \frac{\Delta q}{\Delta t} = \frac{dq}{dt} \quad \textit{denotes the \underline{instantaneous} rate of change} \right).$$

Biot and Savart measured the magnetic field caused by an electric current I (20.1.1). They found that the magnetic field $\vec{B}$ at a point **p** is the sum of the contributions of all the differential $\Delta\vec{B}$ elements, each of which is generated by the current I passing through a differential length $\vec{\Delta l}$ of the wire. Each tiny $\Delta\vec{B}$ is perpendicular to both the small wire length $\vec{\Delta l}$ and the vector $\vec{r}$ from $\vec{\Delta l}$ to **p**. (*See the left panel of Figure* (20.1.2).)

(20.1.2) FIGURE: *The Biot-Savart Law and the Right Hand Rule*

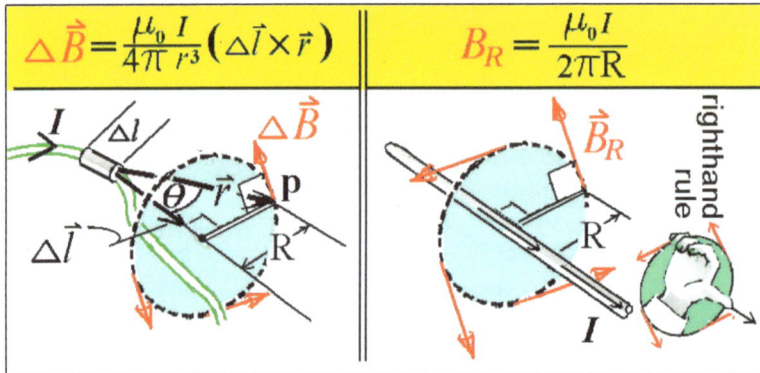

$$\Delta\vec{B}=\frac{\mu_0 I}{4\pi\, r^3}\left(\vec{\Delta l}\times\vec{r}\right) \qquad\qquad B_R=\frac{\mu_0 I}{2\pi R}$$

(20.1.3) NOTE: *The 2-dimensional cross-section of the magnetic field vectors shown in the right hand panel of Figure* (20.1.2) (*consistent with the wheel-axle right hand rule* (17.4.5)), *shows concentric* <u>*circles*</u> *of constant magnitude around an infinite, straight wire. In actual 3-space, the magnetic circles "add up" to form concentric* <u>*cylinders*</u> *of constant magnitude that sheathe the infinite wire.*

(20.2) Quantitative Results for Biot-Savart

We discuss two cases for measurements of induced magnetic fields, namely, when a current I flows through

- a small, differential wire element (20.2.1), and

- a string of differential wire elements, that is, an infinite straight wire (20.2.3).

(20.2.1) **Biot-Savart Law for Differential Wire Elements.**
***Left Panel of Figure* (20.1.2):**   A small, differential magnetic field vector $\Delta\vec{B}$ is induced by a current I flowing through a tiny piece of wire with length Δl. The 3-dimensional vector $\vec{\Delta l}$ represents the 1-dimensional length Δl of the wire segment (*see left panel of Figure*

(17.3.3)). The distance vector $\vec{r}$ runs from the wire segment to the point **p** in space where $\Delta\vec{B}$, the magnetic field vector, is measured. The distance from the wire segment to **p** is $\|\vec{r}\| = r$. Finally, the differential magnetic field vector $\Delta\vec{B}$ is given by

(20.2.1a) $$\Delta\vec{B} = (\frac{\mu_0 I}{4\pi r^3})(\Delta\vec{l} \times \vec{r}),$$ **Biot-Savart Law**

so that its magnitude is

(20.2.1b) $\|\Delta\vec{B}\| = \Delta B = (\frac{\mu_0 I}{4\pi r^2})\Delta l \sin\theta,$ $\begin{cases} \text{since, from (17.6.3),} \\ \|\Delta\vec{l} \times \vec{r}\| = \Delta l(\sin\theta)r \end{cases}$

where θ is the angle in the plane formed by the vectors $\Delta\vec{l}$ and $\vec{r}$. Equation (20.2.1a), a statement of the Biot-Savart law, shows that the differential magnetic field vector $\Delta\vec{B}$ is the cross product (17.6.3) of vectors $\Delta\vec{l}$ and $\vec{r}$. Hence, $\Delta\vec{B}$ is always perpendicular to the plane formed by the vectors $\Delta\vec{l}$ and $\vec{r}$.

(20.2.2)

> **DEFINITION:** *The constant*
>
> **(20.2.2a)** μ_0
>
> *of* (20.2.1a) *is called the magnetic* **permeability constant**.

Permeability, μ_0 (with the 0 subscript), depends on the *vacuum* in which $\Delta\vec{B}$ of (20.2.1a) is measured. Measurements of $\Delta\vec{B}$ taken in other environments — nitrogen, air, water, etc. — produce their own permeability constants, μ.

(20.2.3) Biot-Savart Law for an Infinite Wire.
Right Panel of Figure **(20.1.2):** In this panel, the Biot-Savart law (20.2.1a) is extended from a single differential wire segment with length Δl to an infinite *sum* of such wire lengths, i.e., to an infinite, straight wire carrying the same current I.

The resulting magnetic field vector, $\vec{B}_R = \sum \Delta\vec{B}$, depends on the radial distance, R, from the wire, and has magnitude

(20.2.3a) $$\|\vec{B}_R\| = B_R = \frac{\mu_0 I}{2\pi R}.$$

In (20.2.3a), the vector $\vec{B}_R$ is tangent to the circle with radius R and points clockwise or counterclockwise, in accordance with the wheel-axle orientation of the right hand rule (17.4.5).

(20.2.4) ‖ NOTE: *Equation (20.2.3a) is not trivial. Exercise (20.8.1) takes you through the integration process to prove its validity.*

At all points on the circumference of the dotted circle with radius R (perpendicular to the wire segment vector, $\Delta \vec{l}$), $\|\vec{\Delta B}\| = \Delta B$, the length of the magnetic field vector, is constant — only the *direction* of $\Delta \vec{B}$ changes as it remains tangent to the dotted circle.

We can use this special, simple case to motivate Ampère's Law which is developed in the following section.

(20.3) Ampère's Law

We will need to relate A, a surface area, with L, the boundary, edge, or loop around that area. Closed "potato" surfaces do not have a boundary, but "pancake" surfaces, floating in space, will have an outside boundary. Our surfaces will have no holes or inside edges.

The following notation reinforces the connection between a surface and its boundary:

(20.3.1)

NOTATION: *For a surface in* $\mathbf{R}^3$ *with area A, we use the symbol*

(20.3.1a) $bdry(A)$

to denote the boundary of enclosed area A.

Note that for a closed surface, like a sphere, there is no boundary. But if a 2-dimensional boundary $bdry(A)$ does exist for an area A, then A will not be a unique area — a single boundary can accommodate infinitely many areas. For example, a circle can be the boundary (*edge*) of a flat disk or it can be the rim of a concave bowl.

(20.3.2) $\|$ NOTE: *Similar to our dual usage of the symbol q for both a charged*
$\|$ *particle and its electrical charge (18.2.1), we use the same symbol,*
$\|$ *A, for the physical surface and for the numerical measure of its*
$\|$ *area. Similarly, bdry(A) denotes both the boundary and the length*
$\|$ *of the boundary.*

(20.3.3) THEOREM: (***Ampère's Law for Steady State*** I_{net}***.***)   *Given a*
closed loop bdry(A) defining area A, through which there flow sev-
eral steady-state currents, I_1, I_2, $\cdots$, I_n.   Then the net current,
$I_{net} = I_1 + I_2 + \cdots + I_n$, flowing through the area A (enclosed by
loop bdry(A)), induces a magnetic field $\vec{B}_l$ whose values at a point l
on bdry(A) are constrained by the closed line integral

(20.3.3a) $$\oint_{bdry(A)} \langle \vec{B}_l, d\vec{l} \rangle = \mu_0 I_{net}.$$ **Ampère's Law**

PROOF: We take the simplest case, which is illustrated by the right
hand panel of Figure (20.1.2). The more general non-circular boundary is beyond the scope of this book but an exposition can be found
in the book *Classical Electrodynamics* [21] by J. D. Jackson. We assume the boundary loop, $bdry(A)$, is the circumference of a circular
area A with radius R. Assume that there is only one infinite current-
carrying wire that passes through the center of the circle A. With
the symmetry of the circle, the norm (*but not the direction*) of the
magnetic field vector at any point, l, on the circle can be denoted by

(20.3.3b) $B_l = B_R,$

since B_i depends only on the radius R. Let Δl_i denote the length
of the ith circumference element, $\Delta \vec{l}_i$ on the circumference, $bdry(A)$,
and let l_i be a point in the interval Δl_i. Then

(20.3.3c) $\langle \vec{B}_{l_i}, \Delta \vec{l}_i \rangle = \underbrace{B_{l_i}(\sin 90°)\Delta l_i}_{\text{from Def. (17.6.1)}} = B_R \Delta l_i.$ *from (20.3.3b)*

Using (17.8.5a) to compute the left hand side of (20.3.3a), we obtain

$$\oint_L \langle \vec{B}_R, \Delta \vec{l} \rangle = \lim_{\substack{\max \Delta l_i \\ \to 0}} \left[\sum_{\Delta l_i} B_R \Delta l_i \right]$$ *Definition (17.8.5a)*
and (20.3.3c)

$$= B_R \cdot \left[\lim_{\substack{\max \Delta l_i \\ \to 0}} \sum_{\Delta l_i} \Delta l_i \right] \qquad \begin{array}{c} \textit{factor out constant} \\ \textit{number, } B_R \end{array}$$

$$= B_R \cdot \left[2\pi R \right]. \qquad \begin{array}{c} \lim \sum \\ = \textit{circle circumference} \\ = 2\pi R \end{array}$$

Replace B_R above with its equivalent, $\mu_0 I_{net}/2\pi R$ from (20.2.3a), (*which is proven in Exercise* (20.8.1)), and we see that hypothesis (20.3.3a) results. This proves the theorem. ■

(20.3.4) NOTE: *The use of* (20.2.3a) *in the last line of our proof tells us that the Biot-Savart Law* ((20.2.3) *for a wire*) *implies Ampère's law. The reverse implication is also true.*

. . . .

(20.3.5) FIGURE: *Ampère's Law in the General Case.*

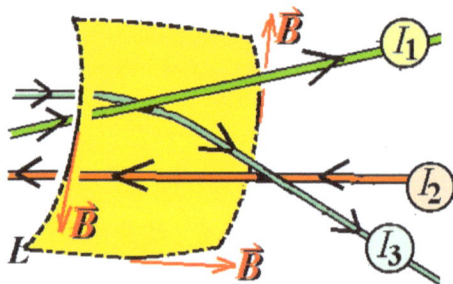

Although we shall consider the case of a *single* wire passing through a region, non-trivial mathematics (*involving closed Jordan curves* (17.4.2)), allow Ampère's law (20.3.3a) to remain valid with more general closed loops L (17.4.3) that may surround several wires. If the kth wire carries current I_k, then the total current through the loop L is $I = \sum_k I_k$, the net (*arithmetic*) sum of all enclosed currents.

(20.4) Maxwell Adds to Ampère's Law

Around 1865, Maxwell noted a deficiency in Ampère's law,

$$\int \langle \vec{B}_l, \vec{dl} \rangle = \mu_0 I_{net} \qquad (20.3.3a),$$

which says that a steady-state current, I_{net}, induces a steady-state magnetic field, $\vec{B}$.

(20.4.1) **Here is the problem:** A CAPACITOR or CONDENSER consists of a pair of conducting surfaces separated by an insulator. In our diagram, a parallel plate capacitor is initially uncharged. As electrons begin to flow clockwise through the wire (which means that the current I flows *counter-clockwise*), the left hand plate accumulates negatively-charged electrons that produce an electric field that repels the like-charged negative electrons from the right hand plate. This "chasing away" of electrons acts as if there is a current passing through the space between the plates where an excess positive charge is created on the right hand plate.

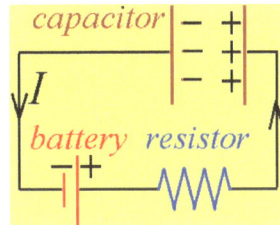

(20.4.2) **There is no physical current or charge** traveling between the plates, yet a magnetic field *is* induced around the capacitor. Although Ampère's law described the effects of charges moving along a wire (that induces a magnetic field), Maxwell realized that a consistent treatment of time-varying (non steady-state) situations can be reached by introducing a term in the right hand side of (20.3.3a) that takes into account the effects of one capacitor plate on the other across the insulting medium. He labeled the term the DISPLACEMENT CURRENT not because it represents a motion of charges, but because it plays a similar rôle to the usual current in the equations. With this deceptively simple modification, the equations no longer suffered from the problems described above, but in addition satisfied other consistency requirements such as conservation of charge. More importantly, the modified equations predicted a wealth of new phenomena whose practical applications have defined many aspects of our current electronic-dependent society.

Maxwell's idea was a momentous discovery indeed.

(20.4.3)
Calc
What term should be added to (20.3.3a) in order to fix Ampère's steady-state law? Maxwell's ingenious solution was to describe the induced magnetic field in terms of the physical, time-varying current, $I(t)$, flowing through the wire in the circuit. He reasoned as follows:

Gauss' law (19.3.1a), pg. 260, tells us that $\Phi_{\vec{E}}(A) = q/\varepsilon_0$. Differentiating both sides of this equality with respect to time t yields

(20.4.3a) $\qquad \dfrac{d}{dt}\Phi_{\vec{E}}(A) = \dfrac{1}{\varepsilon_0} \cdot \dfrac{dq}{dt} = \dfrac{1}{\varepsilon_0} \cdot I(t).$ $\qquad$ *from* (20.1.1)

To match the $\mu_0 I_{net}$ form of Ampère's law (20.3.3a), we solve for $I(t)$ in (20.4.3a) to obtain

(20.4.3b) $\qquad \mu_0 I(t) = \mu_0 \varepsilon_0 \dfrac{d}{dt}\Phi_{\vec{E}}(A)$

which, when added to Ampère's law, yields

(20.4.3c) $\qquad \boxed{\oint_{bdry(A)} \langle \vec{B}_l, d\vec{l} \rangle = \mu_0 I_{net} + \mu_0 \varepsilon_0 \dfrac{d}{dt}\Phi_{\vec{E}}(A).}$ $\qquad$ **Ampère-Maxwell Law**

The Ampère-Maxwell Law (20.4.3c), is the final (corrected) version of Ampère's law (20.3.3a) which now conserves charge. (*Recall A is an area through which the changing circuit flows, and $\vec{B}_l$ is the induced magnetic field measured at points l on the boundary, bdry(A).*)

This ingenious addition of the "virtual" current term (20.4.3b) by Maxwell into (20.4.3c), is a deceptively simple, yet consequential and robust idea.

(20.5) Faraday's Law: Electricity from Magnetism

(20.5.1) **It was Michael Faraday** who first noticed that a moving magnet through a wire loop induced a current corresponding to a non-zero electromagnetic force.

As it turns out, the wire and the current are not necessary. The electric field exists with or without the wire. That is, a changing magnetic field $\Delta\vec{B}$ in any bounded area will induce a driving emf, or electromotive force, $\vec{E}_l$, at points l on any boundary of that area, whether or not that boundary is formed by a physical wire.

To quantify the relation between $d/dt\left(\Phi_{\vec{B}}(A)\right)$, the *change* in a magnetic field through any closed area, A, and $\vec{E}_l$, the induced electrical pressure at point l on $bdry(A)$ (the boundary of A), Faraday devised the following closed line integral equation:

$$\textbf{(20.5.1a)} \oint_{bdry(A)} \langle \vec{E_l}, d\vec{l} \rangle = -\frac{d}{dt}\Phi_{\vec{B}}(A). \qquad \textbf{Faraday's Law}$$

The minus sign in Faraday's law (20.5.1a) is the mathematical expression of Lentz's Law (20.6.1) following, which will state that the direction of the magnetic field produced is always opposite to the direction of the change in the original magnetic field.

(20.6) Lentz's Law: The Positive Side of Negativity

We know that electric fields and currents induce magnetic fields (20.1), (20.4.3c). Faraday's Law (20.5.1a) describes the converse — how changes in magnetic fields induce electric fields or currents.

In this section, we describe LENTZ'S LAW, that describes the "self-canceling" direction of the induced current. Here are the details.

(20.6.1) **Statement of Lentz's Law:** Imagine an area A has closed boundary $bdry(A)$ that encloses a magnetic field $\vec{B}$. Suppose that at one time, the magnetic field changes by a small amount ΔB. Then

> **First**, $\underline{\Delta \vec{B} \to \Delta \vec{E}}$. From Faraday's Law (20.5.1a), we see that the magnetic field change, $\Delta \vec{B}$ within area A, induces an electric field change, $\Delta \vec{E}$, along $bdry(A)$. If a wire loop lies on the closed $bdry(A)$, then a current ΔI is produced in the closed wire circuit due to the electromotive force generated by $\Delta \vec{E}$.

> **Then**, $\underline{\Delta \vec{E} \to \Delta \vec{B'}}$. From the Maxwell-Ampère Law (20.4.3c), an electric field change $\Delta \vec{E}$ (or loop current I) along $bdry(A)$ induces its own magnetic field change, $\Delta \vec{B'}$. The minus sign in Faraday's Law ensures that the induced magnetic field, $\Delta \vec{B'}$, is *opposite* to the original field change $\Delta \vec{B}$. This opposition principle is **Lentz's Law** — although one might question giving a mere minus sign the title of "law."

(20.6.2) Illustrating Lentz's Law

The following figure, which illustrates Lentz's law, analyses *changes* in the number of leftward-directed magnetic field lines enclosed by the ring as it changes positions between ① and ②, and between ③ and ④, in Figure (20.6.3). The loops in positions ① and ④, the most remote from the magnet, are shown capturing only one leftward-directed magnetic filed line. (*All magnetic field lines have a <u>direction</u> as we saw in Fig.* (18.4.2).) The loops closest to the magnet, in Positions ② and ③, capture three leftward-directed field lines.

(20.6.3) FIGURE: *Illustrating Lentz's Law*

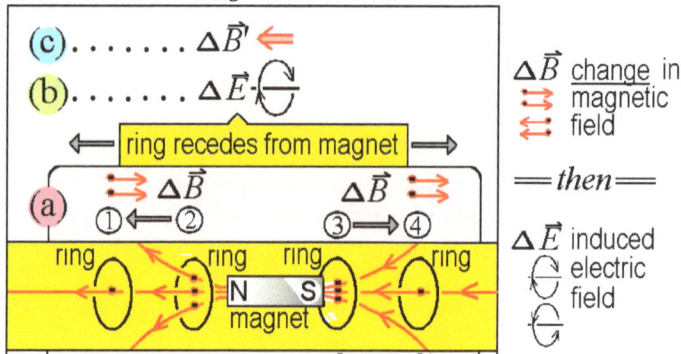

———————— *Ring Recedes from Magnet* ————————

Line (a), Fig. (20.6.3).   *Whenever the ring recedes from the magnet, whether moving leftward,* ① ⇐ ②, *or rightward,* ③ ⇒ ④, *there is a (rightward)* <u>change</u>, $\Delta B = $ ⇄, *in the number of leftward-directed magnetic field lines that are encircled by the wire loop.* In both cases, this rightward-directed change, ⇄ causes a two-line *decrease* in the number of leftward-directed magnetic field lines enclosed by the wire loop — from three down to one.

Line (b), Fig. (20.6.3).   This magnetic field change, $\Delta B = $ ⇄, induces an electric force $\Delta \vec{E}$ in the wire loop, which leads to Line (c) following:

Line (c), Fig. (20.6.3). $\Delta \vec{E}$ induces the magnetic field, $\Delta \vec{B}' = $ ⇐, as defined by the right hand rule (17.4.6). This change always *opposes* the direction of the original magnetic field change, $\Delta B = $ ⇄.

These statements (*along with Exercise* (**20.8.3**)) complete the explanation of Figure (20.6.3), which fully illustrates Lentz's law. ■

(20.7) Maxwell's Four Equations

(**20.7.1**)

HISTORY

Maxwell's four equations first appeared in 1873, in his text *A Treatise on Electricity and Magnetism*. It would be three years later, in October 1876, that ALEXANDER GRAHAM BELL and THOMAS WATSON developed a TELEPHONE that employed the electromagnetic properties articulated by Maxwell. The first voice signal was carried between telephones connected by a two-mile wire.

(**20.7.1a**) **The function of the telephone** depends on the following sequence of frequency-preserving events:

- *A voice will vibrate the air* (***a***)

- *and induce corresponding and varying electrical conductivity of a medium, e.g., carbon particles in the mouthpiece* (***b***),

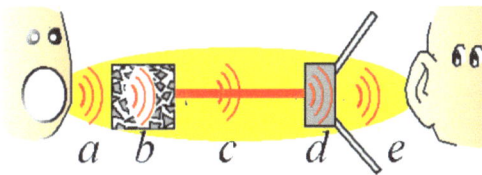

- *that conducts a current with a corresponding varying voltage through a transmission line* (***c***).

- *This fluctuating voltage induces corresponding varying magnetic forces at the receiver's end* (***d***)

- *that "push and pull" on a speaker cone in the receiver's earpiece* (***e***) *with corresponding frequencies that agree with the original voice frequencies. The cone vibrations are transmitted through the air to reproduce the final sound waves that finally arrive at the listener's ear.*

(**20.7.2**) The esthetic and simplifying power of Maxwell's equations impelled MAX BORN to note [3], pg. 181-2:

> The four symbolic formulae [of Maxwell's equations] show wonderful symmetry. Formal agreement of this kind is by no means a matter of indifference. It exhibits the underlying simplicity of phenomena in nature, which remains hidden from direct perception because of the limitations of our senses and

reveals itself only to our analytic faculty.

(20.7.3) **Maxwell's Four Equations:** We combine Gauss' laws of electricity and magnetism, along with Faraday's law (*magnetism induces electricity*), and the Maxwell-Amp`ere law (*electricity induces magnetism*) to obtain the following quartet of equations, known as MAXWELL'S EQUATIONS for electricity and magnetism.

Gauss' law for electricit [*from* (19.3.1a), *pg. 260*]. An electric charge, q_{enc} (a collection of electric "monopoles"), within a closed surface A is computed from the values of the electric field $\vec{E}$ at points on that surface according to the closed surface integral

(20.7.3a)
$$\oint_A \langle \vec{E}, d\vec{A} \rangle = \frac{q_{enc}}{\varepsilon_0}.$$

....

Gauss' law for magnetism [*from* (19.4.1a), *pg. 262*]. (No magnetic monopoles exist) The net magnetic flux (positive field lines in, minus field lines out) through any closed surface A is zero. This quantity is computed from the values of the magnetic field $\vec{B}$ at points on that surface according to the closed surface integral,

(20.7.3b)
$$\oint_A \langle \vec{B}, d\vec{A} \rangle = 0.$$

....

Faraday's law. [*from* (20.5.1a), *pg. 277*] A changing magnetic field, $\vec{B}$, passing through a surface, A, produces an electric field, $\vec{E}_l$, whose values at points l on $bdry(A)$ are constrained by the line integral

(20.7.3c)
$$\oint_{bdry(A)} \langle \vec{E}_l, d\vec{l} \rangle = -\frac{d}{dt} \Phi_{\vec{B}}(A).$$

The minus sign is the mathematical expression of Lentz's law (20.6.3) which imposes a left hand rule (instead of the standard right hand rule) on the orientation of the enclosing loop, $bdry(A)$.

....

Ampère-Maxwell law [*from* (20.4.3c), *pg. 276*]. A changing electric field $\vec{E}$ through a surface A produces a magnetic field $\vec{B}_l$ whose values at points, l, on $bdry(A)$, the closed boundary of surface A, are constrained by the line integral

(20.7.3d) $\oint_{bdry(A)} \langle \vec{B}_l, d\vec{l} \rangle = \mu_0 I_{net} + \varepsilon_0 \mu_0 \dfrac{d}{dt} \Phi_{\vec{E}}(A).$

(20.8) Exercises

(20.8.1)
Exercise

Extending Biot-Savart to an Infinite Wire.   The Biot-Savart law (20.2.1a) is given for a current I passing through a differential wire segment $\Delta \vec{l}$. Extend this result by replacing segment $\Delta \vec{l}$ with a *sum* of such differential lengths, that is, with an infinite, thin, straight wire and show that the induced magnetic field vectors $\vec{B}$ have constant magnitude

(20.8.1a) $\|\vec{B}\| = \dfrac{\mu_0 I}{2\pi R}$

on concentric cylinders (with the wire as central axis) with radius R.

Hint:

Equation (20.2.1b) tells us that the magnitude of the magnetic field vector, induced at point **p**, *by current I through a Δl-length wire is $\Delta B = (\mu_0 I/4\pi r^2)\Delta l \sin\theta$, where r is the distance from Δl to point* **p**.

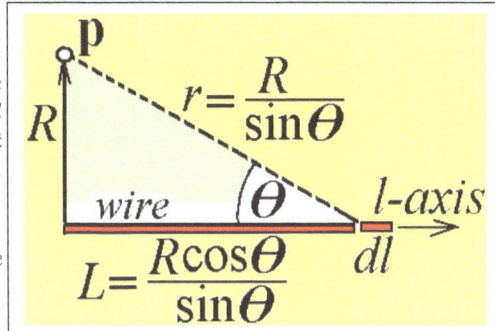

For the purposes of integration, set $\Delta B = dB$ and $\Delta l = dl$, so that the accumulated magnetic field values, B, over <u>half</u> *the wire, from 0 to infinity, is given by the integral*

(*) $\displaystyle \lim_{L\to\infty} \int_{l=0}^{l=L} dB = \int_{l=0}^{l=\infty} dB = \dfrac{\mu_0 I}{4\pi} \int_{l=0}^{l=\infty} (\sin\theta / r^2)\, dl\,.$

Note that the figure shows a right triangle with constant side, R, and variable sides, r and L, both of which are expressed in terms of a single variable, θ. Accordingly, convert the integral (*) to a single-variable integral, $\int_{\theta=\pi/2}^{\theta=0} \cdots d\theta$, where, necessarily, angle θ goes from $\pi/2$ to 0 as the variable L goes from 0 to ∞. This integral over a half ray $[0,\infty)$ will give you only half of the desired result.

(20.8.2)   **The Biot-Savart Law Agrees with the General Ampère Law.** In
Exercise   the infinite wire case of Exercise (20.8.1), show that Ampère's law gives
Calc   the same result, $B_R = \mu_0 I/2\pi R$, for the magnitude of the magnetic field
Trig   vector $\vec{B}_R$ at a point R units from an infinite wire.

integrate along wire   integrate along circumference

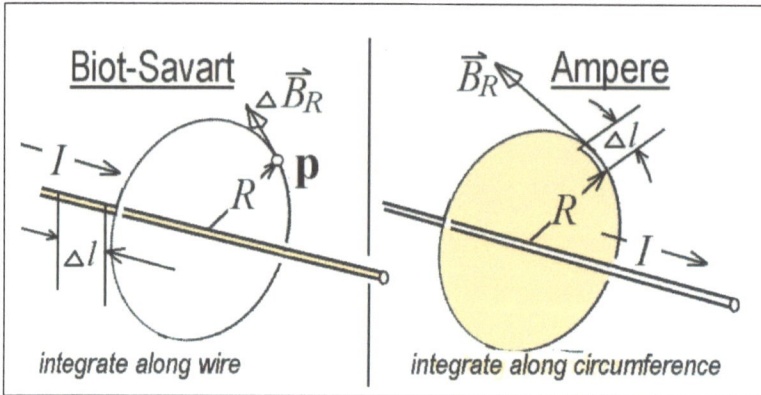

Hint: <u>*Left Panel*</u>*: The Biot-Savart integration proceeds along the $\Delta \vec{l}$ elements of the infinite wire which hosts a current I while the point **p** is
fixed. The value of the differential magnetic vectors $\Delta \vec{B}_R$ add together as
the integration advances along the wire.*

<u>*Right Panel*</u>*:Apmère's Law requires that a current I passes through the cir-
cular area and integration proceeds along the $\Delta \vec{l}$ elements of the circumference, which has length $2\pi R$. The length B_R (but not the direction) of
the magnetic vector $\vec{B}_R$ is constant at every point on the circle with the
current passing perpendicularly through the center.*

(20.8.3)   **Use Figure (20.6.3) to show how Lentz's Law** applies when the ring
Exercise   approaches the magnet.

Hint: *Reason similarly to the argument following Figure (20.6.3) which
describes the case of the magnet receding from the magnet.*

(21) Electromagnetism: A Qualitative View

We show how Maxwell's equations (20.7) and some accurate laboratory measurements lead to an expression for c, the speed in a vacuum of electromagnetic (*EM*) waves. Maxwell did not know that light was a subset of his electromagnetic waves. Other electromagnetic waves include infrared, and ultraviolet radiation, microwaves, radio waves, and X and gamma rays (*see Figure* (21.3.3)).

(21.1) Magnetic Waves from an Infinite Wire

Electromagnetic waves are time-dependent. One of the simplest examples is ordinary household alternating current (AC) in a wire where electrons flow — alternately and repeatedly — in one direction, and then in the opposite direction.

Figure (21.1.1) shows a vertical wire with an alternating current. On the circle perpendicular to the wire (in the X-Y plane) with radius r measured from the wire, the magnetic field vectors are tangent to the r circle and have constant value B_r. As the notation suggests, B_r is constant on circles of radius r but changes as the radius r changes.

Since the current $I(t)$ in the "axle" repeatedly reverses its direction, the right hand rule (17.4.6), (17.4.5), assures us that the orientation of the magnetic ripples — clockwise or counterclockwise — must also reverse in time (*see Figure* (21.1.1)).

Note: Although alternating current occurs in the wires attached to our home appliances, there is no measurable alternating magnetic field. This is because home appliances use two wires that are close enough to each other so that they may be considered to share the same "axle" while their currents run in opposite directions. Hence, their induced alternating magnetic fields cancel each other.

(21.1.1)   FIGURE: *Vertical Alternating Current Propagates*
an Electromagnetic Wave.

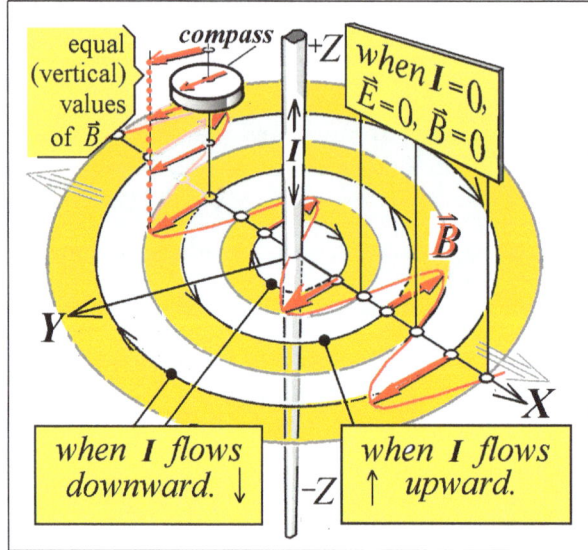

The Biot-Savart law applied to an infinite wire tells us that an upward *current* ↑ *along the Z-axis induces a (cylindrical) magnetic ripple with field vectors oriented in the* counterclockwise *direction. Conversely, a* downward *current* ↓ *along the Z-axis induces a (cylindrical) magnetic ripple with field vectors oriented in the* clockwise *direction* (see Figure (20.1.2), pg. 270).

(21.1.1a)   The compass in Figure (21.1.1) is shown during an "instantaneous snapshot." Its south-seeking needle is tangent to the (cylindrical) ripple and is aligned counterclockwise. Note that the compass would give the same reading at *any* point on a vertical line passing through the original point since the values of the magnetic field $\vec{B}$ are uniform on concentric cylinders surrounding the wire.

(21.1.1b)   A horizontal cross-section of the uniform cylindrical magnetic fields produces circles. In this horizontal plane, as shown in Figure (21.1.1), the vectors of the magnetic field $\vec{B}$ are plotted only on a single line, the X-axis.

At each point x on the X-axis, a vector of length $\|\vec{B}(x)\| = B(x)$ is drawn with its tail at x and tangent to the magnetic field, (i.e., perpendicular to the X-axis). The orientation of the B arrow imitates the behavior of a compass needle (*see Figure* (18.4.2), pg. 248) that might be placed at the point x on the X-axis.

(21.2) **Wave Propagation**

(21.2.1) **The Mechanics of EM Wave Propagation: a preview.** In Figure (21.1.1) we saw how alternating current produced an outward-propagating, time-dependent, alternating *magnetic* field $\vec{B}$. In what follows, we shall see how a companion (and perpendicular) electric field $\vec{E}$ is produced by a varying magnetic field $\vec{B}$.

In fact from Maxwell's equation (20.7.3c) we see that the integral around a boundary that surrounds an <u>electric</u> field mathematically produces an expression that depends on <u>magnetic</u> flux. Similarly, from Maxwell's equation (20.7.3d), the integral around a boundary that surrounds a <u>magnetic</u> field produces an expression that depends on <u>electric</u> flux.

As hypothesized by Maxwell, and as confirmed by countless experiments, a changing electric field induces a changing magnetic field under all circumstances, and *vice versa*.

In (21.2.2), we will look at a snapshot of an alternating current $I(t) \downarrow$ as it increases downward along with the cascade of induced magnetic and electrical fields.

(21.2.2) **F**IGURE: *Electromagnetic Propagation Sequence*

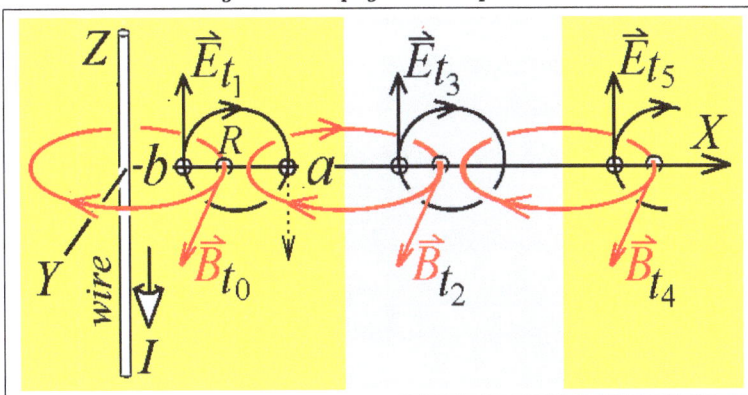

Along the intersection of the X-axis and the circles, $\vec{E}$ vectors are always vertical, while $\vec{B}$ vectors are always horizontal. Also, since magnetic vectors $\vec{B}$ are changing with time, then so are the induced electrical vectors, $\vec{E}$, and vice versa.

Step 0. *At Time t_0, $\vec{B}_{t_0}$ is generated.* The downward electric current, $I \downarrow$ at time t_0, induces the (red) clockwise circle of magnetic field vectors, $\vec{B}$, only one vector of which, $\vec{B}_{t_0}$, is plotted in Figure (21.2.2). The magnetic field vector, $\vec{B}_{t_0}$ is tangential to the red circle. At the point R it:

- intersects the X-axis,
- lies in the X-Y plane,
- is perpendicular to the X-axis, and
- points in the positive Y direction.

At every point on this circle, the magnetic field has value $\|\vec{B}_{t_0}\| = B_R = \mu_0 I(t)/2\pi R$ (20.2.3a), which tells us that if the magnitude of the downward current $I(t) \downarrow$ increases, then so will the magnitude of the scalar B_R. That is,

(21.2.2a) $|I(t)|$ *increases, implies* $\Delta B/\Delta t > 0$.

....

Step 1. *At time $t_1 > t_0$, $\vec{E}_{t_1}$ is generated.* The magnetic field $\vec{B}_{t_0}$ at time t_1 is increasing in time (*see* (21.2.2a)), and, thanks to Faraday's Law (20.7.3c), induces the (black) clockwise circle of electric field vectors, $\vec{E}$, (where only one vector of which is displayed). The minus sign in (20.7.3c) is due to Lentz' law (20.6.3) and determines the direction of induced electric field vector $\vec{E}_{t_1}$. As previously stated, magnetic vectors $\vec{B}$ that are changing with time induce electrical vectors, $\vec{E}$, that are also changing in time and vice versa.

In sum, the electric field vector $\vec{E}_{t_1}$ is tangential to the black circle centered around point R. At point b it:

- intersects the X-axis,
- lies in the X-Z plane,
- is perpendicular to the X-axis, and
- points in the positive Z direction.

It is Lentz' Law (20.6.1), pg. 277, that guarantees the *clockwise* orientation of the electric field vectors around this circle since, by the right hand rule (17.4.5), the magnetic field it induces will oppose the increasing $\vec{B}_{t_0}$.

....

Step 2. *At time $t_2 > t_1 > t_0$, $\vec{B}_{t_2}$ is generated.* The vertical circle of electric field vectors at time t_1 induces the (red) clockwise magnetic field vectors, $\vec{B}$, only one vector of which, $\vec{B}_{t_2}$, is plotted.

Note that it is the Ampère-Maxwell Law (20.7.3d), pg. 280, that guarantees existence of a magnetic field vector $\vec{B}$ that is induced by the changing electric field vector $\vec{E}$. This is due to Maxwell's modification of Ampère's Law (20.3.3a) where he added the crucial $\vec{B}$-dependent term, $\varepsilon_0 \mu_0 \, d[\Phi_B(A)]/dt$.

(The dotted electric field vector at point a, acting like the downward current, $I \downarrow$, determines the clockwise orientation of the induced circle of magnetic field vectors.)

....

(Steps 1 and 2 repeat.) The cycle continues when the horizontal ripple of the previous magnetic field vectors, $\vec{B}$, induces another vertical ring of electric field vectors $\vec{B}$ (apply Step 1), which, in turn, induces a new horizontal loop of magnetic field vectors (apply Step 2), and so on.

Note that we get a propagating *pair* of electric and magnetic field vectors that oscillate as they move away from the (vertical) wire carrying the source alternating current. This linking of electromagnetic pairs is possible only because of Maxwell's modification of Ampère's law.

There is much we can say qualitatively, based on Figure (21.2.2).

....

(21.2.3) Summation: The anatomy of electromagnetic wave propagation induced by an alternating current through a vertical wire:

(21.2.3a) $\vec{B}$ horizontal, $\vec{E}$ vertical for every point on the X-axis.
Note that only two horizontal magnetic field vectors result from the intersection of the horizontal circle of magnetic vectors and the X-axis. Similarly, only two vertical electric field vectors are produced from the intersection of the vertical circle of electric field vectors and the X-axis. Thus, as shown in Figure (21.2.2), magnetic vectors $\vec{B}$ are always perpendicular to the wire, while electric field vectors $\vec{E}$ are always parallel to it. Both vectors emanate from the X-axis and are perpendicular to the X-axis.

(21.2.3b)   Direction of the velocity vector, $\vec{v}$. Figure (21.2.2) also shows that the vectors $\vec{E}$ and $\vec{B}$ are perpendicular to each other, and the direction of the propagated electric-magnetic pair of waves is perpendicular to both $\vec{E}$ and to $\vec{B}$. That is, Nature gives us a physical configuration that coincides with the mathematical cross product (17.6.3), pg. 229, call it $\vec{s}$, where

$$\vec{s} = \vec{E} \times \vec{B}.$$

(21.3) The Geometry of Electromagnetism

(21.3.1)   Fɪɢᴜʀᴇ: *Snapshot of an electromagnetic (EM) wave*

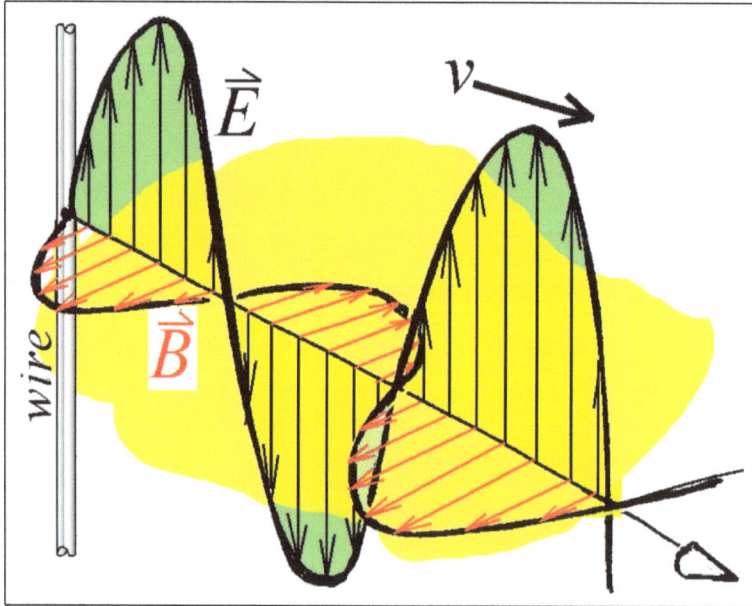

The figure takes one horizontal axis of Figure (21.1.1) and completes Figure (21.2.2). At any fixed point in space and time, the vertical electric field vectors, $\vec{E}$, and horizontal magnetic field vectors, $\vec{B}$, are always perpendicular to each other and their values propagate away from the vertical wire (the source of the alternating current) in the $\vec{E} \times \vec{B}$ direction.

If the current $I(t)$ varies in time as a sinusoidal-like function, then, as (20.2.3a) suggests, the induced magnetic field vectors, $\vec{B}_t$, follow along and take on the same sinusoidal pattern as $I(t)$.

Fields E and B are proportional in the sense of (17.2.1a). This will be shown in (22.3.4c), pg. 302 where we will find that $E = cB$, from which we will also conclude that the constant of proportionality is c, the speed of light. Proportionality guarantees that the graph of the electric field vectors has the same shape as the (perpendicularly drawn) graph of the magnetic field vectors, which, in the case of Figure (21.3.1), is sinusoidal.

....

(21.3.2) **Light: a member of the family.**

- Intuition would dictate that waves, including electromagnetic waves, need a medium — some kind of physical "stuff" to travel through.

After all, we observe water as a carrier for tidal waves and air is a carrier for sound waves. Without water or air there would be no tidal waves or sound. But electromagnetic waves are different — they do not need air or water to propagate, since they travel in the vacuum of space (we do see the stars, after all!).

Now if one does assume that electromagnetic waves are like other waves and need a medium to propagate, then that medium must have rather unique properties: it must fill all the visible cosmos, and yet be so tenuous that it generates negligible friction as the Earth moves around its orbit (otherwise the Earth would have spiraled into the Sun long ago). This hypothesized material was called the luminiferous aether (or aether, for short).

Although Maxwell himself chose to assume the existence of an aether, his equations make no reference to it. Einstein, on the other hand, assumed there was no aether necessary to carry light, a postulate (1.5.2) that greatly simplified his theory of special relativity.

• By the time Maxwell produced his equations that united electricity and magnetism, in which electromagnetic waves consist of traveling electric and magnetic fields that oscillate perpendicular to each other and to the direction of propagation, it had been well accepted (*due to the experiments of Huygens and others*) that light also behaved as a wave, and that its speed (*in a vacuum*) was c. When Maxwell predicted that the speed of electromagnetic waves was also equal to c, he hypothesized that this was no coincidence — light must be a special case.

But how to prove it?

The proof came through the experiments of HEINRICH HERTZ (1857-1894) who observed that if an electromagnetic wave were to encounter a wire aligned with the waves electric field, that field would force the electrons in the wire to oscillate, generating a current that could be measured. The successful implementation of this idea confirmed all aspects of Maxwells predictions and confirmed lights character as an electromagnetic wave. Ironically, Hertz did not perceive the enormous importance of his experiments which he considered to have no practical applications whatever.

. . . .

The following diagram illustrates the electromagnetic spectrum and shows the various wavelengths in comparison with familiar objects.

(21.3.3) Figure: *Comparison of Wavelengths and Objects.*

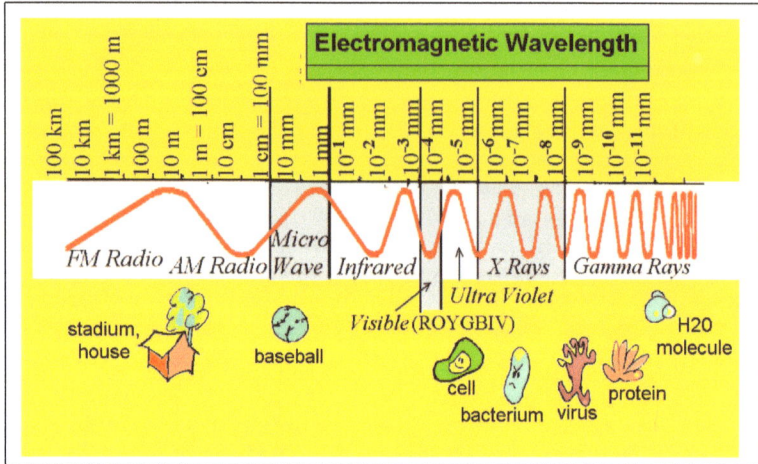

The electromagnetic spectrum includes radio waves (longest) to gamma rays (shortest). Notice the tiny fraction of the spectrum that is taken up by visible light — from about 7.5×10^{-4} mm (red) to 3.5×10^{-4} mm (violet).

Until very recently, Roy G Biv, the R)ed O)range Y)ellow G)reen B)lue I)ndigo V)iolet colors of the visible rainbow, represented the totality of observable reality. With devices such as compasses, radio receivers, and photographic film, we extend our five senses so that we can derive information from a wider range of the electromagnetic spectrum .

(22) Electromagnetism: A Quantitative View

The proper linking of mathematics to reality is not always clear, although the power of mathematical modeling is rarely in doubt. (*For an example of a model based on a false assumption, see (23.6), pg. 322.*) A befuddled character, Miss Withers, articulates this point of view in the 1934 book, *The Puzzle of the Silver Persian* [26], by Stuart Palmer, where we read

> Miss Withers...sat and stared at the red coals of her fire... they presented innumerable, fantastic pictures, but never did they suggest the clear and definite answers to this complex puzzle which the spinster wanted to see. Finally, she took up a piece of hotel stationery. "I might do it by algebra," she thought. "X equals — that's the trouble. X doesn't equal anything. Nor, for that matter, does Y or Z." She swept away her meaningless figures and wished wistfully that she understood relativity.

Theory Meets Observation: In this chapter, we shall see how the theoretically derived wave equation (*Lemma (16.3.7), pg. 206, and Theorem (16.4.1)*) which describes the speed v of a traveling time-dependent wave $f(x, t)$ in one dimension, describes Maxwell's experimental observations exactly for the propagated electrical and magnetic waves, $E(x, t)$ and $B(x, t)$. We will then see how Maxwell then calculated each of their speeds to be $v = c$, the speed of light in a vacuum.

(22.1) Quantitative Preliminaries

(22.1.1)
Calculus
needed

Notations for Fields $B(x,t)$, $E(x,t)$, and Their Arithmetic Differences. We will be interested in the case where the electric and magnetic fields propagate along a direction that we will choose as the X-axis of a reference system. In this case, the magnetic and electric field values, $B(x, t)$, and $E(x, t)$, depend on the *two* variables: distance, x, and time, t. The following table lists compressed notations we will use when one of the variables is held constant (and, hence, suppressed).

(22.1.1a)   **Compressed Notations for $B(x,t)$, $E(x,t)$**

x varies, t fixed	t varies, x fixed
$B_x \overset{def}{=} B(x,t)$	$B_t \overset{def}{=} B(x,t)$
$E_x \overset{def}{=} E(x,t)$	$E_t \overset{def}{=} E(x,t)$

(22.1.1b) Orientation of $\vec{B}_x, \vec{E}_x$: As per Figure (22.2.2), pg. 295, E_x and B_x represent the values of the electric and magnetic fields that are <u>perpendicular</u> to the X-axis at point x — the vector $\vec{B}_x$ is parallel to the Y axis and vector $\vec{E}_x$ is parallel to the Z axis. (*Sometimes E_x is used to denote the component of the electric field along the x direction, a convention we will not follow.*)

With the compressed notation of (22.1.1a), we have an economical way to write the differences between values of $B(x,t)$ and $E(x,t)$ at pairs of points, $\{x, x + \Delta x\}$ when time, t, is fixed; and between values $\{t, t + \Delta t\}$ when distance, x, is fixed. The following notation will be used for differences such as $B_{x+\Delta x} - B_x$.

(22.1.1c)   Notations for ΔB, ΔE

x varies, t fixed	t varies, x fixed
$\Delta B_x \stackrel{def}{=} (B_{x+\Delta x} - B_x)$	$\Delta B_t \stackrel{def}{=} (B_{t+\Delta t} - B_t)$
$\Delta E_x \stackrel{def}{=} (E_{x+\Delta x} - E_x)$	$\Delta E_t \stackrel{def}{=} (E_{t+\Delta t} - E_t)$

Role of Partial Derivatives, $\partial/\partial x$, $\partial/\partial t$, in (22.1.1c). With sufficient (and reasonable) continuity conditions on the function B, we can always make Δx small enough to bring $\Delta B_x/\Delta x$ as close as we like to the partial derivative, $\partial B_x/\partial x$. We say this symbolically by writing

(22.1.1d) $\dfrac{\Delta B_x}{\Delta x} \approx (\dfrac{\partial B_x}{\partial x})$ or $\Delta B_x \approx (\dfrac{\partial B_x}{\partial x})\Delta x.$ $\left\{ \begin{array}{l} \textit{Partial} \\ \textit{derivatives} \\ \textit{are defined} \\ \textit{in (16.3.4).} \end{array} \right.$

A similar argument justifies the other $\approx$ relationships of (22.1.1c) relative to ΔB_t, ΔE_x, and ΔE_t.

(22.1.2) **Mathematicians and physicists** often approximate complicated objects with simple objects that are better understood.

For example, oddly-shaped areas can be approximated by rectangles that are "close enough" to the original area. (*This is a key idea in calculus.*)

In general, reality is often approximated or simplified with idealized models as when, for example, we claim that a rectangular, uniformly dense wooden board with equal weights at either end, and with its pivot in the exact center (*a see-saw*), will remain in a balanced, horizontal position. In the real world, however, the wood density may vary (contrary to our assumption), and our measurements are never *exact* down to the molecular level. However, simplified models are considered valid (enough) when they work effectively.

Figure (22.1.3) illustrates similar simplifications that will be used when we apply the laws of Faraday and Ampère-Maxwell to calculate the speed of light.

(22.1.3) FIGURE: *Approximately Parallel & Approximately Equal Lines.*

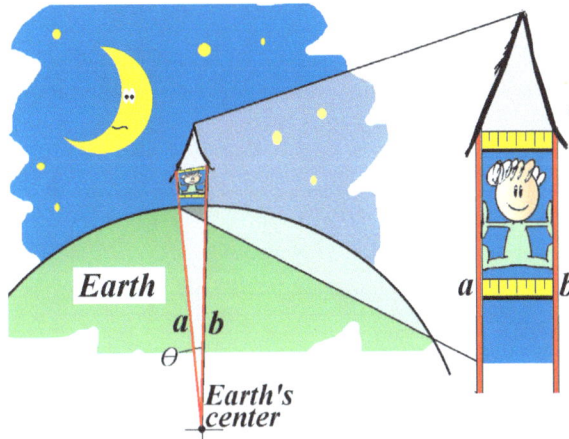

(22.1.3a) **Parallel Lines:** The vertical pie-shaped segment (with the roof on top) has its vertex at the center of the earth. Sides a and b, which form the walls of a building at the surface, are APPROXIMATELY PARALLEL if the vertex angle, θ, is small enough. In the real world, we act as if lines a and b of Figure (22.1.3) really *are* parallel, as when we confidently use a plumb line to align "vertical" walls of a building.

(22.1.3b) **Arcs are approximately straight.** In our everyday experience, we pretend that the surface of a lake is horizontal and straight even though it follows the arc of the earth's circumference. Similarly, when θ in Figure (22.1.3) is very small, it is virtually impossible to know whether the floor and ceiling are true horizontal straight lines, or whether they follow the circular arc of the earth's surface.

But with such small θ, who can tell the difference?

(22.1.3c)   Equal Lines: Often, a non-constant function f is considered to be "approximately constant" if neighboring x and x' produce values, $f(x)$ and $f(x')$, that do not differ significantly. That is, if x is close to x', then $f(x)$ is close to $f(x')$, or $f(x) \approx f(x')$.

For example, suppose sides $a = f(x)$ and $b = f(x')$ in Figure (22.1.3) are values for some function f. Suppose side a is 20 feet shorter than side b so that from the surface of the earth, the house wall a would be a full two stories shorter than wall b. But this inequality begins to disappear when we consider the full lengths of segments a and b. In fact, if the length $a = 4,000\ miles$, and $b = 4,000\ miles + 20\ feet$, then sides a and b are APPROXIMATELY EQUAL in the sense that their lengths differ, relatively speaking, by a tiny amount — about one part in a million, or $1/10,000$ th of 1%.

(22.2) A Quantitative View of Propagation

(22.2.1) The Alternating Current Setup: We focus attention only on the X-axis of Figure (21.1.1), which is a single ray that threads its way through the magnetic field $\vec{B}$ generated by the alternating current.

(22.2.1a)   Graphing B Vectors.   The visual bookkeeping method of recording values of $\vec{B}_x$ at each point x is to imitate the needle of a south-seeking compass placed at that point.

Specifically, at each point x on the X-axis, draw a vector $\vec{B}_x$
- in the X-Y plane,
- perpendicular to the X-axis,
- with tail at the point x, pointing in a north→south orientation, and
- with length $\|\vec{B}_x\| = B_x$, equal to the magnitude of the magnetic field measured at point x.

To aid in the proofs of following theorems, we set down Figure (22.2.2).

(22.2.2) FIGURE: *A Changing Magnetic Field in an Area Induces an Electrical Field on its Boundary*

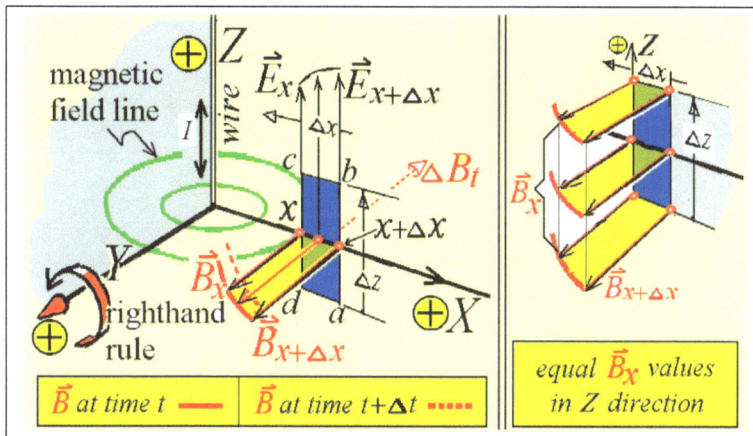

The magnetic $\vec{B}_x$ vectors depend on the radial distance x from the vertical wire and are tangent to the green circles shown in the left hand panel.

The electrical vectors $\vec{E}_x$ also depend on the radial distance x from the vertical wire. However, due to the right-hand rule about the Y-axis, vectors $\vec{E}_x$ are perpendicular to the X-axis, pointing in the Z direction, which is parallel to the wire. (See Note (22.1.1b) describing the configurations of $\vec{E}_x$ and $\vec{B}_x$.)

(22.2.3)
Calculus
needed

THEOREM: *(**Faraday's Law. Magnetic Field Induces Electric Field.**) Given a vertical wire with alternating current, $I(t)$ $\updownarrow$. Then the changing horizontal magnetic field vectors $\vec{B}$ induce vertical electric field vectors, $\vec{E}$, such that at time t and a radial distance x from the wire, we have*

(22.2.3a)
$$-\frac{\partial B_t}{\partial t} = \frac{\partial E_x}{\partial x}.$$

PROOF: *See Figure* **(22.2.2).** Equation (22.2.3a) is the general Faraday's Law of Induction (20.7.3c) as applied to a differential rectangular area $\Delta A = \Delta x \Delta z$ shown in the right hand panel of Figure (22.2.2).

• **The orientation of the boundary of ΔA.** Since the positive direction of the Y-axis coincides with the positive direction of the axle (17.4.5), the right hand rule tells us that the rectangular boundary of the area $\Delta A = \Delta x \Delta z$ in Figure (22.2.2) is oriented *counterclockwise*. That is, orientation of $bdry(\Delta A)$ is described by

(22.2.3b) $bdry(\Delta A) = a \rightarrow b \rightarrow c \rightarrow d \rightarrow a$,   *in that order!*

• **The magnetic flux,** from (19.2.2a), is defined over the differential area $\Delta A = \Delta x \Delta z$ from Figure (22.2.2), to be

(22.2.3c) $\Phi_B(\Delta A) \approx B_t \cdot \Delta A = B_t \cdot \Delta x \Delta z.$

Since we consider the magnetic field at a fixed distance x but at different times, we use B_t in (22.2.3c) to represent $B(x,t)$ (22.1.1c). The philosophy of (22.1.3c) allows us to assume that B is approximately constant over the sufficiently small differential area ΔA. That is, the value of $B_t = B(x,t)$ is as good a choice as $B(x',t)$ when x' is close to x.

• **The right hand side of Faraday's law,** (20.7.3c), requires that we compute $-d/dt\,\Phi_{\vec{B}}(\Delta A)$, the negative of the time derivative of the magnetic flux, $\Phi_{\vec{B}}(\Delta A)$ (19.2.2a). Accordingly, when the position is fixed,

(22.2.3d) $-\dfrac{d}{dt}\Phi_B(\Delta A) = -\dfrac{d}{dt}\left(B_t \cdot \Delta A\right)$ *from (22.2.3c)*

$$= -\left[\dfrac{d}{dt}B_t\right]\cdot\underbrace{\Delta x \Delta z}_{\Delta A} \qquad \begin{array}{l}\textit{factor out the}\\ \textit{constant } \Delta A\end{array}$$

$$= -\dfrac{\partial B_t}{\partial t}\Delta x \Delta z\,. \qquad \begin{array}{l}\dfrac{d}{dt}\textit{ above is actually the}\\ \textit{partial } \dfrac{\partial}{\partial t}\textit{ with x fixed}\end{array}$$

• **The left hand side of Faraday's law,** (20.7.3c), asks us to compute the closed line integral of the induced electric field, $\vec{E}_l$, at points l on $bdry(A)$, namely,

(22.2.3e) $\displaystyle\oint_{bdry(\Delta A)} \langle \vec{E}_l, d\vec{l}\rangle = \oint_{a\,\to\,b\,\to\,c\,\to\,d\,\to\,a} \langle \vec{E}_l, d\vec{l}\rangle$ *counterclockwise,*
                                                                              from (22.2.3b)

$$=\underbrace{\int_{z=a}^{z=b}\langle \vec{E}_{x+\Delta x}, d\vec{z}\rangle}_{(1)}+\underbrace{\int_{x=b}^{x=c}\langle \vec{E}_x, d\vec{x}\rangle}_{(2)}+\underbrace{\int_{z=c}^{z=d}\langle \vec{E}_x, d\vec{z}\rangle}_{(3)}+\underbrace{\int_{x=d}^{x=a}\langle \vec{E}_x, d\vec{x}\rangle}_{(4)}.$$

Each of the integrals above is of the class given by (17.8.6a), where integration occurs along the four counterclockwise oriented chain of segments, $\Delta\vec{z}_1$, $\Delta\vec{x}_2$, $\Delta\vec{z}_3$, and $\Delta\vec{x}_4$ (*see figure at right*). Therefore, the values of integrals (1), (2), (3), and (4) of (22.2.3e), are, respectively,

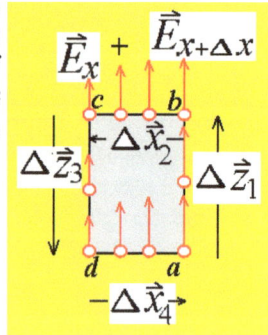

(22.2.3f)

$$\underset{(1)}{\langle \vec{E}_{x+\Delta x}, \Delta\vec{z}_1\rangle} + \underset{(2)}{\langle \vec{E}_x, \Delta\vec{x}_2\rangle}$$

$$+ \underset{(3)}{\langle \vec{E}_x, \Delta\vec{z}_3\rangle} + \underset{(4)}{\langle \vec{E}_x, \Delta\vec{x}_4\rangle}.$$

Using the fact that $\|\Delta x_2\| = \|\Delta x_4\| = \Delta x$ and $\|\Delta z_1\| = \|\Delta z_3\| = \Delta z$, along with Definition (17.6.1), we obtain

(1) $\langle \vec{E}_{x+\Delta x}, \Delta \vec{z}_1 \rangle = E_{x+\Delta x} \underbrace{\cos \theta_1}_{1} \Delta z,$ $\qquad$ $E_{x+\Delta x}$ constant on $\Delta \vec{z}_1$

(2) $\langle \vec{E}_x, \Delta \vec{x}_2 \rangle = E_x \underbrace{\cos \theta_2}_{0} \Delta x,$ $\qquad$ $\vec{E}_x \perp \Delta \vec{x}_2$

(3) $\langle \vec{E}_x, \Delta \vec{z}_3 \rangle = E_x \underbrace{\cos \theta_3}_{-1} \Delta z,$ $\qquad$ E_x constant on $\Delta \vec{z}_3$

(4) $\langle \vec{E}_x, \Delta \vec{x}_4 \rangle = E_x \underbrace{\cos \theta_4}_{0} \Delta x,$ $\qquad$ $\vec{E}_x \perp \Delta \vec{x}_4$

since the angles θ_i between electric field vectors, $\vec{E}$, and the line segments $\Delta \vec{z}$ and $\Delta \vec{x}$ have values $\theta_1 = 0$, $\theta_2 = 90°$, $\theta_3 = 180°$, and $\theta_4 = 0$. Therefore, we have the sum of the inner products (22.2.3f)

$$(1) + (2) + (3) + (4) = (E_{x+\Delta x} - E_x)\Delta z$$

$$= \Delta E_x \Delta z. \qquad \text{from notation of } (22.1.1c)$$

Equate this term, $\Delta E_x \Delta z$ (left hand side of Faraday's law (20.7.3c)), to $(-\partial B_t / \partial t) \Delta x \Delta z$ of (22.2.3d), (the right hand side of Faraday's law (20.7.3c)), and solve for $(-\partial B_t / \partial t)$ to obtain

(22.2.3g) $\qquad -\dfrac{\partial B_t}{\partial t} = \dfrac{\Delta E_x}{\Delta x}.$

Since $\dfrac{\Delta E_x}{\Delta x} \to \dfrac{\partial E_x}{\partial x}$ as $\Delta x \to 0$, equality (22.2.3g) transforms to (22.2.3a), which proves the theorem. ∎

Now that the vertical values of electric field vectors $\vec{E}_x$ and $\vec{E}_{x+\Delta x}$ are established in (22.2.3f), we use the Ampère-Maxwell law (20.7.3d) to construct their newly-induced horizontal magnetic field vectors, $\vec{B}_x$, and $\vec{B}_{x+\Delta x}$.

‖ **NOTE:** These are not the same $\vec{B}_x$ and $\vec{B}_{x+\Delta x}$ magnetic vectors we
‖ started with — these vectors appear Δt time units later.

(22.2.4) Figure: *Electrical Field Change Induces Magnetic Field*

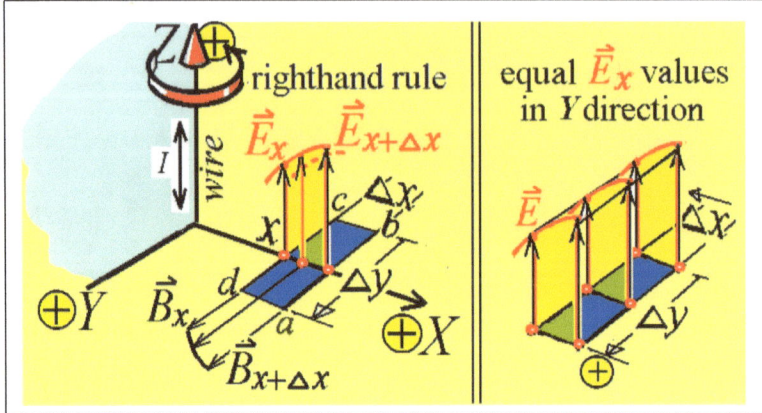

(22.2.5)
Calculus
needed

Theorem: (*Ampère-Maxwell Law. Electric Field Induces Magnetic Field.*) *Given a vertical wire with alternating current, $I(t)$ ↕. Then changing vertical electric field vectors $\vec{E}$ induce horizontal magnetic field vectors $\vec{B}$ such that at time t and at radial distance x from the wire, we have*

(22.2.5a)
$$\mu_0 \varepsilon_0 \frac{\partial E_t}{\partial t} = -\frac{\partial B_x}{\partial x}.$$

Proof:
• **The orientation of the boundary of $\Delta\mathbf{A}$.** Since the Z-axis defines the positive vertical direction, the right hand rule (17.4.5), with the Z-axis as its axle, defines the positive orientation of the rectangular boundary with area $A = \Delta x \Delta z$ in Figure (22.2.2) to be *counterclockwise*. That is, orientation of $bdry(A)$ is described by

(22.2.5b) $bdry(A) = a \to b \to c \to d \to a,$ *in that order!*

....

• **The electric flux,** from (19.2.1a), is defined over the differential area $\Delta A = \Delta x \Delta z$ to be

(22.2.5c) $\Phi_B(\Delta E) \approx E_t \cdot \Delta A = E_t \cdot \Delta x \Delta y.$

Since time t is varying while distance x is fixed, we use E_t in (22.2.5c) to represent $E(x,t)$ (22.1.1c). The philosophy of (22.1.3c) allows us to assume that E is essentially constant over the sufficiently small differential area,

ΔA. That is, the value $E_t = E(x, t)$ is as good a choice as $E_{t'} = E(x, t')$ in the evaluation of E when t' is close to t.

....

• **The right hand side of Ampère-Maxwell law**, (20.7.3d), requires that we compute $\mu_0 \varepsilon_0 \cdot d/dt \Phi_{\vec{E}}(\Delta A)$, which is $\mu_0 \varepsilon_0$ times the time derivative of the electric flux $\Phi_{\vec{E}}(\Delta A)$ (19.2.1a). Accordingly,

(22.2.5d) $\mu_0 \varepsilon_0 \dfrac{d}{dt} \Phi_E(\Delta A) = \mu_0 \varepsilon_0 \cdot \dfrac{d}{dt}(E_t \cdot \Delta A)$ *from (22.2.5c)*

$$= \mu_0 \varepsilon_0 \left[\frac{d}{dt} E_t \right] \cdot \underbrace{\Delta x \Delta y}_{\Delta A}$$ *factor out the constant ΔA*

$$= \mu_0 \varepsilon_0 \frac{\partial E_t}{\partial t} \Delta x \Delta y \, . \quad \begin{cases} \dfrac{d}{dt} \text{ above is actually the} \\ \text{partial } \dfrac{\partial}{\partial t} \text{ with } x \text{ fixed} \end{cases}$$

....

• **The left hand side of Ampère-Maxwell's law**, (20.7.3d), asks us to compute the closed line integral of the induced magnetic field $\vec{B}_l$ at points l on $bdry(A)$, namely,

(22.2.5e) $\displaystyle\oint_{bdry(\Delta A)} \langle \vec{B}_l, d\vec{l} \rangle = \oint_{a \to b \to c \to d \to a} \langle \vec{B}_l, d\vec{l} \rangle$ *counterclockwise, from (22.2.3b)*

$$= \int_{z=a}^{z=b} \langle \vec{B}_{x+\Delta x}, d\vec{y} \rangle + \int_{x=b}^{x=c} \langle \vec{B}_x, d\vec{x} \rangle + \int_{z=c}^{z=d} \langle \vec{B}_x, d\vec{y} \rangle + \int_{x=d}^{x=a} \langle \vec{B}_x, d\vec{x} \rangle .$$
$$\qquad (1) \qquad\qquad (2) \qquad\qquad (3) \qquad\qquad (4)$$

....

Each of the integrals above is of the class given by (17.8.6a), where integration occurs along the four counterclockwise oriented chain of segments, $\Delta \vec{y}_1$, $\Delta \vec{x}_2$, $\Delta \vec{y}_3$, and $\Delta \vec{x}_4$ (*see figure at right*). Therefore, the values of integrals (1), (2), (3), and (4) of (22.2.3e), are, respectively,

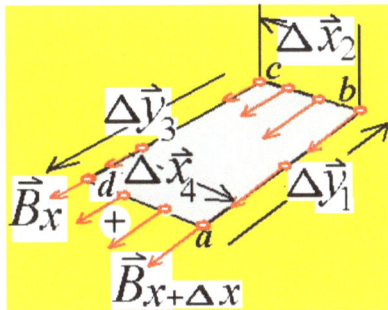

(22.2.5f) $\underset{(1)}{\langle \vec{B}_{x+\Delta x}, \Delta \vec{y}_1 \rangle} + \underset{(2)}{\langle \vec{B}_x, \Delta \vec{x}_2 \rangle} + \underset{(3)}{\langle \vec{B}_x, \Delta \vec{y}_3 \rangle} + \underset{(4)}{\langle \vec{B}_x, \Delta \vec{x}_4 \rangle} .$

Using the fact that $\|\Delta x_2\| = \|\Delta x_4\| = \Delta x$ and $\|\Delta y_1\| = \|\Delta y_3\| = \Delta y$, along with Definition (17.6.1), we obtain

(1) $\langle \vec{B}_{x+\Delta x}, \vec{\Delta} y_1 \rangle = B_{x+\Delta x} \underbrace{\cos \theta_1}_{\text{-1}} \Delta y,$

(2) $\langle \vec{B}_x, \vec{\Delta} x_2 \rangle = B_x \underbrace{\cos \theta_2}_{0} \Delta x,$

(3) $\langle \vec{B}_x, \vec{\Delta} y_3 \rangle = B_x \underbrace{\cos \theta_3}_{1} \Delta y,$

(4) $\langle \vec{B}_x, \vec{\Delta} x_4 \rangle = B_x \underbrace{\cos \theta_4}_{0} \Delta x$

since the angles θ_i between the magnetic field vectors, $\vec{B}$, and the line segments $\Delta \vec{x}$ and $\Delta \vec{y}$ have values $\theta_1 = 180°$, $\theta_2 = 90°$, $\theta_3 = 0°$, and $\theta_4 = 90°$. Therefore, we have the sum of the inner products (22.2.5f)

$$(1) + (2) + (3) + (4) = -(B_{x+\Delta x} - B_x)\Delta y$$

$$= -\Delta B_x \Delta y . \qquad \textit{from notation of (22.1.1c)}$$

Equate this term, $-\Delta B_x \Delta y$ (left hand side of Ampère-Maxwell law (20.7.3d)), to $(\mu_0 \varepsilon_0 \partial B_t / \partial t)\Delta x \Delta y$ of (22.2.5d), (the right hand side of the Ampère-Maxwell law (20.7.3d)), and solve for $\mu_0 \varepsilon_0 (\partial B_t / \partial t)$ to obtain

(22.2.5g) $\qquad \mu_0 \varepsilon_0 \dfrac{\partial B_t}{\partial t} = -\dfrac{\Delta B_x}{\Delta x}.$

Since $\dfrac{\Delta B_x}{\Delta x} \rightarrow \dfrac{\partial B_x}{\partial x}$ as $\Delta x \rightarrow 0$, equality (22.2.5g) transforms to (22.2.5a), which proves the theorem. ∎

(22.3) Theoretical Speed of Wave Propagation

`**We restate and summarize** the results of Faraday's law and the Ampère-Maxwell law. Their differential forms, (22.2.3a) and (22.2.5a), appear in the first row of the following table as entries ① and ②. The second row, with entries ③ and ④ contains equation (16.3.7a), which relates the time derivative, $\partial / \partial t$ and the distance derivative $\partial / \partial x$, for any function traveling rightward with speed v.

To allow for all possibilities, we at first assume the two waves, $\vec{B}$ and $\vec{E}$, might have different speeds, which we denote by v_B and v_E, respectively. (*As we shall see in (22.3.3a), v_B necessarily equals v_E.*)

(22.3.1)

①	$-\dfrac{\partial B}{\partial t} = \dfrac{\partial E}{\partial x}$	②	$\mu_0 \varepsilon_0 \dfrac{\partial E}{\partial t} = -\dfrac{\partial B}{\partial x}$
	Faraday Law **(22.2.3a)**		*Ampère-Maxwell Law* **(22.2.5a)**
③	$\dfrac{\partial B}{\partial x} = -\dfrac{1}{v_B}\dfrac{\partial B}{\partial t}$, (16.3.7a)	④	$\dfrac{\partial E}{\partial x} = -\dfrac{1}{v_E}\dfrac{\partial E}{\partial t}$, (16.3.7a)
		Mixed-Partials Property **(16.3.7a)**	

(22.3.2) **Consequences of** equations ①, ②, ③, and ④ of Table (22.3.1) are further distilled and summarized as follows:

(22.3.2a) $\qquad \dfrac{\partial E}{\partial t} = v_E \dfrac{\partial B}{\partial t}$, $\qquad\qquad$ *from* ①, ④, *eliminate* $\dfrac{\partial E}{\partial x}$

(22.3.2b) $\qquad \dfrac{\partial B}{\partial t} = v_B \mu_0 \varepsilon_0 \dfrac{\partial E}{\partial t}$, $\qquad$ *from* ②, ③, *eliminate* $\dfrac{\partial B}{\partial x}$

(22.3.2c) $\qquad \dfrac{\partial E}{\partial x} = v_B \dfrac{\partial B}{\partial x}$, $\qquad\qquad$ *from* ①, ③, *eliminate* $\dfrac{\partial B}{\partial t}$

(22.3.2d) $\qquad \dfrac{\partial B}{\partial x} = v_E \mu_0 \varepsilon_0 \dfrac{\partial E}{\partial x}$. $\qquad$ *from* ②, ④, *eliminate* $\dfrac{\partial E}{\partial t}$

We now have the tools to show how Maxwell computed c, the speed of electromagnetic (*EM*) waves in a vacuum. Light was not known to be in the family of electromagnetic waves in Maxwell's time, although he was impressed by the "coincidental" fact that his (invisible) electromagnetic waves, and visible light, had exactly the same speed.

(22.3.3) Calculus needed

> **THEOREM:** (***Wave speed in terms of permeability and permittivity.***) *Given an electromagnetic wave with an electric field component E having speed v_E, and a magnetic field component, B, having speed v_B. If the electrical permittivity is ε_0 and magnetic permeability is μ_0, then, necessarily,*
>
> **(22.3.3a)** $\qquad v_E = v_B = c = \dfrac{1}{\sqrt{\varepsilon_0 \mu_0}}$.

PROOF: Since the order of partial derivatives is irrelevant, we may equate $\partial/\partial x$ of Equation (22.3.2a) with $\partial/\partial t$ of Equation (22.3.2c) to obtain $v_E = v_B$ for both the speed v_E of the magnetic wave $\vec{E}$ and the speed v_B of the electric wave $\vec{B}$. Denoting this common speed by

(22.3.3b) $\qquad v_E = v_B = c$

establishes the first two equalities of (22.3.3a). The equal speeds of (22.3.3b) are intuitively correct since, from Figure (21.2.2), the time-dependent field E generates a time dependent field B which, in its turn, generates an electric field farther from the source. In particular, both E and B propagate together, neither lags behind the other, which implies $v_E = v_B$.

. . . .

If we eliminate any one of the four partial derivatives, $(\partial E/\partial x)$, $(\partial E/\partial t)$, $(\partial B/\partial x)$, $(\partial B/\partial t)$, from the equations, (22.3.2a)-(22.3.2d), we will obtain the formula $c = 1/\sqrt{\mu_0 \varepsilon_0}$.

For example, eliminate $(\partial B/\partial t)$ from (22.3.2a) and (22.3.2b), to obtain $\partial E/\partial t = (c^2 \mu_0 \varepsilon_0)\partial E/\partial t$, from which it follows that $1 = c^2 \mu_0 \varepsilon_0$. This completes the third equality of Equation (22.3.3a) and the proof is done.∎

NOTE: ***Wave equation and Maxwell's equations.*** *With (22.3.3a) in hand, we can show that Faraday's law and the Ampère-Maxwell law imply that each of the fields, $\vec{B}$ and $\vec{E}$, satisfies the wave equation (16.4.1a). See Exercise (22.6).*

Now that we know that the speed c (22.3.3a) applies to both components, $\vec{B}$ and $\vec{E}$ of the electromagnetic wave, we can relate their magnitudes at any point x_0 in space.

(22.3.4)
Calculus
needed

> THEOREM: (***Linking E and B.***) *Given an electromagnetic wave with electric field component $\vec{E}$ and magnetic field component $\vec{B}$. Assume the null initial condition,*
>
> **(22.3.4a)** $E(0,0) = B(0,0) = 0$.
>
> *Then the electromagnetic wave propagates with speed c (22.3.3a) in the direction*
>
> **(22.3.4b)** $\vec{E} \times \vec{B}$
>
> *where the magnitudes of the electric and magnetic components, E and B, are proportional as indicated by the equation*
>
> **(22.3.4c)** $E(x_0, t_0) = c \cdot B(x_0, t_0)$
>
> *for all distances x_0 and times t_0.*

PROOF: The direction (22.3.4b) of the propagated wave was established in (21.2.3b).

To validate (22.3.4c), integrate (22.3.2c) over x (holding t_0 fixed). Replacing the symbol $\partial/\partial x$ with d/dx, since the second variable, t_0, is fixed, we have

$$\int_{x=0}^{x=x_0} \left(\frac{dE}{dx}\right) dx = \int_{x=0}^{x=x_0} c \cdot \left(\frac{dB}{dx}\right) dx$$

whose left and right sides compute to be, respectively,

(22.3.4d) $[E(x_0, t_0) - E(0, t_0)] = c \cdot [B(x_0, t_0) - B(0, t_0)].$

Similarly, integrating (22.3.2a) over t, holding $x = 0$ fixed, we obtain

$$\int_{t=0}^{t=t_0} \left(\frac{dE}{dt}\right) dt = \int_{t=0}^{t=t_0} c \cdot \left(\frac{dB}{dt}\right) dt$$

whose left and right sides are, respectively,

(22.3.4e) $[E(0, t_0) - E(0, 0)] = c \cdot [B(0, t_0) - B(0, 0)].$

Now once we add Equations (22.3.4d) and (22.3.4e) and use fact (22.3.4a), which holds that $B(0,0) = E(0,0) = 0$, we see that (22.3.4c) results. This ends the proof. ∎

(22.4) Maxwell's Calculation of c

(22.4.1) **List of Units.** Here follows a brief list, in the MKS (meter-kilogram-second) system of units, that we will use:

- mtr: The **meter**, a unit of distance.
- sec: The **second**, a unit of time.
- C: The **coulomb**, the unit of charge on a particle q.
- A: The **ampere** $= C/\text{sec}$, defined as one coulomb per second. Amperes measure the coulombs/second (an electric current) passing through a wire which is analogous to measuring gallons/second (a liquid current) of water flowing through a conducting pipe.
- N: The **newton**, a unit of force.

From this list, we take away the following: If one ampere A equals one coulomb per second, i.e., $1A = 1C/1\text{sec.}$, then, consistent with methodology of dimensional analysis (Appendix (??)), this equality translates to the unitless quotient

(22.4.2) $1 = \dfrac{A}{C/\text{sec}}.$

Recall that permittivity (18.2.5a) ε_0 appears in Coulomb's Law (18.2.4b), pg. 244, and can therefore be measured in electrostatic experiments. Similarly, permeability (20.2.2) μ_0 appears in the Biot-Savart Law (20.2.1a), pg. 271, and can be measured in magnetostatic experiments. Accepted values are

(22.4.3)
$$\varepsilon_0 = 8.854\ 187\ 817\ 62 \times 10^{-12} \dfrac{C^2}{N \cdot \text{mtr}^2} \qquad and$$

$$\mu_0 = 1.256\ 637\ 061\ 43 \times 10^{-6} \dfrac{N}{A^2}$$

whose product is

$$\varepsilon_0 \mu_0 \approx 11.1265 \times 10^{-18} \dfrac{C^2}{\text{mtr}^2 A^2} \qquad\qquad newtons\ \text{``}N\text{''}\ cancel.$$

$$= 11.1265 \times 10^{-18} \dfrac{C^2}{\text{mtr}^2 \cdot A^2} \underbrace{\left(\dfrac{A}{C/\text{sec}}\right)^2}_{1} \qquad\qquad from\ (22.4.2)$$

$$= 11.1265 \times 10^{-18} \dfrac{\cancel{C}^2}{\text{mtr}^2 \cancel{A}^2} \left(\dfrac{\cancel{A}^2 \quad \text{sec}^2}{\cancel{C}^2}\right) \qquad\qquad unit\ cancellation$$

$$= 11.1265 \times 10^{-18} \dfrac{\text{sec}^2}{\text{mtr}^2}. \qquad\qquad final\ form$$

With $\varepsilon_0 \mu_0$ in hand (*at the end of this chain of equalities*), we finally calculate $(\sqrt{\varepsilon_0 \mu_0})^{-1}$, which is c, the speed of light in a vacuum. In so doing, we discover that

(22.4.4)
$$c = \dfrac{1}{\sqrt{\varepsilon_0 \mu_0}} \approx 2.99792 \times 10^8 \ \dfrac{\text{mtr}}{\text{sec}}$$

where the final units, $\dfrac{\text{mtr}}{\text{sec}}$ (*usually denoted* $\dfrac{m}{s}$ *or* m/s) are, in fact, units of speed.

(22.5) Mathematical Hits

(22.5.1) **Hit:** *Unification.* Maxwell organized the known mathematical models of electricity and magnetism and then made them consistent by adding the "virtual current" or displacement current term $\varepsilon_0 \mu_0 d\Phi_{\vec{E}}(A)/dt$ to Ampère's law (20.4.3c). Without this correction, there would be no conservation of charge and the electrical and magnetic fields would not satisfy the wave equation.

This quartet of equations (20.7.3), commonly known as Maxwell equations, unified the theory of electricity, optics, and magnetism.

(22.5.2) **Hit:** *Predicting Wave Structure.* Maxwell's equations predicted the perpendicularity of the electric and magnetic components of electromagnetic waves (*Figure* (21.3.1)). Moreover, Maxwell's equations show that the electric and magnetic components fields each satisfy the wave equation (22.6.1).

Wave, what wave? This theoretical mathematical wave structure was finally confirmed experimentally by HEINRICH HERTZ (1857-1894) twenty years later, in 1888. (*See* (21.3.2) *for more detail.*)

(22.5.3) **Hit:** *Inspiring Einstein.* Einstein's two basic assumptions of special relativity, especially the assumption that c is constant for all observers (1.5.2b), regardless of their inertial frame, can be traced back directly to Maxwell's work.

In fact, in the very opening lines of his seminal 1905 paper on special relativity [12], Einstein cites the "asymmetries" in electromagnetic theory (*See the quote in* (22.5.3b).)

The asymmetries mentioned by Einstein are these:
The current induced as a magnet moves through a stationary wire loop is determined by the Faraday-Lentz law. If, however, the magnet is held stationary and the wire loop is moved through it, the induced current is determined by a different law, the Ampère-Maxwell law. Despite the difference in the responsible laws, the resulting current is the same and depends only on the relative motion of the wire loop (frame) and the magnet (frame).

(22.5.3a) F<small>IGURE:</small> *Two Frame-dependent Explanations*
for Current I

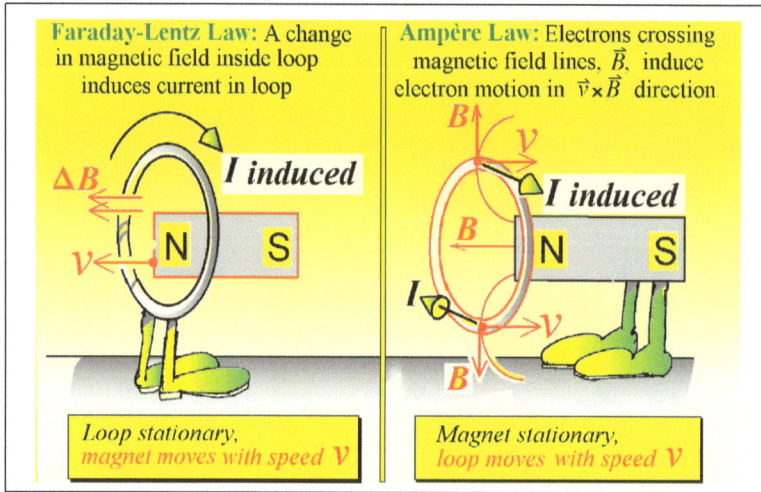

- L<small>EFT</small> P<small>ANEL:</small>   *When the wire is fixed and the magnet moves, current I is induced (in the direction shown) due to Faraday's Law (20.7.3c).*
- R<small>IGHT</small> P<small>ANEL:</small> *When the magnet moves and the wire is stationary, then the <u>same</u> current I is induced, as predicted by the Ampère-Maxwell Law (20.7.3d).*

To quote Einstein from *On the electrodynamics of moving bodies* [12]:

(22.5.3b)   *It is known that Maxwell's electrodynamics — as usually understood at the present time — when applied to moving bodies, leads to asymmetries which do not appear to be inherent in the phenomena. Take, for example, the reciprocal electrodynamic action of a magnet and a conductor. The observable phenomenon here depends on the relative motion of the conductor and the magnet, whereas the customary view draws a sharp distinction between the two cases in which either the one or the other of these bodies is in motion.*

Endorsing Maxwell's work, while taking Galileo's relativity one step further (1.5.2a), Einstein postulated that all laws of physics, *including* those of electromagnetism, shall be deemed valid in <u>all</u> inertial frames of reference.

In addition, Einstein postulated that c, the speed of light in a vacuum, is constant for all observers (*in all frames of reference*) independent of whether the light source, or the observer, is in motion). One might argue that this postulate is unnecessary since the it follows from Maxwell's equations (22.3.3a).

(22.5.4) Hit: *Predicting relation of energy and frequency.* The wave equation defies intuition in the following sense: A rope is tied at one end as shown in Figure (16.1.3) and a wave is propagated in the rope by the vertical oscillation of its loose end which is being pulled with constant "pulling" force T. (*This perceived model is the reality phase of the analysis.*) Intuition would have it that if we hold the pulling force constant, and invest the rope with more energy by oscillating the loose end faster, then the speed of the propagated wave should increase. But it doesn't. Let the traveling wave be modeled by Figure (16.1.6). (*This graphic represents the cartoon phase of the analysis.*) Then, analyzing the mathematical structure — vector decompositions, etc. — we arrive at Equation (16.1.6e) which tells us that as long as the pulling force T is constant, the speed v does not change. In fact, it is the frequency of the traveling wave that will change. (This is the mathematical component of the analysis.)

Although the propagated wave in the rope is a mechanical phenomenon, there is a parallel property with respect to light waves or electromagnetic waves — light that is invested with higher energy does not travel faster (its speed in a vacuum is always c), rather, its *frequency* increases which is evidenced in the visible spectrum by a color change from red (lower frequency) up to violet (higher frequency).

(22.6) Exercises

(22.6.1)
Exercise
Calculus
required

Maxwell equations imply the wave equation. Show that the differential forms of the Faraday and Ampère-Maxwell laws, items ① and ② of (22.3.1), imply that each of the fields, $B(x,t)$ and $E(x,t)$, satisfies the wave equation (16.4.1a).

Hint: *Assume the functions $B(x,t)$ and $E(x,t)$ are sufficiently smooth so that mixed partials are equal (independent of the order of differentiation). That is,*

$$\frac{\partial^2 B}{\partial x \partial t} = \frac{\partial^2 B}{\partial t \partial x} \quad and \quad \frac{\partial^2 E}{\partial x \partial t} = \frac{\partial^2 E}{\partial t \partial x}.$$

The figure shows a graph (drawn horizontally in the X-Y plane) of values of $\vec{B}_x$ on the X-axis over a differential width of Δx.

. . . .

(22.6.2)
Exercise

Prove the following

THEOREM: *The function $B(x,t)$ is real-valued depending on the space variable x and time t that travels to the right as t increases. That is, in the X-Y plane, the graph of $B_x^{(t+\Delta t)}$ is to the right of the earlier $B_x^{(t)}$ graph. Then:*

If $B(x,t)$ increases in x with t fixed ($\Delta B_x > 0$), then $B(x,t)$ decreases in t with x fixed ($\Delta B_t < 0$).

Conversely,

If $B(x,t)$ decreases in x with t fixed ($\Delta B_x < 0$), then $B(x,t)$ increases in t with x fixed ($\Delta B_t > 0$).

Formally, in the notation of (22.1.1a), we have, for any (x,t),

(22.6.2a)
$$\Delta B_x > 0 \ (t \text{ is fixed}) \quad implies \quad \Delta B_t < 0 \ (x \text{ is fixed}),$$
$$\Delta B_x < 0 \ (t \text{ is fixed}) \quad implies \quad \Delta B_t > 0 \ (x \text{ is fixed}).$$

Hint: *Properties (22.6.2a) can be seen in two ways — graphically and analytically: Figure (22.6.2b) shows two panels in which the function $y = B(x,t)$ moves to the right as t increases.*

(22.6.2b)   FIGURE: *Function $y = B(x,t)$ Traveling to the Right as Time Increases*

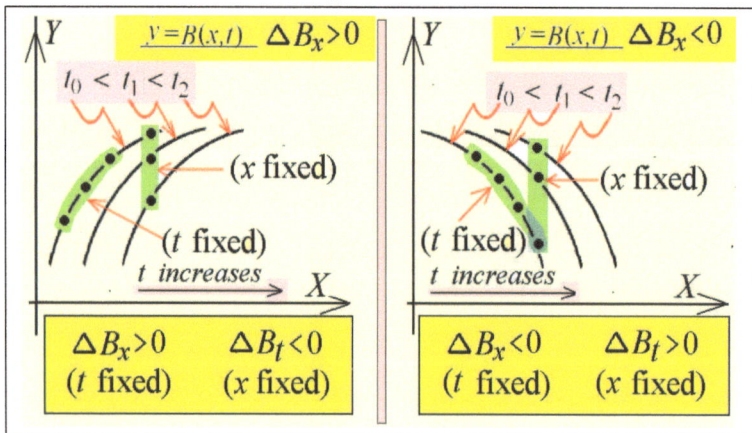

---————————————— Graphically —————————————---

Left panel of (22.6.2b).
Fix *t*. The function $B(x, \cdot)$ in the X-Y plane has positive slope $\underline{(\Delta B_x > 0)}$. This is shown by the points in the slanted grey area where increasing x implies increasing $y = B(x, \cdot)$.

Fix *x*. The sequence of points in the vertical grey area shows that *as t increases*, the values of $y = B(x, t)$ *decrease* $\underline{(B_t < 0)}$.

Right panel of (22.6.2b).
Fix *t*. Function $B(x, \cdot)$ in the X-Y plane has negative slope $\underline{(\Delta B_x < 0)}$. This is shown by the points in the slanted grey area where increasing x implies decreasing $y = B(x, \cdot)$.

Fix *x*. The sequence of points in the vertical grey area shows that *as t increases*, the values of $y = B(x, t)$ *increase* $\underline{(B_t > 0)}$.

The underlined quantities establish (22.6.2a).

---————————————— Analytically —————————————---

In (16.3.7), substitute B for f to see that the signs of $\partial B_x / \partial x$ and $\partial B_t / \partial t$ are always opposite to each other. That is, (16.3.7a) tells us that B_x always increases (decreases) in x as B_t decreases (increases) in t. Briefly put,

$$\text{sign}(\Delta B_x) = -\text{sign}(\Delta B_t).$$

IX. Final Thoughts

Part (X) Final Thoughts. *Some historical context is given along with perspectives on the process of knowing, its applicability, and reliability.*

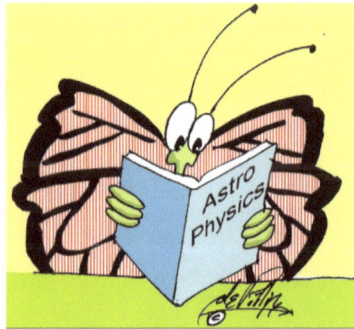

(23) Epilogue: Final Thoughts

(23.1) A Coming of Age

(23.1.1) Old Mathematics. Thousands of years of mathematics have produced more than mere lists of formulas and deduced results. One component of our intellectual heritage that has been defined and nourished by mathematical progress is the *process of proof*.

The theorem of PYTHAGORAS OF SAMOS (*circa* 580-500 B.C.) generalized a property of right triangles that had been known for a few special cases. (*a* RIGHT TRIANGLE *is any triangle in which two sides form a* 90° *angle*.) For example, it was known that for the $(3, 4, 5)$, right triangle, $3^2 + 4^2 = 5^2$, and for the $(5, 12, 13)$ triangle, we have $5^2 + 12^2 = 13^2$. Pythagoras' theorem says that for *all* right triangles with sides (a, b, c), we have $a^2 + b^2 = c^2$. The converse is also true — if the sides (a, b, c) of some triangle satisfy the equation $a^2 + b^2 = c^2$, then it is necessarily a *right* triangle.

Today, there are dozens of independent proofs of the Pythagorean theorem[1].This theorem has come to transcend the formula. Today, mathematicians have produced generalizations that go well beyond triangles in $\boldsymbol{R}^2$. The far-reaching nature of the Pythagorean theorem is nicely described by Robert Crease in his article, *Pythagoras* [7].

Over 2,300 years ago, EUCLID (325-265 B.C.) wrote his *Elements* [15] in which he compiled and organized a body of geometric theorems. More importantly, he provided proofs that rendered the results accessible to and testable by any reader.

Then, ARCHIMEDES (287-212 B.C.), with his use of *infinitesimals*, was on the verge of discovering calculus before he was slain by a Roman soldier at Syracuse as he was drawing figures in the sand during the Second Punic War. The PALIMPSESTS of Archimedes, documents of Archimedes' works that were copied by monks in the tenth century, are the only source of his *The Methods* and *Stomachion* (*a 14-piece version of the 7-piece tangram*) along with the only known Greek text of his work, *On Floating Bodies*.

---

[1] See the online animation (`www.special-relativity-illustrated.com`) or [8], pg. 249 for a non-computational proof.

(23.1.2)
In the twelfth century, the Palimpsest parchment would be washed and cut to be recycled into liturgical texts. The Palimpsests were rediscovered in 1906 and only in 1998, were they made available for further study when a private collector gave them over to scholars at the WALTERS ART MUSEUM *in Baltimore, Maryland. Fortunately, recovery of the erasures are being made there with the use of multispectral imaging.*
(*See* www.archimedespalimpsest.org.)

(23.1.3) Earliest Mathematical Modeling. Early on, the non-concrete, or abstract operations of *addition, subtraction,* and *multiplication* were used to model real, concrete phenomena.

For example, addition and subtraction would be used to count numbers of villagers, sheep, or bushels of grain. The product, or multiplication of numbers, $a \cdot b$, would serve as the mathematical model of the area inside a rectangle with sides a and b.

Measurement of motion, however, was limited to *averages* over long time intervals — a two-hour journey of 60 cubits, say, defined an *average speed* (not *instantaneous speed*) of 30 cubits per hour.

(23.1.4) Newer Mathematics: *The Recent Blossoming.* If four hundred years ago can be called *recent*, then mathematics has only recently been able to accurately model *instantaneous* changes in motions of objects. This achievement came almost simultaneously with the development or invention of integral and differential calculus by GOTTFRIED WILHELM LEIBNIZ (1646-1716) and ISAAC NEWTON (1642-1727). , which finally allowed for the mathematical representation of speeds and accelerations at arbitrary *instantaneous points* of time, as opposed to *extended intervals* of time. (*For more, see The Triumph of Numbers by I. B. Cohen* [6].)

It was about two centuries after the invention of the calculus, when JAMES CLERK MAXWELL (1831-1879) used these tools to develop a mathematical model — four elegant equations — that united the fields of electricity, magnetism and optics and established the wave property of electromagnetic (*EM*) transmissions. As if that weren't enough, Maxwell's equations also predicted the frame-independent speed c (*in a vacuum*) for electromagnetic waves of all frequencies. It would be decades before *visible light* was confirmed to be a special case of Maxwell's electromagnetic waves. (*see Chapter* 20.)

....

(23.1.5) The Scientific Method. The Scientific Method (*see Chapter* (**??**)), a product of the Age of Enlightenment, requires claims of one individual to be experimentally verified and tested by others. One feature of the scientific method that makes it unique in the course of history is that truth is accepted by the persuasion of such testing and *not* by the power of authority or other coercions. Unfortunately, this methodology had not prevailed through the ages. How else can one explain ARISTOTLE'S (384-322 B.C.) apparently unchallenged claim than *men have more teeth than women do*? (*See* (??)*, pg.* ??*.*)

(23.2) Einstein's Annus Mirabilis

(23.2.1)

HISTORY

In 1903, WILBUR AND ORVILLE WRIGHT invent the first gasoline-powered and manned airplane. In 1905, during Einstein's miracle year, MARY ANDERSON invents windshield wipers. In this year, COCA COLA, which was marketed as a health tonic, contained caffeine along with some natural COCAINE from the kola nut. (By 1931 the cocaine was removed.)

(23.2.2) Enter Einstein. The year is 1905, the so-called ANNUS MIRABILIS, (*miracle year*) for Albert Einstein in which he publishes not one, but *four* landmark papers, all in the journal, *Annalen der Physik*. He is a 26-year old clerk in the Swiss Patent Office in Bern and has just earned his doctoral degree in physics from the Zürich Polytechnic Institute of Physics.

(23.2.3) The Photoelectric Effect. In March of 1905, EINSTEIN publishes his <u>first</u> paper. It explains the PHOTOELECTRIC EFFECT, in which a light beam directed at a metallic surface will produce the emission of electrons (electrical current). By positing that light, for all its wave properties, is, at the same time, composed of particles, or individual energy packets (PHOTONS, packets of energy, quanta), Einstein explains why higher intensity (brightness) of the light only produces more electrons (current) but does *not* produce more higher-energy electrons (voltage). This is because a higher energy of photons corresponds to higher frequency (bluer color) and not to increased brightness, which corresponds to a larger number of photons. For this paper, which lay the foundations of quantum theory, Einstein won the 1921 NOBEL PRIZE in physics.

(23.2.4) Brownian Motion. In his <u>second</u> paper of 1905, Einstein explains the random motion of small particles in "motionless" liquid solution — for example, pollen in standing water was observed to be in constant motion even though the water was apparently still.[2] Einstein hypothesizes that

---

[2]Some historians count this work as two papers since it draws on Einstein's

motion of the pollen particles results from their being battered by the moving molecules of the liquid. (*Yet, the molecules are too small to be seen by the naked eye.*) By applying statistical techniques to this *Brownian motion*, Einstein calculated the average distance traveled by the pollen under the assumption that this motion is generated by water molecules. His results were in excellent agreement with the observations and provided a very strong argument for the existence of molecules. At this time, prestigious physicists, such as ERNST MACH (1838-1916), would still be opposed to the idea that molecules even exist.

(23.2.5) Special Relativity. In his third paper of 1905, *On the Electrodynamics of Moving Bodies* [12], which comes to be known as his *special relativity* paper, EINSTEIN acknowledges the work of Maxwell [3] as is seen from his choice of title and his opening lines (*see* (22.5.3b)).

The asymmetry Einstein refers to in his opening lines is between the Faraday-Lentz law (20.7.3c) and the Ampère-Maxwell law (20.7.3d), which, somehow, need two separate frames of reference to explain a single phenomenon. Yet, Maxwell's equations (*see Chapter* (20)) also showed that the speed of light in a vacuum, which Maxwell calculated from a pair of laboratory measurements (*see* (22.4)), does *not* depend on any frame of reference.

Einstein responds to this "asymmetry" in Maxwell's theory (*see Figure* (22.5.3a)) by assuming from the outset that all laws of physics are equally valid in all frames of reference and that the speed of light is the same for all observers, regardless of the relative speed of source and observer.

Einstein's simplification eventually leads us, in spite of our dissenting intuition, to the results that moving clocks run slow and moving objects always shrink in the direction of motion.

(23.2.6) Equivalence of Energy and Mass. EINSTEIN publishes his fourth paper of 1905 entitled *Does the Inertia of a Body Depend Upon Its Energy Content?* where the relativistic addition of speeds $v \oplus w$ (7.3.1) leads directly to $E = mc^2$, the famous equation of equivalence between rest mass m and energy E. Chapter (15) is devoted to a derivation of this

---

1905 doctoral thesis in which he used statistical techniques to infer the size of sugar molecules based on the rate at which sugar dissolved in water. The paper on Brownian motion used some of these statistical techniques.

[3]Ironically, the Zürich Polytechnic Institute of Physics did not offer courses or seminars in Maxwell's theory — Einstein had to study it on his own.

equation.

(23.3) Comparing Relativities

(23.3.1) How Similar are the Galilean and Lorentzian Transformations?
For speeds v, Gaussian matrices have form $G_v = \begin{bmatrix} 1 & -v \\ 0 & 1 \end{bmatrix}$ (5.4.1a). while
Lorentzian matrices take the form
$L_v = \gamma_v \begin{bmatrix} 1 & -v \\ -v/c^2 & 1 \end{bmatrix}$ as given in (7.2.2).

Note that if, in the matrix L_v, we let $c \to \infty$ then the Lorentz matrix L_v
approaches the Galilean matrix G_v. Recall that in Galilean physics, the
speed of light is infinite while in relativistic physics, the speed of light in
a vacuum is the fixed and finite quantity c.

When the speed $c \to \infty$ in L_v, the four individual entries of the matrix L_v
approach the respective four entries of the Galilean matrix G_v.

Although the matrices G_v and L_v approach each other entrywise as $c \to$
∞, they are far apart from each other structurally in the sense that G_v
always has only *one* linearly independent eigenvector (A.5.1) in $\mathbf{R}^2$ while
L_v always has *two* linearly independent eigenvectors.

(*Eigenvector structure is explored in Exercises* (23.9.2) *and* (23.9.3).)

Comparing Galilean and Relativistic Speeds

(23.3.2) Addition-of-Speeds formulas. Given frame $\mathcal{F}^A$ sees frame $\mathcal{F}^B$ moving
with speed $-1 < v < 1$ and frame $\mathcal{F}^B$ sees frame $\mathcal{F}^C$ moving with speed
$-1 < w < 1$. Then frame $\mathcal{F}^A$ sees frame $\mathcal{F}^C$ moving with relative speed
$v + w$ in the Galilean case, and with speed $v \oplus w$ in the Lorentzian case.
Symbolically,

(23.3.2a)
$$\mathcal{F}^A \xrightarrow{v} \mathcal{F}^B \xrightarrow{w} \mathcal{F}^C \ implies \begin{cases} \mathcal{F}^A \xrightarrow{v+w} \mathcal{F}^C & is\ Galilean, \\ \mathcal{F}^A \xrightarrow{v \oplus w} \mathcal{F}^C & is\ Lorentzian. \end{cases}$$

The sums, $v + w$ and $v \oplus w$, which are virtually the same for small values, are each derived in two ways — analytically and geometrically.

(i) **Analytically:** In the classical, Galilean/Newtonian case, the addition of speeds $v + w$ of frame $\mathcal{F}^C$ as seen from frame $\mathcal{F}^A$ is calculated with the Galilean matrix (5.3.4) (*see Theorem* (5.5.2), *pg.* 91). In the relativistic Lorentzian case, the addition of speeds $v \oplus w$ of frame $\mathcal{F}^C$ as seen from frame $\mathcal{F}^A$ is calculated with the Lorentz matrix (7.1.3a) (*see Theorem* (7.4.1), *pg.* 111).

(ii) **Geometrically:** In the classical, Galilean/Newtonian case, we have the following graphic representation of relative speeds v and w in (23.3.2a): There exist certain triangles $\triangle(AB)$ and $\triangle(BC)$ with respective areas r and s (*see Theorem* (5.6.2), *pg.* 93), so that

(23.3.2b) *Speed* $v = \underbrace{\text{Area } \triangle(AB)}_{r}$, *Speed* $w = \underbrace{\text{Area } \triangle(BC)}_{s}$.

From (23.3.2a) and (23.3.2b), we then deduce

(23.3.2c) *Speed* $\mathcal{F}^A \to \mathcal{F}^C$ *is* $\underbrace{\text{Area}[\triangle(AB) + \triangle(BC)]}_{r + s} = v + w.$

Following the wording of (23.3.2b) and (23.3.2c), we show the relativistic property of the addition of speeds.

In the non-classical, Lorentzian case, we have the following graphic representation of relative speeds v and w in (23.3.2a): There exist certain <u>triangle-like</u> regions which we loosely denote as $\triangle(AB)$ and $\triangle(BC)$, with respective areas r and s (*see Figure* (B.3.4), *pg.* 355) so that

(23.3.2d)

Speed $v = \underbrace{\tanh(\text{Area } \triangle(AB))}_{\tanh(r)}$,

Speed $w = \underbrace{\tanh(Area \ \triangle(BC))}_{\tanh(s)}$.

From (23.3.2a) and (23.3.2d), we then deduce

(23.3.2e)

$$Speed\ \mathcal{F}^A \to \mathcal{F}^C \text{ is } \underbrace{\tanh(\text{Area}\,[\triangle(AB) + \triangle(BC)])}_{\tanh(r+s)} = v \oplus w.$$

The constant speed $v \oplus w$ of frame $\mathcal{F}^C$ as seen from frame $\mathcal{F}^A$ is obtained as the hyperbolic tangent of the addition of areas r, s of certain triangle-like regions in the Minkowski spacetime graph. (*See Theorem (7.5.6), pg. 115.*) That is,

(23.3.2f)   speed $v \oplus w = \tanh(r+s)$.

We see in Figure (23.3.3) that for small v and w, $v + w$ and $v \oplus w$ are practically the same. When v and w are small, the T_A and T_B axes (in both Galilean and Lorentzian spacetime graphs) are virtually vertical. Hence, the hyperbola in the right hand panel is very close to the horizontal line of the left hand panel and the graphs come closer to agreement.

----------   **How Similar are the Area Triangles?**   ----------

(23.3.2g)   When Areas Conflate. Another way to force the right hand Lorentzian panel of Figure (23.3.3) to closely approximate the left hand Galilean panel, is to draw the Minkowski diagrams so that the photon speed increases from 1 to ∞. (*In Galilean/Newtonian physics, light has infinite speed.*) That is, take the photon's 45° worldline (*indicating photon speed $v = 1$*) and rotate it clockwise about the origin until it approaches the horizontal X-axis (*indicating photon speed $v = \infty$*). Since the photon's speed is determined by the inverse of the slope of its worldline, a decreasing worldline slope $\downarrow 0$ implies an increasing photon speed $\uparrow \infty$.

Here are the details: The speed of light is infinite in Galilean physics, but is finite in relativistic physics. This is reflected graphically in the spacetime X-T diagrams as follows:

- In Galilean physicis, simultaneous events (fixed value of t) are horizontal lines.

- In relativistic physics, simultaneous events are tangent lines to the hyperbola

$$(t')^2 = t^2 - (x/c)^2 \text{ where } t' \text{ is a fixed time.}$$

(*See Theorem (8.4.1), pg. 123 and the diagram in the body of the proof.*) In fact, $\pm 1/c$ are the inverses of the slopes of the two asymptotes to this hyperbola.

(23.3.3) FIGURE: *Comparing Galilean and Lorentzian Spacetime Graphs: v = the speed of $\mathcal{F}^B$ relative to $\mathcal{F}^A$*

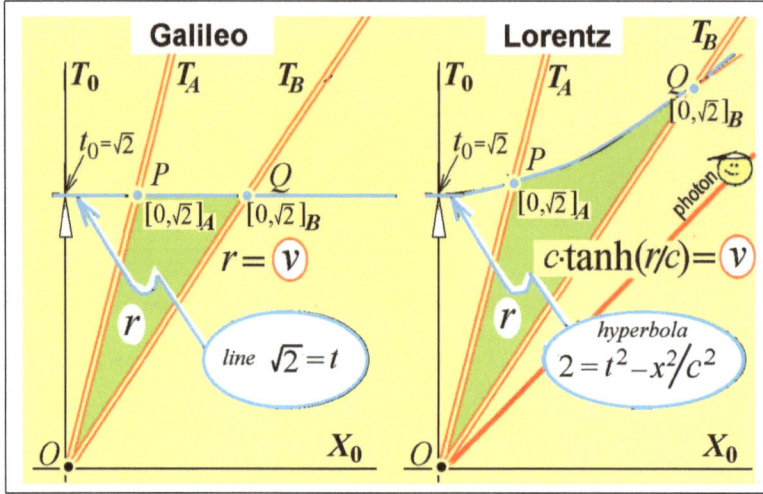

- LEFT PANEL OF FIGURE (23.3.3): *For the classical Galilean case, when the speed of light is essentially infinite, the area $r = v$ of the triangle POQ is exactly equal to the speed v of $\mathcal{F}^B$ as seen from $\mathcal{F}^A$. This was illustrated in the diagram of Theorem (5.6.2), pg. 93.*

- RIGHT PANEL OF FIGURE (23.3.3): *In the relativistic Lorentzian case, the area r of the triangular-like region POQ plays a role similar to that in the Galilean case. When the speed of light for all observers is c, then the speed of $\mathcal{F}^B$ as seen from $\mathcal{F}^A$ is $v = c(\tanh(r/c))$. Now when r/c is tiny, then $\tanh(r/c) \approx r/c$. Hence, when v is tiny, so is area r, in which case $v = c(\tanh(r/c)) \approx c(r/c) = r$, which is the Galilean case of the left panel.*

(23.4) Against Conventional Wisdom

Special relativity tells us that each of us carries our own personal clock which, from each individual's point of view, never runs slower than the personal clock of any other individual.

In spite of this truth — that clocks in motion run slower — the difference in the ticking rates of the clocks is so tiny that we never perceive this difference in our day-to-day experience.

If clock rates vary and the variances are not apparent, then it is easy to behave as if clock rates do *not* vary. This is similar to saying that since the earth's curvature is not apparent (*is a lake flat or does it bulge in the middle?*) then it is easy to behave in day-to-day activities as if the earth is flat.

Hence, we believe the opposite of the relativistic truth. Experience tells us it is "obvious" that all clocks, regardless of their motion, run at the *same* rate.

(23.4.1)

> This Galilean (*non-relativistic*) interpretation of physics says it is possible to construct a single theoretical reference clock by which we may all synchronize our personal watches, regardless of our relative speeds.

Here are two examples:

(23.4.2) **I. A Successful Universal Clock:** *Measuring Longitude.* It was JOHN HARRISON (1693-1776) who responded to the challenge announced in 1714 by Isaac Newton, in the name of the *Royal Society of London for the Promotion of Natural Knowledge*, to develop an accurate means for ships at sea to determine their east-west distance (longitude) from England. A prize of £20,000 was offered. While other scientists thought that certain astronomical observations might lead to a solution, Harrison decided to build a *clock* of extreme accuracy that would withstand the continual lurchings of a ship at sea. (*Every* 15° *interval of east-west longitude corresponds to one hour's difference between noon at each end. See Exercise* (23.9.5).)

After almost half a century of trials, Harrison's fourth generation clock, called H4, was tested in 1764 at sea and successfully met the conditions of the Royal Society's challenge. In spite of Harrison's documented success, influential scientists of the Royal Society refused to authorize the award to a lower-class "tinkerer." Only after the intervention of KING GEORGE III in 1772 did Harrison finally receive his £20,000 award. See the book *Longitude* by Dava Sobel [31] for a nice overview of these events.

In sum, the success of Harrison's clock confirmed that it is useful and even necessary to adopt the Galilean view of time (23.4.1) where arbitrary observers can synchronize their clocks to a single and *accurate* universal clock.

(23.4.3) **II. A Successful Use of Universal Time:** *Train Schedules and Time*

Zones. Another example of effective use of the Galilean view of time (23.4.1) which re-enforced conventional wisdom that all clocks run at the same rate, is seen in the history of railroad timetables, all of which are, indeed, synchronized to a single, theoretical clock. In addition, TIME ZONES were created for the very purpose of facilitating the synchronization of railroad schedules over wide regions to the single theoretical Galilean reference clock.

Less than 200 years ago, there were no time zones. Each locality determined noon to be that time of day when the sun was at its highest point in the sky. The sun dial, with noon being marked by the shortest shadow, was an adequate timepiece — as long as coordination was not necessary with sundials of other communities.

In England, The Great Western Railway, as part of an industry-wide attempt to counteract the inconsistencies of local times, started to run on fixed London time in November of 1840. Following suit, most public clocks were on the same London time by 1855.

In the United States and Canada, it was 1883 when railroads created synchronized, regional time zones. However, the general population had not yet warmed to the notion. Local pride was a factor causing Detroit to maintain two or more time standards until 1905 (the year of Einstein's paper [12]) when they finally adopted Central Standard Time.

(23.4.4) **These examples** show how EINSTEIN'S thinking in 1905, where any observer sees moving clocks running slower, went against conventional wisdom and "common sense" which held that any observer sees all clocks running at the same rate.

(23.5) Some Experimental Results

(23.5.1) **Earliest Experiments.** We have already encountered experiments whose results are consistent with the theory of special relativity.

For example, light was first confirmed to have finite speed with the experiment of Rømer who inferred this property from observations of Io, a moon of Jupiter (??). (*In Galilean physics, the speed of light is infinite.*) This finite speed was measured, to reasonable accuracy, by Fizeau via his ingenious cogs-and-mirrors experiment (??).

Later, de Sitter showed that the speed of light did not depend on the speed of its source when he tracked the light from binary stars (*Figures* (**??**) *and* (**??**)). This is consistent with our skipping stone model of wave transmission which predicts the observed wave speed is independent of the speed of its source. (*see the Skipping Stone model* (16.5) *and Figure* (16.5.4)).

On the other hand, the speed of the transmitting medium *is* communicated to (added to) the speed of the transmitted wave (16.5.5). This is why the "failure" of the Michelson-Morley experiment (**??**) eventually convinced scientists that light, unlike any other known wave, did *not* need a transmitting medium or aether (**??**).

Finally, using data from laboratory experiments (*Chapters* (18)-(22)), Maxwell calculated c, the theoretical speed of an electromagnetic wave in a vacuum. Moreover, the mathematics did not *require* a transmitting medium for electromagnetic waves (*although Maxwell thought there must be one*) nor did Maxwell's model require a particular frame of reference for these waves (22.4).

Confirming the Theory of Special Relativity

(23.5.2) *Can We Observe Moving Clocks Running Slow?* A Mu Meson, or Muon for short, is an elementary particle with a negative charge and a mass about 20 times that of the electron. Statistically, half of any group of stationary muons will decay after about 3 microseconds (*one microsecond is* 10^{-6}, *a millionth, of a second*). In any muon population, some muons live more than 3 microseconds and others live less but on average, the life span is

$$\text{life}_{stationary} = 3 \text{ microseconds.}$$

In nature, high-velocity muons with speeds v_{moving} close to the speed of light are produced when cosmic rays — subatomic charged particles that originate in outer space — hit Earth's atmosphere. Experimental observation shows that the average life span of these high-speed communities of muons is

$$\text{life}_{moving} = (\gamma_v) \cdot \text{life}_{stationary},$$

where $\gamma_v = 1/\sqrt{1 - (v_{moving}/c)^2}$ is the familiar Lorentzian factor (*see* (1.8.2) *and* (7.1.3b)) and v_{moving} is the observed muon speed.

As predicted by special relativity, the clock in the moving frame of the

streaming muons runs slower than Earth clocks by exactly a factor of
$\sigma_v = \sqrt{1 - (v_{moving}/c)^2} < 1$ *(recall* (1.8.4)*)*.

(23.5.3) *GPS Satellites and Relativity.* From the dozen or so GPS (GLOBAL
POSITIONING SYSTEM) satellites circling Earth, accurate calculations of
Earth positions are based on the accuracy of the onboard atomic clocks.

Special relativity tells us that all moving clocks run slow. Hence onboard
clocks of GPS satellites, which are in motion relative to Earth, will run
slower than Earth clocks. The slower rate is about 7 microseconds each
day.

General relativity says that nearness to mass will also cause clocks to run
slow. In particular, the tug of Earth's gravitational field slows the onboard
GPS clocks. But when the GPS clocks are placed in orbit, further from
Earth, the gravitational effect is reduced. This causes the GPS clocks to
run *faster* than they did when assembled on Earth. The faster rate is about
45 microseconds each day.

As a result, onboard GPS atomic clocks, when measured on Earth, are designed to run 38 microseconds slower each day than accurate Earth clocks.
Without this adjustment, accumulated clock errors would cause GPS calculations of global positions to be off by more than 10 kilometers (6 miles)
each day.

(23.5.4) **The formula $v \oplus w$ for the relativistic addition of speeds** v **and** w,
was closely approximated by measurements from the experiment of Fizeau
(*see Section* (13.2)).
....

(23.5.5) ‖ NOTE: **For a comprehensive overview** *of experimental results that*
           ‖ *confirm properties of special relativity, see the website of John Baez at*
           ‖ www.math.ucr.edu/home/baez/physics/Relativity/SR/experiments.html.

(23.6) Bad Assumption, Good Result

(23.6.1) **A remarkable dispute.** In a conversation with DAVID ELLERMAN, it
was noted that the book, *The Best of All Possible Worlds* pp. 48-50 [14]
by IVAR EKELAND describes the disagreement between RENÉ DESCARTES
(1596-1650) and PIERRE FERMAT (1601-1665) over the following:

(23.6.1a) | **Question:** Does light travel faster through air
              | than it does through water?

Descartes answered this question in the negative, maintaining that the speed of light *increases* as it passes from air to water, while Fermat believed the opposite — that the speed of light *decreases* after entering water from air. Ironically, with opposite and contradictory hypotheses, these two men derived the valid result known as **Snell's Law**, one form of which is:

(23.6.2)

SNELL'S LAW. *A ray of light passes through air at speed v_1 with* ANGLE OF INCIDENCE θ_1 *(see diagram on the right). The light ray then passes through water at speed v_2 and* ANGLE OF REFRACTION θ_2. SNELL'S LAW *states that there exists a constant m, (the* REFRACTIVE INDEX *or* INDEX OF REFRACTION*), that is independent of the angle θ_1 such that the following relation holds true:*

(23.6.2a) $\quad \sin \theta_1 = m \cdot \sin \theta_2.$

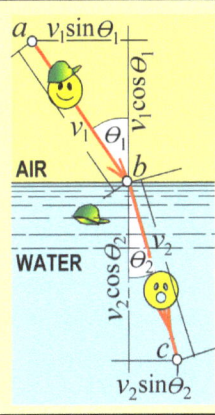

Although the Dutchman WILLEBRORD SNELL discovered the relationship (23.6.2a) in 1620, he did not prove it. Descartes was the first to actually derive it.

How did Descartes do it? He assumed (*incorrectly*) that once the light ray entered water, its vertical speed component increased while the horizontal speed remained unchanged.

Fermat, on the other hand, assumed (*correctly*) that in the Figure of (23.6.2) with fixed points a and c, light would travel in the least time over all a-b-c paths as the position of entry point b varied horizontally.

Both Descartes and Fermat arrived at the same form (23.6.2a) of Snell's Law and both would agree that, in the case of air-to-water travel, the value of the refractive index[4] of (23.6.2a) should be $m \approx 1.33$. But

(23.6.2b) $\quad m = \begin{cases} (v_2/v_1) & \text{\textit{according to Descartes,}} \\ (v_1/v_2) & \text{\textit{according to Fermat.}} \end{cases}$

Since the speed of light would not be known for some two hundred years[5], it was impossible at this time to decide which of the two ratios of (23.6.2b) was valid. (**Note:** *Exercise* (23.9.7), *will guide you through the derivations*

---

[4] Index of refraction for various materials is discussed in (13.2.2c).

[5] In 1849, Fizeau devised a mechanism for measuring c in air. (*See* (**??**), pg. **??**). In 1851, Fizeau measured the speed of light in water (*see* (13.2), pg. 164).

of (23.6.2a) and (23.6.2b).)

(23.7) A Limited Reality

(23.7.1) **Reality as Source + Perceived Effect.**   It is interesting to note how, in one form or another, philosophers through the ages have addressed reality in terms of *effect* and *cause*, or *perception* and *source* of the perception.

For example, PLATO (427-347 B.C.) likened our reality to that of beings imprisoned in a cave. From birth, these individuals can only perceive the shadows and sounds (*effects*) of the true and real world outside the cave (*source*). Anyone managing to find his or her way out of the cave would be met by the blinding richness of the outside reality.

Plato goes further with this metaphor and introduces a moral component — the enlightened escapee is obliged to return to the cave in order to inform the inhabitants about the outside reality.

What did SOCRATES[6] (469-399 B.C.) have to say about this moral obligation? In his work *The Republic, Book VII* [28], Plato devised an imaginary dialogue between his teacher, Socrates, and a character named Glaucon. During this discourse, Socrates — not the most optimistic person in the world — says,

> Men would say of him that [ascended up to the light] ... that it was better not even to think of ascending; and if any one tried to ... lead [any of us] up to the light, let them only catch the offender, and ... put him to death.

DAVID HUME (1711-1777) claimed that we determine our reality from our collection of individual perceptions (*effects*) resulting from a much larger reality (*source*), just as we perceive individual railroad cars (*effects*) of a larger train (*source*). Hume was not so sure that the full train of reality could be accurately known from the separate cars alone.

IMMANUEL KANT (1724-1804) was so positive about the reality of objects that he invented the new (*clarifying?*) term, "noumenal." To say objects are NOUMENAL is to say they are real — outside and independent of the mind, as opposed to a phenomenon (*perceived effect*), also called a "thing-in-itself."

---

[6]There are no known writings due to Socrates. It is mostly through the writings of his student, Plato, that we know of his ideas.

(23.7.2) Beyond Dimension. EDWIN A. ABBOTT (1838-1926), in his book *Flatland* [1], not only endorses the existence of an external reality, but goes further by suggesting that much, if not most, of true reality is *beyond perception*. In Abbott's book, the universe of the Flatlander is a table-top land where there are only two space dimensions. Inhabitants of this two-dimensional table-top universe have no conception of up or down. They only know left-right and backward-forward dimensions.

(23.7.2a) In Flatland, a creature from the third dimension — a Roundlander, say — can intervene and produce results that seem downright magical to the Flatlanders. For example, when a Roundlander pokes a finger into Flatland space, it is perceived as a horizontal line, as is every other object in Flatland. But when the finger is pulled perpendicularly upward — a direction unknown to the Flatlanders — the Flatlanders see the finger-line mysteriously shrink and then disappear into thin (*flat?*) air!

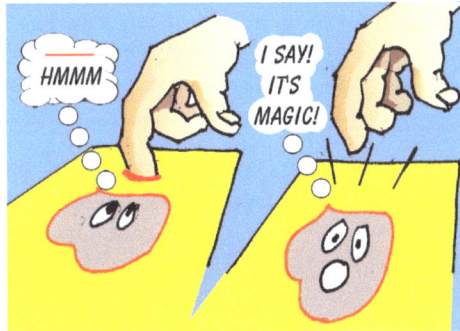

(23.7.2b)   The third space dimension is as confounding to the Flatlanders as a fourth space dimension is to us. Suppose two Flatlanders traveling along separate perpendicular roads meet at the intersection. Like two cars at a four-way stop, one Flatlander must wait for the other to pass before resuming its (his/her?) own journey. Now suppose a voice from the third dimension proclaims, *"Build a bridge so that no traveler has to wait for the other!"* The awe-struck Flatlanders listening to this disembodied voice would have no idea of what a bridge would be in three-space, nor how it could possibly help them to avoid one traveler having to wait for the other. (*Although its practicality may be in doubt, Exercise* (23.9.6) *asks you to mathematically construct a bridge in four-dimensional space so that two objects at the same point* (x, y, z) *in* $\mathbf{R}^3$ *will be separated in* $\mathbf{R}^4$.)

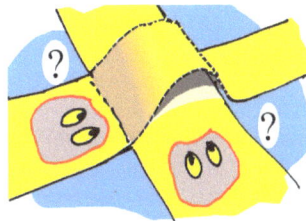

....

(23.7.2c) Our tunnel vision of reality. Abbott's two-dimensional Flatlanders are limited in the sense that they are unaware of and cannot perceive our richer three-dimensional reality. But we are also limited since much, if not all, of what we believe to be real, lies *beyond* what we perceive with our five input ports (senses).

For example, we do not perceive radio waves until a radio receiver transforms them to sound. Similarly, a Geiger counter converts radioactivity (*the spontaneous emission of invisible matter or energy from the nucleus of an unstable atom*) into sound or a visible graph.

(23.7.2d)   Gadgets expand our reality in the sense that instruments convert (or reduce) extra-sensory phenomena to sights, sounds, smells, tastes, or sensations that we then interpret as (effects of) "reality."

We saw this, for example, in the experiments of Rutherford (**??**), pg. **??**. Invisible (*to humans, at least*) alpha particles ricochet off invisible atoms of gold and hit a screen whose coating transforms each collision into a visible blip. After Rutherford applied his *intellect* and *imagination*, we expanded reality.

In another case, experimenters applied their intellect and imagination to the readings of certain physical instruments so that today we know that c, the speed of light in a vacuum, is invariant for all observers. (1.5.2b). In a remarkable *tour de force*, EINSTEIN took this single observation about c and applied *his* intellect and imagination in a grand thought experiment to produce the theory of special relativity.

(23.7.2e)   We are butterflies of the Universe. The butterfly, with a life-span of only a few weeks or months, is ignorant of our realities[7] — of   mathematics,   music,   humor,   emotion.   On   the   other hand, as clever as we may believe ourselves to be, we suffer the anguish of acknowledging the incompatibility of three successful theories of physical reality: (i) special and (ii) general relativity (*for large scale physics*), and (iii) quantum theory (*for small scale physics*). Special and general relativity both assume continuous motion and deterministic causality while quantum theory describes   discontinuous   motion

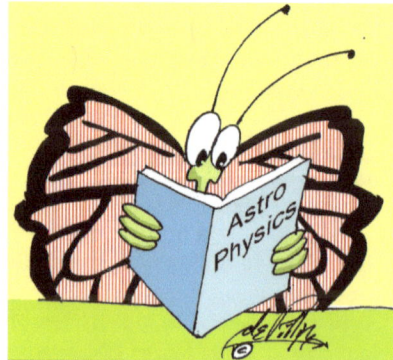

---

[7]For all its limitations, the Monarch butterfly does have abilities that exceed our own. Apart from knowing how to fly, every last one of them can navigate their individual migration route over the thousands of miles between North America and Mexico.

that is influenced by laws of probability that are inconsistent with deterministic causality.

As long as our limitations bar us from seeing the larger reality in which all these theories are mutually consistent, we remain butterflies of the Universe.

(23.8) PIES Reality

(23.8.1) **A Useful Oversimplification.** Reality is comprehended in stages: (i) *Physical* perception is massaged or processed by our (ii) *intellect*. The (iii) *emotional* pleasure of discovery impels further exploration after which, we may seek (iv) the *meaning* of it all.

(23.8.2) TABLE: *Four Facets of Reality*

P)physical	*the physical world.*
I)intellect	*thought, imagination, and logic.*
E)motion	*sensations and feelings.*
S)piritual	*the structure that provides meaning.*

Interrelations. These four suggested facets of Reality are not independent of each other. For example, emotions can affect our physical state and poor physical health can affect our thinking abilities.

(23.8.3) EXAMPLES: *Emotion versus Intellect.*
- Our intellect may accept the notion that moving clocks run slow, but is this idea emotionally acceptable?
- Our engineers may understand the aerodynamics of flight, but some of them might still feel uncomfortable, emotionally, with the fact that jumbo jets, weighing several tons, are being held aloft only by thin air.
- One may feel emotionally distressed walking under a ladder even though the intellect sees no cause for concern.

(23.8.4) **Many spiritualities** can be defined since there are many beliefs about our Meaning or Reason-for-Being in the Universe. Meaning will be interpreted by some, as non-existent — it is only the physical world that truly exists. Meaning, for others, will be understood by recognizing an external force or entity, and this entity may be highly impersonal (Nature) or highly personal (God).

(23.8.5) **Do scientists share a common view of Reality and Meaning?** Suffice it to quote the following interesting statements of the eminent scholars and physicists, Steven Weinberg and Freeman Dyson:

Steven Weinberg [32] proclaimed

> With or without RELIGION, you would have good people doing good things and evil people doing evil things. But for good people to do evil things, that takes religion.

But Freeman Dyson [10] added

> Weinberg's statement is true as far as it goes, but it is not the whole truth. To make it the whole truth, we must add an additional clause: "And for bad people to do good things — that takes religion."

(23.8.6) **How Much Uncertainty Will a Scientist Tolerate?** Given that scientists studying the physical universe attempt to minimize *uncertainty* (*see Section* (**??**)), one might assume that scientists must not tolerate uncertainty in any other discipline. Consider, then, these comments of Freeman Dyson [11]:

> Both as a scientist and as a religious person, I am accustomed to living with uncertainty. Science is exciting because it is full of unsolved mysteries, and religion is exciting for the same reason. The greatest unsolved mysteries are the mysteries of our existence as conscious beings in a small corner of a vast universe. Why are we here? Does the universe have a purpose? Whence comes our knowledge of good and evil? These mysteries, and a hundred others like them, are, as yet, beyond the reach of science.

(23.8.7) **Finally,** this book has attempted an explanation of the processes by which we may come to understand a very small part of our Reality — the physical part. This is the beginning.

(23.9) Exercises

(23.9.1)
Exercise

Using Definition (A.5.1), show the following properties:

(23.9.1a) If $\vec{X}$ is an eigenvector for a matrix A, then so is any scalar multiple, $a\vec{X}$ for any scalar $a \neq 0$.

(23.9.1b) Suppose a matrix A has two linearly independent eigenvectors, $\vec{X}_1$ and $\vec{X}_2$ with the *same* eigenvalue, λ. Then show that the sum $\vec{X}_1 + \vec{X}_2$ is also an eigenvector with the same eigenvalue, λ.

(23.9.1c) Suppose a matrix A has two linearly independent eigenvectors, $\vec{X}_1$ and $\vec{X}_2$ with the *same* eigenvalue, λ. Then show that the linear combination $a_1\vec{X}_1 + a_2\vec{X}_2$, for scalars a_1 and a_2, is also an eigenvector with the same eigenvalue λ.

Hint: *Instead of showing this result from scratch, use the results of* (23.9.1a) *and* (23.9.1b).

(23.9.2)
Exercise

Using Definition (A.5.1), find eigenvectors and eigenvalues for the following matrices:

(23.9.2a) Show that for the $n \times n$ identity matrix, I_n, *any and every* vector $\vec{X} \neq \vec{0}$ is an eigenvector. Moreover, the eigenvalue is always $\lambda = 1$.

(23.9.2b) Show that for any 2×2 triangular matrix $A_2 = \begin{bmatrix} a & b \\ 0 & c \end{bmatrix}$ with real entries, a, b, and c, the vector $\vec{X}_0 = \begin{bmatrix} 1 \\ 0 \end{bmatrix}$ and all its multiples $\lambda\vec{X}_0$ is the only set of eigenvectors of A_2. Moreover, a is the eigenvalue associated with $\vec{X}_0$.

(23.9.2c) Given the Galilean matrix $G_v = \begin{bmatrix} 1 & -v \\ 0 & 1 \end{bmatrix}$. Then show that the vector $\vec{X}_0 = \begin{bmatrix} 1 \\ 0 \end{bmatrix}$ and all its multiples $\lambda\vec{X}_0$ is the only set of eigenvectors of G_v. Moreover, 1 is the eigenvalue associated with $\vec{X}_0$.

Hint: *Use* (23.9.2b).

(23.9.3)   Given the Lorentz matrix $L_v = \gamma_v \begin{bmatrix} 1 & -v \\ -v & 1 \end{bmatrix}$ (7.2.2).
Exercise

> **(23.9.3a)**   Using Definition (A.5.1), find eigenvectors and eigenvalues of L_v.
>
> **Hint:** *From Definition* (A.5.1), *obtain two equations in the variables x, t, v, and eigenvalue λ. Put all terms on one side so that both equations equal zero, then show that the two eigenvalues are $\lambda = (1 \pm v)\gamma_v$. Substitute each of these eigenvalues back into the equations to obtain eigenvectors $\{\vec{X}_0, \vec{X}_1\}$, which will be mutually perpendicular (see (A.2.6b), pg. 337). Note from (23.9.1a), that once you find an eigenvector $\vec{X}_i$ for λ, any non-zero multiple $a\vec{X}_i$ is an equivalent eigenvector for the same λ.*

(23.9.4)   Show that the geometric assertions of (23.3.2g) are valid. That is, show
Exercise   that the rotation of the 45° worldline of the photon forces the Lorentzian triangle-like region of the right hand panel to approach the Galilean triangle of the left hand panel.

(23.9.5)   Show how an accurate clock allows for accurate calculation of longitude
Exercise   for ships at sea (*see* (23.4.2)).

> **Hint:** *An explorer carries a clock that accurately reads the time of city A where noon (sun at its maximum height) is observed at 12:00pm. If noon for the remote explorer occurs when her clock reads 11:00am, or one hour previous to noon in city A, then she knows she is exactly 15° longitude west of city A. If she sees noon at time 10:30am, or 1.5 hours previous to noon at city A, then she is 1.5×15° longitude west of city A. In general, if the explorer sees local noon occuring n hours before (after) 12:00pm, then her position is $n \times 15^O$ longitude west (east) of city A.*

(23.9.6)   Show how persons A and B who are approaching a collision at point
Exercise $[x_0, y_0, z_0]$ in three-dimensional space $\boldsymbol{R}^3$, can be saved when one of them veers onto a bridge in four-dimensional space $\boldsymbol{R}^4$.

> **Hint:**   *To reenforce intuition, let us see, analogously, how collision is avoided in a lower dimension. Accordingly, suppose A and B are headed for a collision at point $[x_0, y_0]$ in $\boldsymbol{R}^2$. If we embed $\boldsymbol{R}^2$ into $\boldsymbol{R}^3$ — see the figure to the right — then A and B can each be put into different planes.*

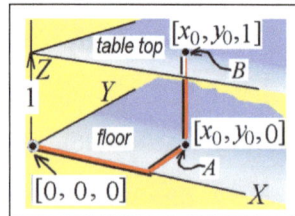

> *The collision point $[x_0, y_0]$ in $\boldsymbol{R}^2$ corresponds to two distinct coordinates in $\boldsymbol{R}^3$, namely, $[x_0, y_0, 0]$ for A on the "floor" of $\boldsymbol{R}^3$ and $[x_0, y_0, 1]$ for B on the "tabletop" of $\boldsymbol{R}^3$. Similarly, formally embed $\boldsymbol{R}^3$ into $\boldsymbol{R}^4$ — without*

the benefit of pictures (1.3.3) — *so that the would-be collision point for A has new coordinates* $[x_0, y_0, z_0, 0]$ *on the "floor" of* $\mathbf{R}^4$ *and the collision point for B has new coordinates* $[x_0, y_0, z_0, 1]$ *on the "tabletop" of* $\mathbf{R}^4$.

Two Derivations of Snell's Law (23.6.2a):

(23.9.7)
Exercise Cal-
culus needed

(23.9.7a) Derive (23.6.2a) and (23.6.2b) using Descartes' (*false*) hypotheses, namely, that the speed of light increases after it passes from air to water.

Hint: *Equate the horizontal components of* v_1 *and* v_2 *in the Figure of* (23.6.2) *and the result is immediate.*

(23.9.7b) Derive (23.6.2a) and (23.6.2b) using Fermat's (*correct*) hypothesis, namely, that in the figure to the right, light travels along path *a-b-c* in the least time.

Hint: *Write the times that light requires to travel along paths a-b and b-c as a function of x (see figure to the right). Find the x value that produces the minimum.*

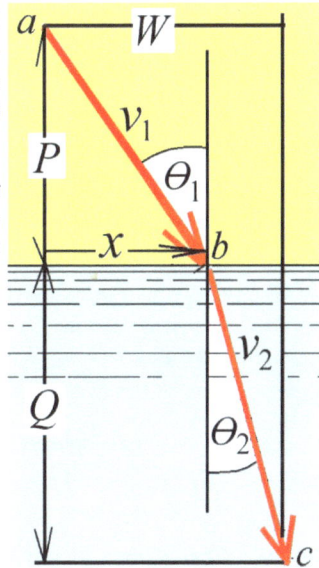

X. Appendices

Appendix A: Basics of Linear Algebra: . *An overview of vector spaces, linear transformations and their matrices, eigenvectors, and eigenvalues. Linear algebraic structure plus a graphical interpretation of the Minkowski diagram lead directly to a derivation of the Lorentz transformation.*

Appendix B: *The hyperbolic geometry behind Minkowski diagrams is the key to explaining the paradoxes and the non-intuitive relativistic addition $(v \oplus w)$ of speeds v and w.*

Appendix C: Deconstruction of a Moving Train: *What might be regarded as a single unit at a single moment (e.g., a train) is analyzed as an infinite sequence of points, each of which has its own clock or*

time. This is how we can know that the <u>front end</u> of a moving train is younger than its <u>back end</u> — albeit by an imperceptible amount.

Appendix ??: Dimensional Analysis *outlines the process of dimensional analysis that is especially popular with engineers for checking consistency in calculations.*

Appendix ??: Rings of Functions and Matrices *isolates the underlying axiomatic structure behind matrix calculations. As a bonus, we use the same structure to see exactly why a **minus times a minus must be a plus** (Exercise (??)).*

Appendix ??: The Scientific Method. *We describe the process of falsification as a key ingredient of the scientific method which becomes a self-correcting process.*

Appendix ??: Mathematical Logic. *Human reasoning is shown to be constrained by rules of logic that are very similar to rules of algebra. Moreover, and contrary to popular belief, we see why Sherlock Holmes almost never deduced anything. (Hint: Holmes' conclusions were almost always inferences, or inductions, not deductions.)*

Appendix ??: Early Measurements of c. *We offer details of some of the earliest and ingenious methods for measuring c, the finite speed of light. Even in the time of Galileo and Newton, it could not be proven that the speed of light was finite. In this chapter, you will see how you can use your home microwave oven to measure c in a 20-second experiment (Exercise (??), pg. ??).*

(A) Linear Algebra Overview

(A.1) Mathematics as a Conduit to Reality

(A.1.1) **This book is for** the reader who is familiar with calculus and matrices (linear algebra). Little or no physics is required.

Sections (A.1.2)-(A.1.4) present a brief overview of the link between linear algebra and perceived reality. We first present a graph of a moving particle where *two* sets of (distance vs. time) axes are used:

(A.1.2) **F**IGURE: X_A-Y_A and X_B-Y_B **axes used to plot a point (dotted line) moving horizontally.**

The horizontal oval shows the red dot moving horizontally as perceived in "observed reality."

The two-dimensional graph, the mathematical model of this reality, shows the

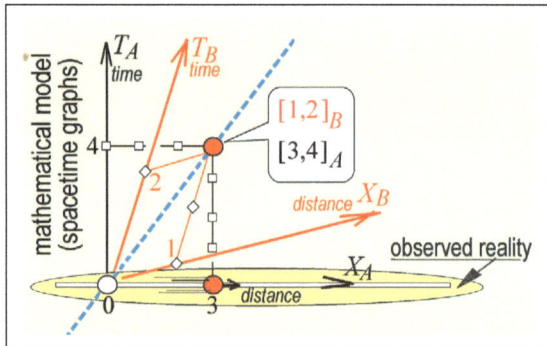

moving dot where the X_A-T_A coordinates are $[3, 4]_A$ (where distance = 3 units and time = 4 seconds) and the X_B-T_B coordinates of the same point are $[1, 2]_B$ (where distance = one units and time = 2 seconds).

(A.1.3) **Linear Algebra:** *A Natural Language for Special Relativity* The following overview uses terminology and definitions of linear algebra whose locations in this text are indicated.

   (A.1.3a) **Constant Speed of Parallel Rulers.** Special relativity, broadly speaking, is the study of Nature's laws using identical rulers that move in parallel tracks at constant speeds relative to each other. Identical clocks are found at each point of each ruler.

   (A.1.3b) **Motion as Straight Lines.** When the trajectory of a point on one of the horizontally-moving rulers is plotted on a two-dimensional distance-versus-time graph (an *X-T spacetime* graph), the result is a straight line. This is due to the constant speed of the ruler. (*See the dotted line in Figure (A.1.2) that we will later discuss* *in detail.*)

(A.1.3c) Inherent Vector Space Structure and Matrices. Every X-T spacetime graph is actually a *vector space* (A.2.1). So when we pass from one set of coordinates to another — from $[\,3\,,4\,]_A$ to $[\,1\,,2\,]_B$ in Figure (A.1.2), for example — then we are talking about a *function* (A.3.2) on a *vector space* (A.2.1) that sends zero to zero and sends straight lines to straight lines. The only functions that do this are so-called *linear functions* (A.4.1) which can always be represented by *matrices* (A.4.3), (A.4.6). These matrices are the *Lorentz transformations* (7.1.3) that we discuss in detail later.

(A.1.4) In sum, constant motion is graphically represented by straight lines in spacetime graphs. Thanks to the vector space structure of spacetime graphs, we may pass from one coordinate system to another (from one ruler to another) using the Lorentz transformation (or Lorentz matrix) from which we gain insight into all the paradoxes of special relativity.

(A.2) Vector Spaces

To fix ideas, we present some mathematical concepts that are used in this book.

(A.2.1)

> **D**EFINITION (informal): A VECTOR SPACE is any set of "objects" called VECTORS that can be added together (*if $\vec{X}$ and $\vec{Y}$ are vectors, then so is $\vec{X} + \vec{Y}$*) and multiplied by a number (*if $\vec{X}$ is a vector and a is a number, then $a\vec{X}$ is also a vector*).

For example, the "objects" (or vectors) of Definition (A.2.1) might be 2×3 MATRICES (2×3 *arrays of numbers*) or FUNCTIONS (*see Definition* (A.3.2)) such as the set of all POLYNOMIALS of the form $a_2 x^2 + a_1 x + a_0$ on the real line.

(A.2.2)

> **D**EFINITION: In the special case that the "objects" or "vectors" of (A.2.1) are lists of ordered k-tuples of real or complex numbers, then the numbers in the ordered list are called COORDINATES or COMPONENTS. These k-tuple vectors belong to a k-dimensional space which we denote $\boldsymbol{R}^k$.

(A.2.3) NOTATION: The k-tuples of Definition (A.2.2) are written in one of two forms, depending on context, namely as a row vector

$\vec{X} = [x_1, x_2, \ldots, x_k]$ or, as a column vector,

$$\vec{X} = \begin{bmatrix} x_1 \\ x_2 \\ \vdots \\ x_k \end{bmatrix}, \text{ which is also written } \quad \text{col}[x_1, x_2, \ldots, x_k].$$

(A.2.4) A SCALAR is a real or complex number.

The addition of *pairs* of vectors and multiplication of a *single* vector by a scalar are defined as follows:

For all scalars (real or complex numbers) a, and all vectors

    (A.2.4a) $\vec{X} = [x_1, x_2, \ldots, x_k], \quad \vec{Y} = [y_1, y_2, \ldots, y_k].$

we define the following vector operations,

    VECTOR ADDITION
    (A.2.4b) $[x_1, x_2, \ldots, x_k] + [y_1, y_2, \ldots, y_k]$ *entrywise*
               $\overset{def}{=} [x_1 + y_1, x_2 + y_2, \ldots, x_k + y_k]$ *addition*

    SCALAR MULTIPLICATION
    (A.2.4c) $a[x_1, x_2, \ldots, x_k] \overset{def}{=} [ax_1, ax_2, \ldots, ax_k].$ *entrywise*
                                            multiplication

For more on (A.2.4c), see Exercise (3.5.1).

(A.2.5)   NOTE: *Strictly speaking, there are two distinct plus signs in* (A.2.4b). *The boldface* **+** *sign is used between pairs of* $\underline{k\text{-tuples}}$ *while the* + *sign only appears between pairs of* $\underline{numbers}$. *This distinction is often ignored where the same plus sign is used for both binary* (**??**) *operations.*

When x_1, x_2, and x_3 are space coordinates and $x_4 = t$ represents time (*hence, all scalars must be real*), then the graphs of these 4-dimensional vector spaces are *spacetime diagrams* that we shall develop in (4.1.2).

(A.2.6) The INNER PRODUCT of two vectors, $\vec{X}, \vec{Y}$ of the form (A.2.4a), is defined to be

(A.2.6a) $\quad \langle \vec{X}, \vec{Y} \rangle \overset{def}{=} \sum_{i=1}^{k} x_i \bar{y}_i$ where for each variable y_i, the symbol

$\bar{y}_i$ denotes its COMPLEX CONJUGATE (A.2.13a). Vectors $\vec{X}, \vec{Y} \in \boldsymbol{R}^k$ are said to be PERPENDICULAR for real coordinates, and ORTHOGONAL for complex number coordinates if

(A.2.6b) $\quad \langle \vec{X}, \vec{Y} \rangle = 0.$

(A.2.6c) Geometry in terms of formulas. Note how the geometric quantity of *perpendicularity* (or 90° between pairs of vectors) is expressed in terms of the *analytic expression* (A.2.6a). That is, $\vec{X}$ is geometrically perpendicular to $\vec{Y}$ if and only if the expression $\langle \vec{X}, \vec{Y} \rangle$ is numerically equal to zero (A.2.6b).

(A.2.6d) EXAMPLE: **Illustrating Definition (A.2.6b)**

Definition (A.2.6b) says that vectors $[1,1]$ and $[a,b]$ are **geometrically** *perpendicular to each other if and only if* **analytically** *we have*

$\quad < [1,1], [a,b] >= 0.$

Definition (A.2.6a) implies that this inner product equals zero if and only if $b = -a$, for any and all a. This defines an infinity of vectors $[a, -a]$ (including $\boldsymbol{0} = [0,0]$) each of which is perpendicular to $[1,1]$.

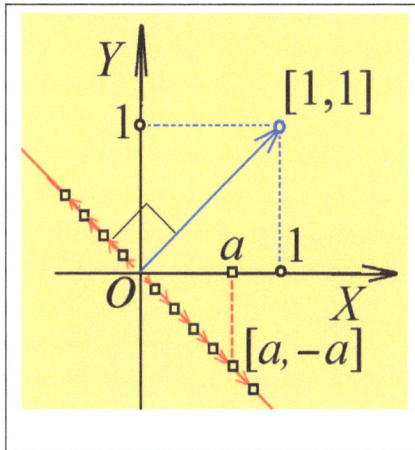

Moreover, with (A.2.6b) we can extend the notion of perpendicularity beyond two or three dimensions. For example, even if $k > 3$, then for vectors $\vec{X}, \vec{Y} \in \boldsymbol{R}^k$ (A.2.3) we have a meaningful definition for $\vec{X}$ being perpendicular to $\vec{Y}$ even though we do not have pictures (e.g., arrows) to represent these vectors.

(A.2.7) Any k-term sum, $r_1 + r_2 + \ldots + r_k$, can be represented with the SIGMA NOTATION which can take any of the following forms: $\sum_{i=1}^{i=n} r_i,$ $\sum_{i=1}^{n} r_i,$ $\sum_{i}^{n} r_i,$ or $\sum_{i} r_i.$

(A.2.8) The LINEAR COMBINATION of vectors $\{\vec{X}_i\}$ in $\boldsymbol{R}^k$ is any sum of the form $\sum_{i} a_i X_i$ where each a_i is a scalar.

(A.2.9) A BASIS for a k-dimensional vector space $\boldsymbol{R}^k$ is a fixed set of k non-zero vectors $\{\vec{X}_i\}$ from which any vector $\vec{X}$ in $\boldsymbol{R}^k$ can be expressed in only one way as a linear combination

> **(A.2.9a)** $\vec{X} = \sum_{i}^{k} a_i \vec{X}_i$ *for k unique scalars $\{a_i : i = 1, 2, \ldots, k\}$.*

(A.2.10) A set of vectors $\{\vec{X}_i\}$ in $\boldsymbol{R}^k$ is LINEARLY INDEPENDENT if the there is only one way to write the zero vector as a linear combination of the vectors $\{\vec{X}_i\}$, namely, by setting each $a_i = 0$ in (A.2.9). That is,

> **(A.2.10a)** $\vec{0} = \sum_{i} a_i \vec{X}_i$ *only if each $a_i = 0$.*

Equivalently, vectors $\{\vec{X}_i\}$ are LINEARLY INDEPENDENT if for any vector $\vec{Y}$,

> **(A.2.10b)** $\vec{Y} = \sum_{i} a_i \vec{X}_i = \sum_{i} b_i \vec{X}_i$ *only if each $a_i = b_i$.*

That is, for any $\vec{Y}$, there is only one way to write $\vec{Y}$ as a linear combination of the vectors $\{\vec{X}_i\}$.

(A.2.11) A STANDARD BASIS for a vector space $\boldsymbol{R}^k$ is the particular set of k vectors, $\{\vec{E}_1, \vec{E}_2, \ldots, \vec{E}_k\}$,

> **(A.2.11a)**
> $$\underbrace{\begin{bmatrix} 1 \\ 0 \\ \vdots \\ 0 \end{bmatrix}}_{\vec{E}_1} \quad \underbrace{\begin{bmatrix} 0 \\ 1 \\ \vdots \\ 0 \end{bmatrix}}_{\vec{E}_2} \quad \ddots \quad \underbrace{\begin{bmatrix} 0 \\ 0 \\ \vdots \\ 1 \end{bmatrix}}_{\vec{E}_k}.$$

(A.2.12) EXAMPLE: To illustrate (A.2.9a) when the basis is the standard basis, any vector $\vec{X}$ in $\boldsymbol{R}^k$ has the form

$$\vec{X} = \begin{bmatrix} x_1 \\ x_2 \\ \vdots \\ x_k \end{bmatrix} = x_1 \underbrace{\begin{bmatrix} 1 \\ 0 \\ \vdots \\ 0 \end{bmatrix}}_{\vec{E}_1} + x_2 \underbrace{\begin{bmatrix} 0 \\ 1 \\ \vdots \\ 0 \end{bmatrix}}_{\vec{E}_2} + \cdots + x_k \underbrace{\begin{bmatrix} 0 \\ 0 \\ \vdots \\ 1 \end{bmatrix}}_{\vec{E}_k} .$$

(A.2.13) The Euclidean LENGTH or NORM of a k-vector of form
$\vec{X} = [x_1, x_2, \ldots, x_k]$ (or $\mathrm{COL}[x_1, x_2, \ldots, x_k]$) is defined by

(A.2.13a) $\underbrace{X = \|\vec{X}\|}_{\substack{equivalent \\ notation}} \overset{def}{=} \sqrt{x_1\,\bar{x}_1 + x_2\,\bar{x}_2 + \cdots + x_k\,\bar{x}_k} \geq 0$

where $\bar{x}_i$ is the COMPLEX CONJUGATE of x_i. That is, if for real numbers a_i and b_i we have $x_i = a_i + ib_i$, then its cojugate is $\bar{x}_i = a_i - ib_i$. Hence, for any scalar α, $\|\alpha\,\vec{X}\| = |\alpha|\,\|\vec{X}\|$ where $|\alpha|$ is the *absolute value* of α.

(A.2.14)

> **DEFINITION:** *In any vector space, a coordinatefree representation of the* STRAIGHT LINE *L generated by two vectors $\vec{X}$ and $\vec{Y}$ ($\vec{X} \neq \vec{Y}$) is the set L of all (λ-dependent) points $\vec{Z}_\lambda$ of the form*
>
> **(A.2.14a)** $L = \{\vec{Z}_\lambda \text{ of the form } \lambda\vec{X} + (1-\lambda)\vec{Y} \text{ for } -\infty < \lambda < \infty\}.$

(A.2.15) **NOTE:** The equation (A.2.14a) is *coordinate free* and is therefore valid for vectors $\vec{X}$ and $\vec{Y}$ in *any* vector space as defined in (A.2.1). This means we can extend (or abstract) the geometric notion of a line of *points* or *vectors* containing the two vectors $\vec{X}$ and $\vec{Y}$ even though, as with vector spaces of *matrices* or *polynomials*, we do not have a picture like that of Figure (A.2.16) following .

(A.2.16) Figure: *Line L equals all points $\lambda \vec{X} + (1-\lambda)\vec{Y}$ as λ varies.*

Any pair of distinct vectors on line L will suffice to generate L in accordance with (A.2.14a). The generating vector pair $\{\vec{X}, \vec{Y}\}$ in Equation (A.2.14a) and Figure on the right is not unique.

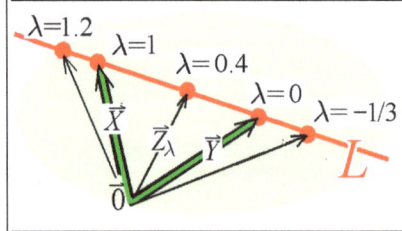

(A.2.17) Interpreting λ in (A.2.14a) and Figure (A.2.16): The scalar (number) λ tells us where the vector $\vec{Z}_\lambda = \lambda \vec{X} + (1-\lambda)\vec{Y}$ of (A.2.14a) lies relative to the line segment running from $\vec{Y}$ and $\vec{X}$. For example,

$\lambda = -1/3 \quad Z_{-1/3}$ lies in front of the line segment from $\vec{Y}$ to $\vec{X}$
$\lambda = 0 \quad\quad\ Z_0 = \vec{Y}$
$\lambda = 0.4 \quad\ \ Z_{0.4}$ lies almost halfway between $\vec{Y}$ and $\vec{X}$
$\lambda = 1 \quad\quad\ Z_1 = \vec{X}$
$\lambda = 1.2 \quad\ \ Z_{1.2}$ lies beyond the line segment from $\vec{Y}$ to $\vec{X}$.

(A.2.18) The LINE SEGMENT between vectors $\vec{X}$ and $\vec{Y}$, denoted by $\overline{XY}$, is the subset of points in line L (A.2.14a) for which $0 \le \lambda \le 1$. The DISTANCE between $\vec{X}$ and $\vec{Y}$, or the LENGTH of $\overline{XY}$ is defined by

(A.2.18a) $dist(\vec{X}, \vec{Y}) \stackrel{def}{=} length(\overline{XY})$
$$= length(\vec{X} - \vec{Y})$$
$$= \|\vec{X} - \vec{Y}\|$$
$$= \sqrt{(x_1 - y_1)^2 + (x_2 - y_2)^2 + \cdots + (x_k - y_k)^2}.$$

(A.3) Functions

Informally, a FUNCTION is a relation between a set in inputs and a set of outputs where each input is related to (associated with) only one output.

(A.3.1) EXAMPLE: $f(x) = x^2 + 1$: Often, but not always, the inputs of a function are *numbers* and *there is a rule or formula* relating input to output. (*Example (A.3.3a) describes a function whose inputs and outputs are not numbers, and there is no rule or formula that pairs input with output.*) When we write $f(x) = x^2 + 1$, we are stating a rule that says "to produce an output, $f(x)$, we square each input x, then add 1". The input-output pair is $(x, f(x))$, or $(x, x^2 + 1)$.

Here follows a more formal definition which defines the function to be a list of the input-output pairings:

(A.3.2)

> **DEFINITION:**
> $f : \mathcal{X} \to \mathcal{Y}$, from set $\mathcal{X}$ to set $\mathcal{Y}$ is a set of pairs,
>
> $$f \stackrel{def}{=} \{[x, y]\} \quad \text{or} \quad f \stackrel{def}{=} \{[x, f(x)]\},$$
>
> where the first element is an <u>input</u> element x from $\mathcal{X}$, and there is only one (second) <u>output</u> element $y = f(x)$ in $\mathcal{Y}$.
>
> If for every second element y in $\mathcal{Y}$, there is only one input element x in $\mathcal{X}$, then the function f is said to be ONE-TO-ONE, also denoted 1-1 or 1-to-1.
>
> We say that the function f takes element x to element $y = f(x)$.

(A.3.3) EXAMPLES: **Functions.** Consider the three sets,

(A.3.3a) $\qquad \mathcal{X} = \{1, \triangle, \spadesuit\}, \quad \mathcal{Y} = \{5, \odot\},$ and $\mathcal{Z} = \{6, 7\}$

each of which has only finitely many elements. Use of the symbols $\triangle, \spadesuit$, and $\odot$ emphasizes the fact that the elements of the sets $\mathcal{X}$ and $\mathcal{Y}$ are not necessarily numbers.

We define the functions $f : \mathcal{X} \to \mathcal{Y}$ and $g : \mathcal{Y} \to \mathcal{Z}$ as the sets of ordered pairs (A.3.2)

(A.3.3b) $\qquad f = \{[1, \odot], [\triangle, 5], [\spadesuit, 5]\}$

$\qquad\qquad g = \{ [5, 6], [\odot, 7]\}$

which, in "$f(x)$" notation, take the forms

(A.3.3c) $\qquad f(1) = \odot, \quad f(\triangle) = 5, \quad f(\spadesuit) = 5.$

$\qquad\qquad g(5) = 6, \quad g(\odot) = 7$

Display (A.3.3c) shows that for each input x from $\mathcal{X}$ (*or* y from $\mathcal{Y}$) there is only one output element $f(x)$ in $\mathcal{Y}$ (*or* $g(y)$ in $\mathcal{Z}$).

(A.3.4)

> **DEFINITION:** A 1-to-1 function $f{:}\mathcal{X} \rightarrow \mathcal{Y}$, the set of ordered pairs (x, y), has an INVERSE FUNCTION $f^{-1}{:}\mathcal{Y} \rightarrow \mathcal{X}$ which is the set of ordered pairs (y, x). Inputs of f become outputs of f^{-1} and outputs of f become inputs of f^{-1}.

(A.3.5) **EXAMPLE: An Inverse Function.** Note that function f (A.3.3b) is not 1-1 because two distinct inputs $\triangle$ and $\spadesuit$ have the same output 5. In fact, $f(\triangle) = f(\spadesuit) = 5$. The function g is 1-1 so we apply Definition (A.3.4) and *reverse the order of the ordered pairs* of function $g{:}\mathcal{Y} \rightarrow \mathcal{Z}$ (A.3.3b), to obtain the inverse function $g^{-1}{:}\mathcal{Z} \rightarrow \mathcal{Y}$. That is,

(A.3.5a) $g(5){=}6, \qquad g(\odot){=}7$

so that $g^{-1}(6){=}5, \quad g^{-1}(7){=}\odot$

(A.3.6)   The   COMPOSITION   of two given functions,

(A.3.6a) $g : \mathcal{X} \rightarrow \mathcal{Y}$ and $f : \mathcal{Y} \rightarrow \mathcal{Z}$,

is a *third* function, denoted by

(A.3.6b) $(f \circ g) : \mathcal{X} \rightarrow \mathcal{Z}$,

that is defined by the separate actions of f and g in the following way:

(A.3.6c) $(f \circ g)(x) \stackrel{def}{=} f(g(x))$ in $\mathcal{Z}$

for all x in $\mathcal{X}$. In the terminology of (A.3.2), the function $f \circ g$ sends element $x \in \mathcal{X}$ to element $f(g(x)) \in \mathcal{Z}$.

(A.3.7) **EXAMPLE: A Composite Function.** Consider functions $f : \mathcal{X} \rightarrow \mathcal{Y}$ and $g : \mathcal{Y} \rightarrow \mathcal{Z}$ defined in (A.3.3b). Then Definition (A.3.6) tells us that the composite function $g \circ f$ is described as follows:

(A.3.7a) $\begin{aligned} g \circ f(1) &= g(f(1)) = g(\odot) = 7 \\ g \circ f(\triangle) &= g(f(\triangle)) = g(5) = 6 \\ g \circ f(\spadesuit) &= g(f(\spadesuit)) = g(5) = 6 \end{aligned}$

which has ordered pair notation $g \circ f = \{[1, 7], [\triangle, 6], [\spadesuit, 6]\}$.

(A.4) Linear Functions and Matrices

(A.4.1) A LINEAR FUNCTION $T{:}\mathcal{X} \to \mathcal{Y}$ for vector spaces $\mathcal{X}$ and $\mathcal{Y}$ is defined by three properties that (i) $T(\vec{0}) = \vec{0}$, (ii) T preserves proportion, and (iii) T takes straight lines (A.2.14a) to straight lines (*proved in Theorem* (4.5.6), *pg.* 78). i.e.,

(A.4.1a) $\quad T(\lambda_1\vec{P_1} + \lambda_2\vec{P_2}) \overset{def}{=} \lambda_1 T(\vec{P_1}) + \lambda_2 T(\vec{P_2})$

for all vectors, $\vec{P_1}$, $\vec{P_2}$ in $\mathcal{X}$ and all scalars, λ_1, λ_2.

Using mathematical induction, the 2-term expression (A.4.1a) that defines the linearity of T can be extended to the n-term expression

(A.4.1b) $\quad T(\sum_{i=1}^{i=n} \lambda_i\vec{P_i}) = \sum_{i=1}^{i=n} \lambda_i T(\vec{P_i})$

for all scalars λ_i and all vectors $\vec{P_i}$, $i = 1, 2, \ldots, n$.

A useful property of any linear function T is that once its values $T(x)$ are known for a <u>small set</u> of input xs, (*a basis*), the values $T(x)$ are known for <u>all</u> possible input vectors x. Here are the details.

(A.4.2)

> **T**HEOREM: *All the values of the linear function $T{:}\mathcal{X} \to \mathcal{Y}$ for vector spaces $\mathcal{X}$ and $\mathcal{Y}$ are **uniquely determined just by the values of T on any basis (A.2.9)** of $\mathcal{X}$.*

PROOF: Let $\vec{X}$ be any vector. Let $\{\vec{X_i}, i = 1, 2, \ldots, n\}$ form a basis (A.2.9) which then guarantees the existence of unique scalars, $\{\lambda_i, i = 1, 2, \ldots, n\}$, such that

(A.4.2a) $\quad \vec{X} = \sum_{i=1}^{i=n} \lambda_i\vec{X_i}$ $\hspace{3cm}$ *from* (A.2.9).

Then

(A.4.2b) $\quad T(\vec{X}) = T(\sum_{i=1}^{i=n} \lambda_i\vec{X_n}) = \sum_{i=1}^{i=n} \lambda_i T(\vec{X_i})$ $\hspace{1.5cm}$ *from* (A.4.1b).

Since each λ_i is known, and each $T(\vec{X_i})$ is known, (*the value of T on the basis vector $\vec{X_i}$*), it follows that the last expression of (A.4.2b) is known, which means the first expression, $T(\vec{X})$, is completely determined by its values on the n basis vectors $\{\vec{X_i}\}$. ∎

MATRICES

(A.4.3)   The $m \times k$ MATRIX A (m rows, k columns) defined by the mk scalars
$a_{ij}, i = 1, 2, \ldots, m, \ j = 1, 2, \ldots, k$ is written as the rectangular array

(A.4.3a) $A = (a_{ij}) = \begin{bmatrix} a_{11} & a_{12} & \cdots & a_{1k} \\ a_{21} & a_{22} & \cdots & a_{2k} \\ \vdots & \vdots & \ddots & \\ a_{m1} & a_{m2} & \cdots & a_{mk} \end{bmatrix} \begin{matrix} \leftarrow r_1(A) \\ \leftarrow r_2(A) \\ \cdot \\ \leftarrow r_m(A) \end{matrix}$

$$\begin{matrix} \uparrow & \uparrow & & \uparrow \\ c_1(A) & c_2(A) & & c_k(A) \end{matrix}$$

where

(A.4.3b)

• $r_i(A) = [a_{i,1}, a_{i,2}, \ldots, a_{i,k}]$ is the ith row of $m \times k$ matrix A and

• $c_j(A) = \text{col}[a_{1,j}, a_{2,j}, \ldots, a_{m,j}]$ denotes its jth column.

With (A.4.3b), we have an alternate notation for (A.4.3a), namely,

(A.4.3c)

$$A = (a_{ij}) = \begin{bmatrix} r_1(A) \\ r_2(A) \\ \vdots \\ r_m(A) \end{bmatrix} = \begin{bmatrix} c_1(A) & c_2(A) & \cdots & c_k(A) \end{bmatrix}.$$

Let the $k \times n$ matrix B (k rows, n columns) defined by the kn scalars
$j = 1, 2, \ldots, k, \ l = 1, 2, \ldots, n$, be written

(A.4.3d) $B = (b_{jl}) \overset{def}{=} \begin{bmatrix} b_{11} & b_{12} & \cdots & b_{1n} \\ b_{21} & b_{22} & \cdots & b_{2n} \\ \vdots & \vdots & \ddots & \vdots \\ b_{k1} & b_{k2} & \cdots & b_{kn} \end{bmatrix}.$

Then the   MATRIX PRODUCT, $A * B$,   is the $m \times n$ matrix

(A.4.3e) $A * B = D = (d_{il}) = \begin{bmatrix} d_{11} & d_{12} & \cdots & d_{1n} \\ d_{21} & d_{22} & \cdots & d_{2n} \\ \vdots & \vdots & \ddots & \vdots \\ d_{m1} & d_{m2} & \cdots & d_{mn} \end{bmatrix}$

(m rows, n columns) which is defined by the mn scalars c_{il} where
$i = 1, 2, \ldots, m, l = 1, 2, \ldots, n$, and

(A.4.3f) $\quad d_{il} = \langle\, r_i(A)\,,\, \overline{d_l(B)}\,\rangle = \displaystyle\sum_{s=1}^{s=k} a_{is}\,\overline{d_{sl}}.\quad see\quad$ (A.2.6) *and* (A.4.3b)

where $\overline{b_{sl}}$ is the complex conjugate (A.2.13a) of b_{sl} in case b_{sl} is complex.

(A.4.4) EXAMPLES: **Matrix Products:** When the matrix A is a 2×2 square matrix and the right hand factor is a column vector matrix $\vec{X}$ or a square 2×2 matrix, then (A.4.3e), (A.4.3f) give the entries of the products as follows:

(A.4.4a) $\quad \underbrace{\begin{bmatrix} a & b \\ c & d \end{bmatrix}}_{A} * \underbrace{\begin{bmatrix} x \\ y \end{bmatrix}}_{\vec{X}} = \underbrace{\begin{bmatrix} ax+by \\ cx+dy \end{bmatrix}}_{D = A * \vec{X}}$

(A.4.4b) $\quad \underbrace{\begin{bmatrix} a & b \\ c & d \end{bmatrix}}_{A} * \underbrace{\begin{bmatrix} p & q \\ r & s \end{bmatrix}}_{B} = \underbrace{\begin{bmatrix} ap+br & aq+bs \\ cp+dr & cq+ds \end{bmatrix}}_{A * B}.$

From (A.4.4b), there is a 2×2 matrix I_2 that is FEEBLE[1] in the sense that after multiplication, I_2 leaves any factor unchanged. Specifically, for any 2×2 matrix A, the products, $I_2 * A$ and $A * I_2$ are still equal to A. Here is the calculation:

(A.4.4c) $\quad \underbrace{\begin{bmatrix} 1 & 0 \\ 0 & 1 \end{bmatrix}}_{I_2} * \underbrace{\begin{bmatrix} a & b \\ c & d \end{bmatrix}}_{A} = \underbrace{\begin{bmatrix} a & b \\ c & d \end{bmatrix}}_{A} = \underbrace{\begin{bmatrix} a & b \\ c & d \end{bmatrix}}_{A} * \underbrace{\begin{bmatrix} 1 & 0 \\ 0 & 1 \end{bmatrix}}_{I_2}.$

It also follows from (A.4.4b) that any 2×2 matrix $A = \begin{bmatrix} a & b \\ c & d \end{bmatrix}$, where $ad - bc \neq 0$, has a unique MULTIPLICATIVE INVERSE

(A.4.4d) $\quad A^{-1} = \begin{bmatrix} a & b \\ c & d \end{bmatrix}^{-1} = \dfrac{1}{ad - bc}\begin{bmatrix} d & -b \\ -c & a \end{bmatrix}$ *see footnote[2]*

which brings us back to the identity I_2 in the sense that

(A.4.4e) $\quad A^{-1} * \begin{bmatrix} a & b \\ c & d \end{bmatrix} = \begin{bmatrix} a & b \\ c & d \end{bmatrix} * A^{-1} = \begin{bmatrix} 1 & 0 \\ 0 & 1 \end{bmatrix} = I_2.$

We show that *any* linear function (or transformation) T can always be

---

[1] Similarly, 0, the number zero is "feeble" with respect to addition since its combination with any other number x leaves it intact. That is, for all x, $0 + x = x + 0 = x$. Formally, feeble elements are called identities.

[2] Scalar multiplication by a matrix (*a vector*) is defined in (A.2.4c).

effected by lefthand multiplication by a certain $m \times k$ matrix MX(T) which we now define. (*The proof that* MX(T) *is the desired matrix follows in Theorem* (A.4.6).)

(A.4.5)   NOTATION: *Suppose* $T: \boldsymbol{R}^k \to \boldsymbol{R}^m$ *is a linear transformation. Let* $\{\vec{E}_1, \vec{E}_2, \ldots, \vec{E}_k\}$ *be the* k *standard basis vectors of* $\boldsymbol{R}^k$, *each of which is written as a* $k \times 1$ *column. Then we define the* $m \times k$ *matrix whose* j*th column equals* $T(\vec{E}_j)$ *in* $\boldsymbol{R}^m$ *where* $j = 1, 2, \ldots, k$. *That is,*

(A.4.5a) $\text{MX}(T) \overset{def}{=} \left[T(\vec{E}_1)\, T(\vec{E}_2)\, \ldots\, T(\vec{E}_j)\, \ldots\, T(\vec{E}_k) \right]$

$$\qquad\qquad\ \uparrow\qquad\ \uparrow\qquad\quad \uparrow\qquad\quad \uparrow$$
$$\qquad\qquad Col\,1\quad Col\,2\qquad Col\,j\qquad Col\,k$$

(A.4.6)

> THEOREM: **Matrix multiplication characterizes all linear transformations.** For any linear $T{:}\boldsymbol{R}^k \to \boldsymbol{R}^m$ as defined in (A.4.1a), $m \times k$ matrix MX(T) given by (A.4.5a) has the property that for all $\vec{X}$ in $\boldsymbol{R}^k$,
>
> **(A.4.6a)** $\underbrace{T(\vec{X})}_{T\ \underline{of}\ X} = \underbrace{\text{MX}(T) * \vec{X}}_{matrix\ \text{MX}(T)\ \underline{times}\ vector\ \vec{X}}$.

PROOF: **Choose any** $\vec{X}$ in $\boldsymbol{R}^m$. Then $\vec{X}$ is represented by the linear combination (A.2.9a)

(A.4.6b) $\vec{X} = a_1\vec{E}_1 + a_2\vec{E}_1 + \ldots + a_j\vec{E}_j + \ldots + a_k\vec{E}_k = \sum_{j=1}^{k} a_j\,\vec{E}_j.$

for unique scalars $\{a_j\}$ and the standard basis vectors (A.2.11) $\vec{E}_j$ in $\boldsymbol{R}^m$ where $j = 1, 2, \ldots, k$.

A useful observation is the fact that from any $m \times k$ matrix, the jth column, an $m \times 1$ sub-matrix, can be isolated by simply multiplying on the right by the jth standard basis vector $\vec{E}_j$. In the notation of (A.4.3b), for any $k \times m$ matrix A, $c_j(A) = A * \vec{E}_j$. In our case, set $A = \text{MX}(T)$, a matrix which, by definition (A.4.5a), guarantees that its jth column is $T(\vec{E}_j)$. Then

(A.4.6c)   *The* jth *column of* MX(T) = MX(T) $* \vec{E}_j = T(\vec{E}_j)$.

Since $\vec{X} = \sum_{j=1}^{k} a_j\vec{E}_j$ as given in (A.4.6b),

(A.4.6d) $T(\vec{X}) = T(\sum_{j=1}^{k} a_j\vec{E}_j) = \sum_{j=1}^{k} a_jT(\vec{E}_j)$ $\begin{cases} linearity \\ of\ T\ \text{(A.4.1b)} \end{cases}$

$$= \sum_{j=1}^{k} a_j \left(\text{MX}(T) * \vec{E}_j \right) \qquad \textit{from (A.4.6c)}, \ T(\vec{E}_j) = \left(\text{MX}(T) * \vec{E}_j \right)$$

$$= \text{MX}(T) * \left(\sum_{j=1}^{k} a_j \vec{E}_j \right) \qquad \textit{factor out MX}(T)$$

$$= \text{MX}(T) * \vec{X} \qquad \textit{since } \vec{X} = \sum_{j=1}^{k} a_j \vec{E}_j \textit{ from (A.4.6b)}.$$

In this string of equalities, equate the first and last terms to show that (A.4.6a) is true. This ends the proof.∎

(A.4.7) EXAMPLE: **The 2×2 MX(T)= $[T(\mathbf{E}_1) \ T(\mathbf{E}_2)]$ for $T : R^2 \to R^2$.**

For all $\vec{X} = \begin{bmatrix} x_1 \\ x_2 \end{bmatrix}$ in R^2, and arbitrary numbers a, b, c, d, we define

(A.4.7a) $T \left(\begin{bmatrix} x_1 \\ x_2 \end{bmatrix} \right) = \begin{bmatrix} ax_1 + bx_2 \\ cx_1 + dx_2 \end{bmatrix}$ *linear function $T : R^2 \to R^2$*

$$= \begin{bmatrix} a & b \\ c & d \end{bmatrix} \qquad\qquad * \begin{bmatrix} x_1 \\ x_2 \end{bmatrix} \qquad \textit{from (A.4.4a)}$$

$$= \left[T \left(\begin{bmatrix} 1 \\ 0 \end{bmatrix} \right) \ T \left(\begin{bmatrix} 0 \\ 1 \end{bmatrix} \right) \right] * \begin{bmatrix} x_1 \\ x_2 \end{bmatrix}$$

$$\textit{from (A.4.7a)}, \ T(\begin{bmatrix} 1 \\ 0 \end{bmatrix}) = \begin{bmatrix} a \\ c \end{bmatrix} \textit{ and } T(\begin{bmatrix} 0 \\ 1 \end{bmatrix}) = \begin{bmatrix} b \\ d \end{bmatrix}$$

$$= \underbrace{\left[T \left(\vec{E}_1 \right) T \left(\vec{E}_2 \right) \right]}_{\text{MX}(T)} * \begin{bmatrix} x_1 \\ x_2 \end{bmatrix} \qquad \left\{ \vec{E}_1 = \begin{bmatrix} 1 \\ 0 \end{bmatrix}, \vec{E}_2 = \begin{bmatrix} 0 \\ 1 \end{bmatrix} \right.$$

which confirms $\text{MX}(T) = \left[T(\vec{E}_1) \ T(\vec{E}_2) \right]$ (A.4.6a) when $m = k = 2$.

(A.4.8)

> NOTATION: *It will often be convenient to replace the symbol* MX(T) *for the $m \times n$ matrix of T with the simpler symbol T. That is,*
>
> $$\text{MX}(\mathbf{T}) \equiv T.$$

(A.5) Eigenvectors and Eigenvalues

Intuitively, an *eigenvector* (sometimes called a ***proper vector***) of a given matrix is a vector that is stretched or contracted by a certain factor known as the *eigenvalue* (also called a ***proper value***) under the action of (*multiplication by*) that matrix. Here are the formal definitions.

(A.5.1)

DEFINITION: Given an $n \times n$ matrix T. Then an $n \times 1$ vector $\vec{X} \neq 0$ in $\mathbf{R}^n$ is called an <u>EIGENVECTOR</u> or <u>PROPER VECTOR</u> of T if

(A.5.1a) $T(\vec{X}) = T * \vec{X} = \lambda \vec{X}.$

The "stretching" scalar λ, which may be real or complex, is called an <u>EIGENVALUE</u> or <u>PROPER VALUE</u> of the matrix T.

(A.5.2) **There is always at least one eigenvector for any matrix T.** A deep theorem, equivalent to the *Fundamental Theorem of Algebra*,[3] guarantees that every square matrix has at least one eigenvector which comes with an associated eigenvalue. Also note that Definition (A.5.1) requires that the eigen*vector* $\vec{X} \neq 0$ while the eigen*value* $\lambda = 0$ is allowed.

(A.5.3) **There are at most n linearly independent (A.2.10) eigenvectors for any $n \times n$ matrix T.** For example, the matrix $\begin{bmatrix} a & 1 \\ 0 & a \end{bmatrix}$ has only one eigenvalue, namely a, and its only eigenvector is $\vec{X} = \text{col}[1,0]$ (*or any multiple* $b\vec{X}$ *of* $\vec{X}$ *where* $b \neq 0$). The matrix $aI_2 = \begin{bmatrix} a & 0 \\ 0 & a \end{bmatrix}$ has only one eigenvalue, a, yet *every* non-zero vector is an eigenvector.

Finding Matrix of T When $T(X_i) = \lambda_i X_i$ for Known Eigenvectors X_i and Eigenvalues λ_i.

The following theorem is essential in the proof of Theorem (7.1.3).

---

[3] *The Fundamental Theorem of Algebra* says that for any nth degree polynomial, $p(z) = \sum_{k=0}^{n} a_k z^k$, where the coefficients a_k are complex numbers, there are exactly n complex roots $\{z_i\}$ where $p(z_i) = 0$ for each $i = 1, 2, \ldots, n$.

(A.5.4)

THEOREM: *Given an $n \times n$ matrix T that has eigenvectors that also form a basis for $\boldsymbol{R}^n$. That is,*

(A.5.4a) $\quad T(\vec{X}_i) = T * \vec{X}_i = \lambda_i \vec{X}_i \quad$ *where $i = 1, 2, \ldots, n$,*

for scalars $\{\lambda_i\}$ and n vectors $\{\vec{X}_i\}$ that are also linearly independent in $\boldsymbol{R}^n$. Then using the $n \times n$ matrices

(A.5.4b)

$$S = \begin{bmatrix} \vec{X}_1 & \vec{X}_2 & \ldots & \vec{X}_n \end{bmatrix} \quad and \quad D = \begin{bmatrix} \lambda_1 & 0 & & 0 \\ 0 & \lambda_2 & & 0 \\ 0 & 0 & \ddots & 0 \\ 0 & 0 & & \lambda_n \end{bmatrix},$$

$$\uparrow \qquad \uparrow \qquad \qquad \uparrow$$
$$Col\,1 \quad Col\,2 \qquad \quad Col\;n$$

we have

(A.5.4c) $\quad T = S * D * S^{-1}.$

PROOF: Note that S is the matrix whose ith column is the eigenvector $\vec{X}_i$. From (A.4.6a), the ith column of S is $\vec{X}_i = S(\vec{E}_i) = S * \vec{E}_i$ for all standard basis vectors $\{\vec{E}_i\}$ (A.2.11) of $\boldsymbol{R}^n$. That is,

(A.5.4d) $\quad S * \vec{E}_i = \vec{X}_i.$ $\quad$ Equivalently, $S^{-1} * \vec{X}_i = \vec{E}_i.$

(A.5.4e) $\quad$ Now for all i, $i = 1, 2, \ldots, n$,

$$(S * D * S^{-1}) * \vec{X}_i = S * D * (S^{-1} * \vec{X}_i) \qquad * \text{ multiplication associativity}$$

$$= S * (D * \vec{E}_i) \qquad\qquad S^{-1} * \vec{X}_i = \vec{E}_i \text{ from (A.5.4d)}$$

$$= \lambda_i S * (\vec{E}_i) \qquad\qquad D * \vec{E}_i = \lambda_i \vec{E}_i \text{ from (A.5.4b)}$$

$$= \lambda_i \vec{X}_i \qquad\qquad\qquad S * (\vec{E}_i) = \vec{X}_i \text{ from (A.5.4d)}$$

$$= T * \vec{X}_i. \qquad\qquad\qquad Hypothesis \text{ (A.5.4a)}$$

From the chain of equalities (A.5.4e), $(S * D * S^{-1}) * \vec{X}_i = T * \vec{X}_i$ (*also denoted by* $T(\vec{X}_i)$) for all basis vectors $\{X_i\}$, $i = 1, 2, \ldots, n$. It follows that $(S * D * S^{-1}) * \vec{X} = T * \vec{X}$ for <u>all</u> $\vec{X}$ in $\boldsymbol{R}^n$ and not just for basis vectors (Theorem (A.4.2)). With (A.5.4c) being established, the proof is done. ■

(B) Hyperbolic Functions

(B.1) Overview

We come to the hyperbolic functions, $\cosh x$, $\sinh x$, and $\tanh x$ (pronounced "coe-sh", "sinsh", and "tansh", respectively). The functions $\cosh x$ and $\sinh x$ are the even and the odd parts of the function e^x where x is a real number. *For more on even and odd functions, see Exercise* (B.4.1), pg. 357. Here are the definitions:

(B.1.1)

> **DEFINITION:** *The* HYPERBOLIC SINE, COSINE, AND TANGENT, sinh, cosh, *and* tanh, *are defined for all real* x *as follows:*

(B.1.1a)
$$\sinh(x) = S_x \overset{def}{=} \frac{e^x - e^{-x}}{2},$$

(B.1.1b)
$$\cosh(x) = C_x \overset{def}{=} \frac{e^x + e^{-x}}{2},$$

(B.1.1c)
$$\tanh(x) = T_x \overset{def}{=} \frac{e^x - e^{-x}}{e^x + e^{-x}}$$
$$= \frac{S_x}{C_x} \overset{def}{=} \frac{\sinh(x)}{\cosh(x)}.$$

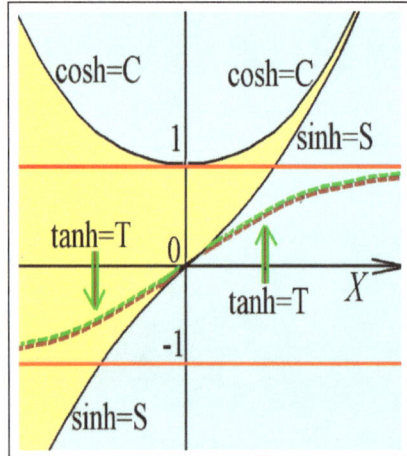

(B.1.2) Ranges of Hyperbolic Functions. Note that for all real x,

$$-1 < \tanh(x) = T_x < 1,$$

$$1 \le \cosh(x) = C_x < \infty,$$

$$-\infty < \sinh(x) = S_x < \infty.$$

(B.1.3) **Some Hyperbolic Identities:** Lines (B.1.3a), (B.1.3c), and (B.1.3d), flow directly from the core definitions (B.1.1).

(B.1.3a) $\qquad C_x^2 - S_x^2 = 1.$ $\qquad\qquad$ *Divide this line by C_x^2 to obtain*

(B.1.3b) $\qquad 1 - T_x^2 = 1/C_x^2.$

(B.1.3c) $\qquad C_{x+y} = C_x C_y + S_x S_y.$ $\qquad$ *Use $x + y$ in Def. (B.1.1b)*

(B.1.3d) $\qquad S_{x+y} = S_x C_y + C_x S_y.$ $\qquad$ *Use $x + y$ in Def. (B.1.1a). Then divide this line by (B.1.3c) to obtain*

(B.1.3e) $\qquad T_{x+y} = \dfrac{T_x + T_y}{1 + T_x T_y}.$ $\qquad$ *Divide (B.1.3d) by (B.1.3c). Then use Def. (B.1.1c)*

(B.2) Even and Odd Functions

(B.2.1)

> **DEFINITION:** *An* EVEN FUNCTION *f_E acts like x^n where n is even. That is, an even function, like x^n will always absorb the minus sign in the sense that*
>
> $$f_E(-x) = f_E(x) \quad for\ all\ x.$$

(B.2.2)

> **DEFINITION:** *An* ODD FUNCTION *f_O, behaves like x^n where n is odd. Specifically, any odd function, like x^n, allows minus signs to factor out in the sense that*
>
> $$f_O(-x) = -f_O(x) \quad for\ all\ x.$$

Any function $f : R \to R$ decomposes uniquely into an even part f_1 and an odd part f_2. The interesting thing is that the forms for f_1 and f_2 are easily obtained. In fact, we have the

(B.2.3)

Theorem: *For any function $f : R \to R$, define f_1 and f_2 by*

(B.2.3a) $f_1(x) = \dfrac{f(x) + f(-x)}{2}$ *and* $f_2(x) = \dfrac{f(x) - f(-x)}{2}$.

Then

(B.2.3b) $f = f_1 + f_2$ *where f_1 is even and f_2 is odd.*

Moreover, decomposition (B.2.3b) is unique in the sense that if there are other functions, g_1 and g_2 that are even and odd respectively, such that

(B.2.3c) $f = f_1 + f_2 = g_1 + g_2$,

then, necessarily, $f_1 = g_1$ *and* $f_2 = g_2$.

Proof: Exercise (B.4.1), part (3).

(B.3) Invariant Areas of Transformed Hyperbolas

(B.3.1) Figure: *Triangular and Trapezoidal Areas are Equal*

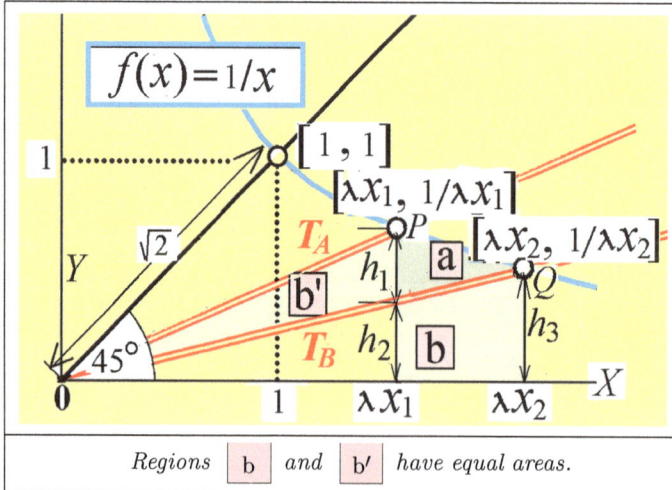

Regions b *and* b′ *have equal areas.*

In Lemma (B.3.2) and Theorem (B.3.3) following, we will show that for fixed x_1 and x_2 on the X-axis in Figure (B.3.1), the areas of all subtended triangular areas $\overline{POQ}$ are constant and independent of the "sliding" scalar, λ.

First, we prove

(B.3.2)

> **LEMMA:** (***Area Equality***) *In Figure* (B.3.1), *the perpendicular X-axis and Y-axis are identically scaled. The points* $0 < x_1 < x_2$ *are arbitrary but fixed on the X-axis. We also have axes* T_A, T_B, *hyperbola* $f(x) = 1/x$, *and arbitrary scalar* $\lambda > 0$. *Then*
>
> **(B.3.2a)**   *Area of triangle* $\boxed{b'}$ = *Area of trapezoid* $\boxed{b}$.
>
> *Moreover, this value is independent of* λ.

PROOF: In Figure (B.3.1), the left hand and right hand sides of the trapezoid $\boxed{b}$ are $h_2 = (x_1/(\lambda\, x_2^2))$ and $h_3 = (1/(\lambda\, x_2))$, respectively. The horizontal base of the trapezoid has length $(\lambda x_2 - \lambda x_1)$. By taking the product, *average height* $\times$ *width* $= \frac{1}{2}(h_2 + h_3)(\lambda x_2 - \lambda x_1)$, we obtain

(B.3.2b)   Trapezoid Area $\boxed{b} = \dfrac{1}{2}\left(1 - \left(\dfrac{x_1}{x_2}\right)^2\right).$ *independent of* λ

Now triangle $\boxed{b'}$ has (vertical) base $h_1 = (1/\lambda)(1/x_1 - x_1/x_2^2)$ and (horizontal) altitude λx_1. Hence, $1/2$ *base* $\times$ *altitude* has value

(B.3.2c)   Triangle Area $\boxed{b'} = \dfrac{1}{2}\left(1 - \left(\dfrac{x_1}{x_2}\right)^2\right),$ *independent of* λ

which proves (B.3.2a). The theorem is done. ∎

. . . .

We come to the theorem that quantifies the triangular area $\overline{POQ}$ in Figure (B.3.1) bounded by the hyperbola $f(x) = 1/x$ and the T_A and T_B axes.

(B.3.3)

> Theorem: (***Calculation of Area of Fig.*** **(B.3.1)**)   *In Figure*
> *(B.3.1), we are given two areas:*
>
> - *Area(* a *+* b *) of the subtended "triangle" bounded by the* T_A
>   *and* T_B *axes and the hyperbola* $f(x) = 1/x$,
>
> - *Area(* a *+* b′ *) of the subtended "trapezoids".*
>
> *Then* <u>*both*</u> *areas are invariant under* λ *and have common value*
>
> **(B.3.3a)** $r_{A,B} = \ln\left(\dfrac{x_2}{x_1}\right).$ *area r independent of* λ.
>
> *Equivalently,*
>
> **(B.3.3b)** $e^{r_{A,B}} = \dfrac{x_2}{x_1}$ *exponential of both*
>   *sides of* (B.3.3a).

Proof: Since (B.3.2a) guarantees that the area of   b   = area of   b′ ,
it follows from Figure (B.3.1) that the area of triangle   a   +   b′   = the
area of trapezoid   a   +   b , which computes to be

(B.3.3c) $\displaystyle\int_{\lambda x_1}^{\lambda x_2} \left(\frac{dx}{x}\right) = \ln(\lambda x_2) - \ln(\lambda x_1) = \ln\left(\frac{x_2}{x_1}\right) = r_{A,B},$

a quantity that is *independent of* λ which we have denoted $r_{A,B}$. The very
last equality of (B.3.3c) is (B.3.3a). Taking the exponential of both sides
of (B.3.3a) gives us (B.3.3b). Having established (B.3.3a) and (B.3.3b),
the theorem is proved. ■

The following theorem uses a counterclockwise 45° rotation of Figure (B.3.1)
to relate the area $r_{A,B}$, the hyperbolic tangent tanh, and the trigonometric
tangent tan.

(B.3.4) F<small>IGURE</small>: *Area-Preserving Rotation of Figure* **(B.3.1)**.

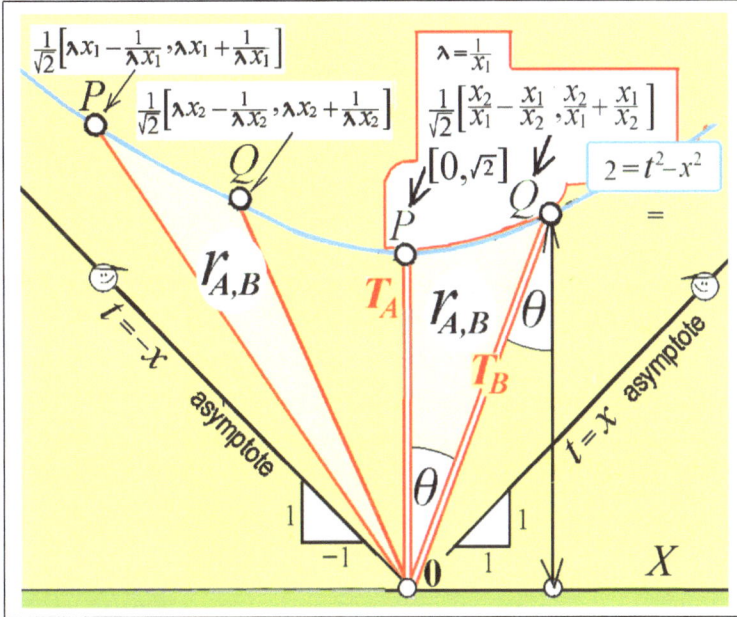

(B.3.5)

> T<small>HEOREM</small>: (***Relating Hyperbolic Area to Speed***): *In the diagram* (B.3.4) *the* T_A *and* T_B *axes intersect the hyperbola* $2 = t^2 - x^2$ *at points* P *and* Q *that are arbitrary except for the requirement that*
>
> - *Area* $r_{A,B}$ *is fixed for triangular regions* $\overline{P0Q}$.
>
> *If the line segment* $\overline{OP}$ (*or* T_A-*axis*) *is perpendicular to the* X-*axis, forming angle* θ *with the* T_B-*axes, then*
>
> **(B.3.5a)** $\tan(\theta) = \tanh(r_{A,B})$.
>
> *If* $\tan(\theta) > 0$ (*if* $\tan(\theta) < 0$), *then* Q *falls to the right* (*to the left*) *of* P.

P<small>ROOF</small>: We rotate Figure (B.3.1) 45° counter clockwise to produce Figure (B.3.4). Then

- $\underline{f(x) = 1/x \text{ rotates into the hyperbola } 2 = t^2 - x^2 \text{ of (B.3.4)}}$: Counter clockwise rotation through an angle θ of any point $\mathrm{col}[x, y]$ in $\mathbf{R}^2$ can be produced by a left hand multiplication by the orthogonal transformation

(B.3.5b) $\begin{bmatrix} \cos(\theta) & -\sin(\theta) \\ \sin(\theta) & \cos(\theta) \end{bmatrix} = \dfrac{1}{\sqrt{2}} \begin{bmatrix} 1 & -1 \\ 1 & 1 \end{bmatrix}$ *when* $\theta = 45°$.

Using the matrix (B.3.5b) on a typical point $\mathrm{col}[x, 1/x]$ of the hyperbola $f(x) = 1/x$ in Figure (B.3.1), we obtain

(B.3.5c)

$$\frac{1}{\sqrt{2}} \begin{bmatrix} 1 & -1 \\ 1 & 1 \end{bmatrix} * \begin{bmatrix} x \\ 1/x \end{bmatrix} = \frac{1}{\sqrt{2}} \begin{bmatrix} x - 1/x \\ x + 1/x \end{bmatrix} = \begin{bmatrix} \tilde{x} \\ \tilde{t} \end{bmatrix}. \qquad \begin{cases} \tilde{x} \stackrel{def}{=} (x - 1/x)/\sqrt{2} \\ \tilde{t} \stackrel{def}{=} (x + 1/x)/\sqrt{2} \end{cases}$$

The output coordinates, $\tilde{x} = (x - 1/x)/\sqrt{2}$ and $\tilde{t} = (x + 1/x)/\sqrt{2}$, have the property $2 = \tilde{t}^2 - \tilde{x}^2$ which defines the hyperbola in the rotated figure (B.3.4).

- *Choose* $\lambda = 1/x_1$. As $\lambda > 0$ varies and points P and Q slide along the hyperbola $2 = t^2 - x^2$, the resulting triangular areas $\overline{POQ}$ have constant area $r_{A,B}$ (B.3.3a). For $\lambda = 1/x_1$, the T_A-axis of triangular region $\overline{POQ}$ is vertical. The T_B-axis of frame $\mathcal{F}^B$ forms an angle θ, say, which tells us that

(B.3.5d) $\tan(\theta) = \dfrac{(x_2/x_1 - x_1/x_2)}{(x_2/x_1 + x_1/x_2)}$ *from Coordinates of* Q *in* (B.3.4)

$$= \frac{(e^{r_{A,B}} - e^{-r_{A,B}})}{(e^{r_{A,B}} - e^{-r_{A,B}})} \qquad \textit{from (B.3.3b)}$$

$$= \tanh(r_{A,B}). \qquad \textit{from Def. (B.1.1c)}$$

This establishes (B.3.5a) for all constant-area regions $\overline{POQ}$ of Figure (B.3.4). Now the speed $v < 0$

   if and only if $\tanh(r_{A,B}) < 0$ *from* (B.3.5d)

   if and only if $r_{A,B} < 0$ *from Figure* (B.1.1)

   if and only if $\ln(x_2/x_1) < 0$ *from* (B.3.3c)

   if and only if $x_2 < x_1$ $\ln(x_2/x_1) < 0$ *means* $(x_2/x_1) < 1$

   if and only if axis T_B is left of T_A *see axes in* (B.3.1)

The proof is done. ◼

(B.4) Exercises

Even and Odd Functions. For the functions f_1 and f_2 given by (B.2.3a), prove that (B.2.3b) is true. That is, show

(1) $f = f_1 + f_2$ and

(2) f_1 and f_2 of (B.2.3a) are even and odd, respectively.

. . . .

(3) Prove (B.2.3c) is valid — there is only one way to decompose a function f into an even and an odd part.

Hint: *Assume f_1, g_1 are even and f_2, g_2 are odd where $f = f_1 + f_2 = g_1 + g_2$. Take the difference to get $(f_1 - g_1) + (f_1 - g_2) = 0$. Since the zero function is the only function that is both even and odd, argue that, necessarily, $f_1 = g_1$ and $f_2 = g_2$.*

Cosh occurs in Nature. The curve of the hyperbolic cosine, $\cosh(x) = C(x)$, shown in the figure of (B.1.1), pg. 350, appears in nature and is called the **catenary curve**. Hanging strings or cables of uniform linear density (ρ *pounds per linear foot, say*) fall into a catenary pattern. Examples include suspension bridge cables and fine, flexible necklaces.

Question: Analyze the Diagram (B.4.2b) to show that the forces produce a catenary.

Hint: *Consider Steps (B.4.2a)-(B.4.2f) following:*

(B.4.2a)   Calculus Preliminaries: hyperbolic derivatives

 If $y(x) = A \cosh(Bx),$

 then $y'(x) = AB \sinh(Bx)$

 and $y''(x) = AB^2 \cosh(Bx).$

(B.4.2b)   Three Forces on a Cable Segment: W_{x_0}, F_0, F_{x_0}

The cable segment in the diagram on the right is stationary. Hence, the sum of the following three forces acting on the cable equals zero.

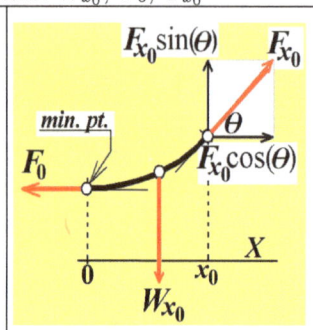

. . . .

W_{x_0} = Weight

$\quad = \rho \times$ arc length, $x = 0$ to $x = x_0$

$\quad = \rho \times \displaystyle\int_{x=0}^{x=x_0} \sqrt{1 + [y'(x)]^2}\,dx$

F_0 = internal horizontal force at cable minimum,

F_{x_0} = internal force at $x = x_0$.

(B.4.2c)   A Hanging Cable Coincides with the Catenary Cosh(x).

Step 1: Note that Diagram (B.4.2b) decomposes F_{x_0} into its horizontal and vertical components. Set the sum of all horizontal forces to zero. Then set the sum of all vertical forces to zero. From these two equations, deduce the unitless expression

(B.4.2d) $y'(x_0) = \dfrac{\rho}{F_0} \displaystyle\int_{x=0}^{x=x_0} \sqrt{1 + [y'(x)]^2}\,dx.$

Step 2: After differentiating (B.4.2d) with respect to x, show that

(B.4.2e) $\left[\dfrac{F_0}{\rho} y''(x)\right]^2 - \left[y'(x)\right]^2 = 1.$

Step 3: Use the differentiation properties (B.4.2a) of $A\cosh(Bx)$ and the identity (B.1.3a) to show that (B.4.2e) implies

(B.4.2f) $y(x) = \dfrac{F_0}{\rho} \cosh(\dfrac{\rho}{F_0}x) +$ const.

(C) Deconstructing a Moving Train

(C.1) Motion Alters Age

(C.1.1) **Consider triplets of identical age,** each having been born at *precisely* the same moment. (*We are inventing a biology in which triplets can be born simultaneously instead of sequentially.*) The triplets are seated inside a railroad car with two of them sitting at opposite ends while the third triplet sits in the middle row (*see Figure* (C.2.1)).

If the railroad car is traveling in the rightward direction, then, when each triplet is exactly opposite an observer on the platform, the triplets' various ages, as measured by the observers will all be different. This is illustrated in Figure (C.2.1) where the time is **6.15** on the platform, and the (*formerly same-age*) triplets in the train get older as we pass from the front (right) to the rear (left) of the railroad car.

Conversely, on a leftward traveling train, the triplets get younger as we pass from right to left. (*Exercise* (C.3.1).)

(C.1.1a) **To resolve this dilemma,** we must leap beyond our intuition and imagine the train (or any connected rod, or ruler) as a continuously connected collection of individual "slices" like beads on a string, where each slice, or bead, possesses its own synchronized clock.

In the following discussion, the train of Figure (C.2.1) can be thought of as three individual "slices," each with its own observer and clock.

(C.2) Minkowski Diagram for a Moving Train

In Figure (C.2.1) following, we consider a train as a set of infinitely many individual points. This is why the left hand panel of Figure (C.2.1) represents all of the train's coordinates as an *area* or *band* (of coordinates) on the spacetime graph.

(C.2.1) FIGURE: *A Moving Train in Several Time Slices*

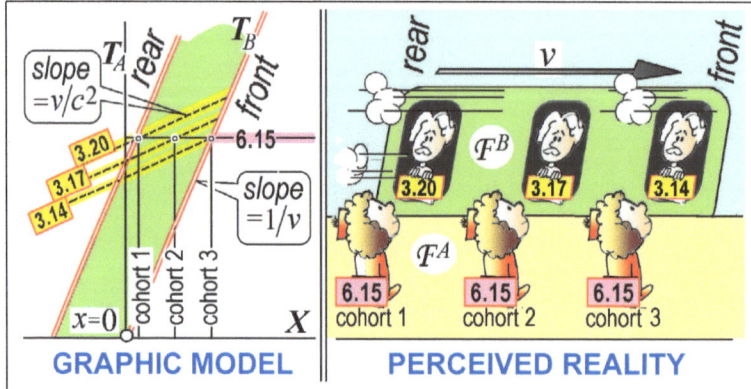

- LEFT PANEL OF (C.2.1) (*Graphic Model*):The stationary frame of the train platform has a vertical T-axis which intersects the horizontal X-axis at $x = 0$.

  WORLDLINES.
  ¬ **The three platform cohorts** have worldlines that are vertical lines parallel to the T-axis.

  ¬ **The front (rightmost) end of the train at point** $x = 0$ in the moving frame has relative speed v. Its worldline radiates from the origin with slope $(1/v)$, i.e., at an angle θ from the vertical which implies $v = \tan\theta$. This worldline coincides with the T'-axis of the moving frame.

  ¬ **The rear (leftmost) end of the train** has its worldline to the left of, and parallel to, the T'-axis. The shaded region between the left and right worldlines represents the infinitely many parallel spacetime worldlines of the points of the train in between.

  LINES of SIMULTANEOUS EVENTS.
  In the stationary frame, lines of simultaneous events are horizontal $(\rightarrow)$. The graph shows simultaneous events at time 6.15. In the moving frame of the train, each line of simultaneous events $(\nearrow)$ has slope (v/c^2) (*Theorem* (8.4.1)). The graph shows parallel (slanted) lines of simultaneous events at respective times, 3.14, 3.17, and 3.20.

- RIGHT PANEL OF (C.2.1)(*Perceived Reality*): Three cohorts (observers) are fixed at equal distances on a railroad platform. The individual cohort clocks are synchronized with each other. A train passes along the platform moving from left to right. When the platform clocks all read 6.15 , each cohort observes the moving clock (on the train) at the other end of her individual line-of-sight corridor.

Although all clocks on the train are synchronized within the train's frame, the readings of the train's clocks from the platform at time 6.15 are (reading from left to right) 3.20 , 3.17 , and 3.14 .

In sum, at time 6.15 on the platform, the rear (leftmost part) of the train is seen to be older than the front (rightmost part) of the train.

(C.3) Exercises

(C.3.1)
Exercise

Shifting Age of Train Segments. Modify Figure (C.2.1) to show that for a leftward moving train, the left end will be younger than the right end as measured by synchronized clocks on the platform.

Bibliography

[1] Abbott, Edwin A. *Flatland: A Romance of Many Dimensions*. Dover (Thrift Edition), 96 pp., 1992.

[2] Born, Max. *Einstein's Theory of Relativity*. Dover, New York, New York, 1965.

[3] Cohen, I. B. *The Triumph of Numbers*. W. W. Norton & Co., 2005.

[4] Crease, Robert. Pythagoras. *Physics World*, January 2006.

[5] de Pillis, John. *777 Mathematical Conversation Starters*. Math. Assoc. of Amer., Washington, D. C., 2002.

[6] Drake, Stillman (trans.). *Galileo Galilei: Dialogue Concerning the Two Chief World Systems — Ptolomaic and Copernican*. University of California, Berkeley, 1957.

[7] Dyson, Freeman. Religion from the outside. *The New York Review of Books, pp. 4-8*, LII(11), June 22 2006.

[8] Dyson, Freeman. *Progress in Religion*. Acceptance speech for the Templeton Prize for Progress in Religion, Washington National Cathedral, May 16, 2000.

[9] Einstein, Albert. Zur elektrodynamik bewegter körper (on the electrodynamics of moving bodies). *Annalen der Physik*, 17:891, 1905.

[10] Einstein, Albert. *Sidelights on Relativity*. Dover, New York, New York, October 1983.

[11] Ekeland, Ivar. *The Best of All Possible Worlds*. University of Chicago Press, 2007.

[12] Euclid (introduction by Thomas Heath). *The Thirteen Books of Euclid's Elements, Books 1 and 2 (paperback)*. Dover, New York, 1956.

[13] Hamming, R. W. On the unreasonable effectiveness of mathematics. *Amer. Math. Monthly*, 87(2), 1980.

[14] Hawking, Stephen W. *A Brief History of Time: From the Big Bang to Black Holes*. Bantam; Rei edition, March 1, 1988.

[15] Jackson, J. D. *Classical Electrodynamics*. Wiley, 3rd ed., 1998.

[16] Palmer, Stuart. *The Puzzle of the Silver Persian*. Bantam Books, ISBN:0-553-25934-2, 1934.

[17] Penrose, Roger. *The Road to Reality*. Knopf, 2004.

[18] Plato (Kamtekar and Lee). *The Republic*. Penguin Classics, 2003.

[19] Sobel, Dava. *Longitude: The True Story of a Lone Genius Who Solved the Greatest Scientific Problem of His Time*. Debate, 1997.

[20] Weinberg, Steven. (interview). *New York Times*, April 20, 1999.

[21] Wigner, Eugene P. On the unreasonable effectiveness of mathematics in the natural sciences. *Comm. on Pure and Appl. Mathematics*, 13(1), 1960.

Index

364

$\mathcal{N}$ _____

O _____

www.ingramcontent.com/pod-product-compliance
Lightning Source LLC
Chambersburg PA
CBHW050802220326
41598CB00006B/91